普通高等教育"十四五"规划教材

Modern Animal Reproductive Technology

Shi Fangxiong
Mao Dagan

China Agricultural University Press
· Beijing ·

Abstract

Modern animal reproductive technology (MART) is offered as a professional course based on animal reproduction science. The topics covered are techniques of estrus detection, ovulation control, fertilization, pregnancy diagnosis, parturition control, oocyte and embryo cryopreservation, sex control, embryo transfer, stem cell, animal cloning, transgenic animal, hormone immunization and genetic immunization, and avian reproductive biotechnology.

It includes many features about and beyond the text materials in each chapter. The text is informative and reader-friendly, with logical structure and excellent illustrations. Emphasis is laid on the connection of the theory to the practice, and the teachability and learnability. There is a summary at the beginning, and assignments at the end of each chapter. Ideological and philosophical elements are integrated into the book. This textbook can be used as a reference book for undergraduate and graduate students of animal husbandry, veterinary medicine and life science in domestic and overseas universities.

图书在版编目(CIP)数据

现代动物繁殖技术 = Modern Animal Reproductive Technology：英文/石放雄，茆达干主编. --北京：中国农业大学出版社，2023.6

ISBN 978-7-5655-2985-6

Ⅰ.①现… Ⅱ.①石… ②茆… Ⅲ.①动物－繁殖－高等学校－教材－英文 Ⅳ.①S814

中国国家版本馆 CIP 数据核字(2023)第 091715 号

书　名 Modern Animal Reproductive Technology

作　者 Shi Fangxiong　Mao Dagan

策划编辑 赵　艳　张苏明　　**责任编辑** 赵　艳

封面设计 中通世奥

出版发行 中国农业大学出版社

社　址 北京市海淀区圆明园西路 2 号　　**邮政编码** 100193

电　话 发行部 010-62733489，1190　　读者服务部 010-62732336

编辑部 010-62732617，2618　　出　版　部 010-62733440

网　址 http://www.caupress.cn　　**E-mail** cbsszs@cau.edu.cn

经　销 新华书店

印　刷 涿州市星河印刷有限公司

版　次 2023 年 6 月第 1 版　2023 年 6 月第 1 次印刷

规　格 185 mm×260 mm　16 开本　18.25 印张　455 千字

定　价 58.00 元

Editorial Committee

Editors in chief:

Shi Fangxiong 石放雄(Nanjing Agricultural University)
Mao Dagan 茆达干(Nanjing Agricultural University)

Associate editors:

Wei Quanwei 魏全伟 (Nanjing Agricultural University)
Hua Guohua 滑国华(Huazhong Agricultural University)
Gong Ting 龚　婷(Guizhou University)

Editorial members: (In alphabetical order)

Huang Pan 黄　攀(Jiangsu University)
Li Juan 李　娟(Nanjing Agricultural University)
Liang Aixin 梁爱心(Huazhong Agricultural University)
Lü Zengpeng 吕增鹏(China Agricultural University)
Muhammad Faheem Akhtar (Liaocheng University)
Wang Changfa 王长法(Liaocheng University)
Wang Chunsheng 王春生(Northeast Forestry University)
Wang Zhengchao 王正朝(Fujian Normal University)
Wang Ziyu 王子玉(Nanjing Agricultural University)
Wu Guoyun 吴国云(Jiangxi Agricultural University)
Yu Minli 于敏莉(Nanjing Agricultural University)
Zhang Wei 张　伟(Henan Institute of Science and Technology)
Zhou Zhengrong 周峥嵘(Jiangsu University)

Reviewers in chief:

Yang Liguo 杨利国(Huazhong Agricultural University)
Hou Guangxu 侯广旭(Nanjing Agricultural University)

Introduction for Authors

Shi Fangxiong, PhD, professor of the Zhongshan Project in Nanjing Agricultural University (NAU). He got his PhD degree of veterinary medicine from Tokyo University of Agriculture and Technology, Japan in 2001 and got his post-doctoral training in cellular and molecular biology in the Green Center of Reproduction, the University of Texas Southwestern Medical Center at Dallas and the Department of Cellular and Molecular Biology, California State University at Los Angeles, USA during 2001-2004. He was the Editor-in-Chief of *Animal Husbandry and Veterinary Science* during 2007-2018.

Mao Dagan, PhD, professor of NAU where she got her PhD degree in animal genetics, breeding and reproduction in 2003. She worked as a visiting scholar in Department of Obstetrics and Gynecology of University of Nebraska Medical Center (UNMC) in 2013. She is one of directors of Animal Reproduction Branch of Chinese Association of Animal Science and Veterinary Medicine (CAAV). She serves as a professional reviewer for several SCI journals and Chinese core journals.

Wei Quanwei, PhD, associate professor of NAU where he got his PhD degree in animal genetics, breeding and reproduction in 2013. He got his post-doctoral training in Department of Microbiology and Physiological Systems in University of Massachusetts Medical School in 2020. He is a member of CAAV, and Society for the Study of Reproduction (SSR). He serves as the editorial board member of *Frontiers in Nutrition* and a professional reviewer for several SCI journals and Chinese core journals.

Hua Guohua, PhD, professor of Huazhong Agricultural University (HZAU) where She got her PhD degree in animal genetics, breeding and reproduction in 2009. She visited UNMC twice during 2013-2015 and 2019-2020. She is a member of CAAV and SSR. She serves as the editorial board member of *Animal Reproduction Science* and *Animals*, and a professional reviewer for a series of SCI journals and Chinese core journals.

Gong Ting, PhD, associate professor of Guizhou University. She got her PhD degree of animal genetic, breeding and reproduction from NAU in 2016, and got a chance of academic exchange with the veterinary medicine department of Tokyo University of Agriculture and Technology. She has been a director of Animal Reproduction Society of CAAV.

Huang Pan, PhD, associate professor of Jiangsu University. He got his PhD degree of animal genetics, breeding and reproduction from NAU in 2012, and got his post-doctoral training in the area of nutrition and gut health in the National University of Singapore during

2022-2023.

Li Juan, PhD, professor of NAU. She got her PhD degree of medicine from Faculty of Human Healthy in Aarhus University, Denmark in 2009, and got her post-doctoral training in the area of reproduction from Faculty of Agriculture in Aarhus University, Denmark during 2009-2011.

Liang Aixin, PhD, associate professor of HZAU where received a PhD degree in animal genetics, breeding and reproduction in 2009. She worked as a visiting scholar at the University of Naples "Federico II" in 2014, and at UNMC in 2019. She was a guest editor for *Frontiers in Genetics* during 2021-2022.

Lü Zengpeng, PhD, associate professor of China Agricultural University where he got his PhD degree of animal nutrition and feed science in 2018, and got his post-doctoral training in animal genetics, breeding and reproduction science in NAU during 2018-2021. He is the secretary of the National Broiler Nutrition Steering Committee.

Muhammad Faheem Akhtar, PhD, lecturer of Liaocheng University. He got his PhD degree in animal genetics, breeding and reproduction from NAU in 2019 and got post-doctoral training in area of molecular and medical biotechnology in Nanjing Normal University during 2019-2021.

Wang Changfa, PhD, professor of Liaocheng University. He got his PhD degree from Ocean University of China in 2003. He is one of the leading talents of Shandong Taishan Industry, chief expert of Donkey Industry Innovation Team of Shandong Modern Agricultural Industry Technology System, chief scientist of Donkey Industry Technology Collaboration Innovation Center of University in Shandong.

Wang Chunsheng, PhD, associate professor of Northeast Forestry University (NEFU). He got his PhD degree of animal genetics, breeding and reproduction from Northeast Agricultural University in 2008 and got his post-doctoral training in the developmental biology in NEFU during 2008-2014. He worked as a visiting scholar at the University of California, Davis during 2017-2018.

Wang Zhengchao, PhD, professor of Fujian Normal University and a distinguished professor of Fujian Province Minjiang Scholars. He got his PhD degree of animal genetics, breeding and reproduction from NAU in 2008 and got his post-doctoral training in Virginia Commonwealth University during 2008-2015. He is an editorial board member *Experimental and Therapeutic Medicine*.

Wang Ziyu, PhD, lecturer of NAU where he got his PhD degree of animal genetics, breeding and reproduction. He went to University of Western Australia to investigate the sheep industry in Western Australia.

Wu Guoyun, PhD, lecturer of Jiangxi Agricultural University. She got her bachelor degree of Animal Science from College of Animal Science and Technology of Jiangxi Agricultural University in 2013 and PhD degree of animal genetics, breeding and reproduction in NAU in 2019.

Yu Minli, PhD, associate professor of NAU. She got her PhD degree in animal genetics, breeding and reproduction from Zhejiang University in 2012 and worked as a visiting scientist in reproductive biology in the Department of Biomedical Sciences, Cornell University during 2016-2017.

Zhang Wei, PhD, associate professor of Henan Institute of Science and Technology. He got his PhD degree of animal genetics, breeding and reproduction from NAU in 2011. He is a director of Animal Reproduction Society of Henan Animal Husbandry and Veterinary Society. His mainly research is physiological regulation of animal reproduction, protection of livestock germplasm resources and molecular breeding and healthy breeding of livestock.

Zhou Zhengrong, PhD, associate professor of Jiangsu University. She got her PhD degree in animal genetics, breeding and reproduction from NAU in 2013. She worked as a visiting scholar in David Geffen School of Medicine, University of California at Los Angeles, USA during 2019-2020.

Preface

The educational internationalization has become an irreversible trend in China's universities. About 20 years ago, we used to have offered a bilingual course of New Animal Reproductive Technology for college students majoring animal sciences. Recently, we have opened some full English courses on new animal reproductive technology for undergraduate, master degree and even doctoral degree students. However, the textbooks of such professional courses in English in line with the syllabuses are not available in China. Therefore, we assembled the professors and young scholars in this field from ten universities to form an editorial board to write this textbook of *Modern Animal Reproductive Technology* (MART). This English textbook on animal reproductive technology is a preliminary attempt to meet the needs of current all-in-English teaching, as well as research and production needs of animal husbandry, veterinary medicine and life sciences.

This textbook is compiled under the guidance of scientific, advanced, systematic, practical and extensive principles, taking into account of the characteristics of teaching in colleges and universities, as well as the needs of the development of modern animal husbandry and biological pharmacy. In teaching material structure, the combination of basic theory and fundamental technology is highlighted and the frontier issues, practicability and maneuverability are focused on. In terms of content, the basic theories and practical technologies related to biotechnology essential for life science are given priority and the scientific nature of theories and advance in methods are emphasized, including techniques of estrus detection, ovulation control, fertilization technique, pregnancy diagnosis, parturition control, oocyte and embryo cryopreservation, sex control, embryo transfer, stem cell, animal cloning, transgenic animal, hormone immunization and genetic immunization, and avian reproductive biotechnology. Therefore, this textbook can be used not only as a learning textbook for animal husbandry, veterinary medicine and life science majors in colleges and universities, but also as a reference book for those engaged in animal husbandry, biopharmaceutical and other industries and product research and development.

This textbook consists of 12 chapters. The assignment of writing tasks for each chapter is as follows: Introduction, Shi Fangxiong; Chapter 1, Wei Quanwei; Chapter 2, Wang Changfa, Gong Ting and Muhammad Faheem Akhtar; Chapter 3, Zhou Zhengrong and Huang Pan; Chapter 4, Wang Zhengchao; Chapter 5, Wu Guoyun; Chapter 6, Wang Ziyu; Chapter 7, Hua Guohua; Chapter 8, Wang Chunsheng; Chapter 9, Li Juan; Chapter 10,

Zhang Wei; Chapter 11, Mao Dagan and Liang Aixin; Chapter 12, Yu Minli and Lü Zengpeng. After the first drafts were submitted, the final drafts were compiled by the deputy editors and the chief editors.

The publication of our textbook went smoothly. Firstly, we would like to express our gratitude to Professor Yang Liguo of Huazhong Agricultural University and Professor Hou Guangxu of Nanjing Agricultural University to read and review our whole book. Secondly, we thank Ms. Zhang Suming and Ms. Zhao Yan, editors of China Agricultural University Press for carefully planning and editing. Thirdly, we thank Professor Huang Ruihua and Dr. Hou Liming of Nanjing Agricultural University, and Professor Fu Chunquan of College of Agriculture, Jinhua Polytechnic for providing some photographes and videotapes. Last but not least, we thank all the members of the Editorial Committee for their hard work and dedication. Our book was financially supported by the Textbook Construction Fund of College of Animal Science and Technology, Nanjing Agricultural University and the Unveiled Projects of Jiangsu Provincial Seed Industry Revitalization (JBGS〔2021〕108), China.

During the compilation, College of Animal Science and Technology of Nanjing Agricultural University and China Agricultural University Press have provided us with their full support and cooperation. We hereby express our deep gratitude to them.

Although the editors have made great efforts in the compilation, the shortcomings of the textbook are unavoidable due to the limitations of the editors and the rapid development of reproduction technology. We sincerely hope that readers and experts give more valuable suggestions for supplementary revision in the second edition.

Editors in Chief

Prof. Shi Fangxiong & Prof. Mao Dagan

2022. 12

Contents

Introduction

As climate warming and frequent occurrence of extreme weather in the world, the survival and developmental environment for animals and human beings become very challenging. However, our longing for a better life has never changed. Animal reproductive technology is a key tool in pursuing that goal. Modern Animal Reproductive Technology (MART) is a comprehensive emerging technical discipline based on the basic theory and principle of animal reproduction, and combining classic and advanced engineering techniques to improve the performance and efficiency of livestock production, or to save endangered and valuable animals.

MART is a major component of animal reproduction sciences, some parts develop from classic techniques like artificial insemination (AI) and embryo transformation (ET), while others are created from new methods like gene engineering, stem cells, gene conversion and animal clone, hormone immunization and gene immunization technology. Along with the rapid development of science and technology, the techniques used in modern animal reproduction is constantly updated and improved, especially AI is an important direction of development.

MART involves many disciplines of life sciences, like molecular biology, cellular biology, immunology, biochemistry, genetics, and almost all sub-disciplines of animal biosciences. Combining cutting-edge basic disciplines outside biological fields such as physics, chemistry, mathematics, computer and informatics, it has formed a comprehensive discipline with mutual penetration of multiple disciplines. Among them, several importance breakthroughs in life sciences have led to milestone achievements in the history of animal reproductive technology, for example, without cell culture and microscopy, there would be no production and application of cloned animals. Clarifying DNA double helix structure and semi-reserved replication mechanism and decoding of the genetic code have provided foundations for transgenic technology. By applying the methodologies and theories of modern microelectronics and computer sciences, animal reproductive technology is going deep into the molecular level and is advancing faster than ever before.

Two centuries ago, our pioneers made contributions to the artificial insemination (AI) in farm animals. This technique used in dairy cattle breeding is extremely successful nowadays, a technique accomplished by improving the selection differential wherein one highly selected male is mated with thousands of females. *In vitro* fertilization (IVF) and embryo transfer were developed in the early to mid-20th century as part of a general awakening of interest in reproduction in mammals. IVF with fresh or frozen-thawed semen has been developed and optimized for laboratory animals and livestock to increase the female genetic contribution to the gene pool. Such techniques and other newly developed/developing techniques will be introduced from mechanisms to clinic application in this textbook.

The picture for animal production in the beginning of the 21st century is far from the scenario in the 20th century. Animal production, and agriculture in general, in the 20th century was influenced by the need to feed and clothes growing populations after a period of war in the developed world. Strategies for providing food and clothes were implemented at any cost, including intensified use of natural resources to the limit and the extensive use of chemical substances (pesticides, antimicrobials, or hormones), which act as xenobiotics. The world population will continue growing in the present century. According to the estimates by the Food and Agriculture Organization (FAO), the global population will be about 8.5 billion in 2030 and 9.7 billion in 2050, which means food production should increase by 70 percent. However, as we all know, natural resources are limited and any action on the environment has long-lasting consequences. Hence, agriculture, especially animal production, needs to adapt to a changing environment by improving its efficiency and sustainability, and reducing its impact via low-carbon economy. Many biotechnological assisted reproductive techniques (ART) are currently used to control the reproductive processes of farm animals, and the main aim is to develop and implement relevant tools for improving fertility of farm animals with easily applied, affordable and effective techniques. Great efforts are made by scientists to achieve the aims of smart animal production, particularly in the field of animal reproduction and specifically in the management of the reproductive activities.

Modern animal reproductive technology helps farmers increase the production of farm animals, and artificial intelligence allows easy data entry on farm records, monitoring farm activities, analyzing economic performance, improving animals' health and improving soil richness. All these features and functions aim to achieve "smart farming". Artificial intelligence uses data to improve the quality and clarity of decision making across all levels of the animal husbandry. It can outperform humans in determining if individual animals meet market specifications through the forecast of individual animal conditions. However, as more farms are intimate with modern technology, artificial intelligence and sensing technologies will begin to play a more important role in helping farmers understand patterns and solutions to stubborn problems in modern livestock farming.

Chapter 1 Estrus and Ovulation Control

Summary

Dynamic changes of ovarian cycle happen in the adult female animal as follicles develop. During these ovarian cyclic changes, especially around ovulation, the females show significant changes in the telltale phenomenon and behavior known as estrus. At estrus, the most important event is ovulation. To examine estrus and predict ovulation time is a basic technique to improve reproduction performance. To be acceptable for routine use in commercial livestock production units, estrus and ovulation control serves for the purpose of breeding animals. With the development of animal genetics, and breeding and reproduction technology, a variety of estrus and ovulation control technologies have been developed, which make animal ovulation and estrus control more and more precise. In this chapter, we briefly introduce the technologies about estrus detection and ovulation control, and further discuss their application methods.

The estrous cycle is the rhythmic changes that occur in the female reproductive system from one estrus to another. When a female animal grows to a certain age, the ovarian follicles develop and secrete estrogen, causing a series of changes in reproductive organs, sexual behavior and sexual desire. Most follicles develop and go atresia and only a few follicles ovulate finally. At peri-ovulation, most female animals present a physiological state, called an estrus stage (estrus). At normal estrus, female animals have obvious sexual desire, are ready to have sex and a series of internal changes occur in the form and function of reproductive organs. The development, maturation, ovulation and estrogen production of ovarian follicles are the essence of estrus, and changes in external reproductive organs and sexual behavior are the external characteristics of estrus, which can be used for estrus identification and timely breeding. Fig. 1-1 shows the phases of the bovine estrous cycle.

Asexuality is a condition in which female animals lack a regular reproductive cycle. As a result, animals become unproductive and cause huge economic losses to farmers or breeders. Therefore, it is very important to solve this problem in the livestock production. For physiological anestrus, reproductive hormones or some animal husbandry management measures can be used to induce estrus, shorten the reproductive cycle, and improve the

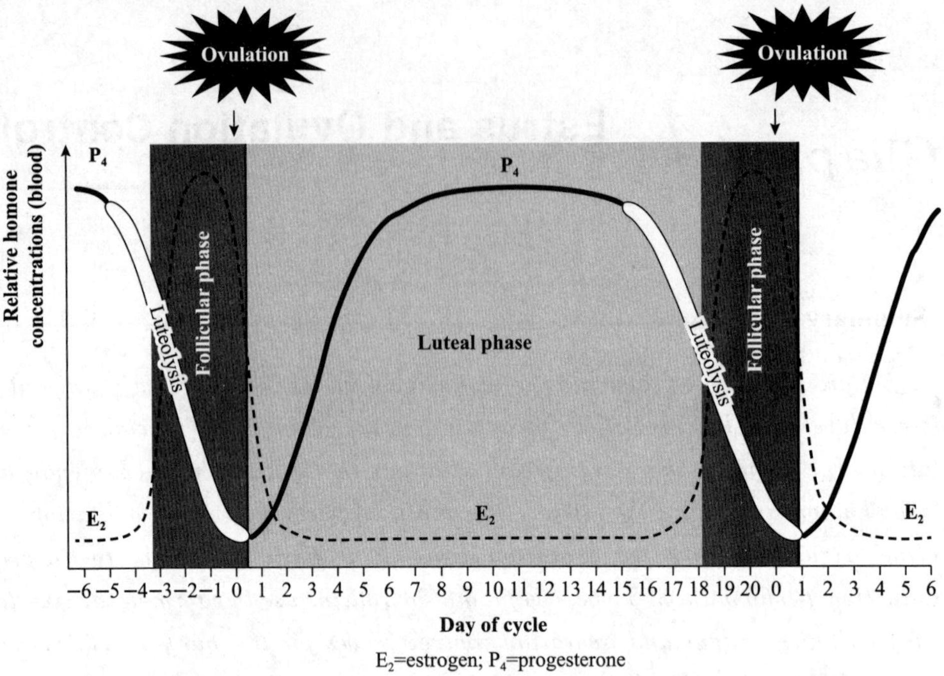

Fig. 1-1 Phases of the bovine estrous cycle (Senger et al. , 2012)

fecundity of livestock. It can also induce a group of female animals to estrus at the same time, that is, to improve the efficiency of modern animal husbandry. There are several established methods to induce estrus in animals, ranging from the use of hormones to the use of bio-stimulants. In addition to inducing estrus, some techniques can also be synchronized on demand. Estrus synchronization facilitates simplified management practices and advanced technologies such as embryo transfer, *in vitro* fertilization, and cloning.

In animal reproduction management, almost everything related to the actual mating must be done when the female is ready. Accordingly, there has been considerable interest in developing the programs that manipulate ovarian activity to regulate ovulation to allow mating at predetermined times. At present, the technology related to ovulation control is still the focus of reproductive research. People are committed to developing effective pharmacological and physical methods to control the ovulation of animals and apply them to animal production.

1.1 Principle of estrus diagnosis

Estrus diagnosis is an important technique in animal reproduction. Estrus identification can determine the estrus stage of the animal and predict the time of ovulation, so as to determine the appropriate mating time, timely mating or artificial insemination, and improve the conception rate and fertility of the female animal. The abnormal estrus or reproductive tract disease of the female animal can be also discovered or identified early for timely treatment.

Estrus in animals is a series of changes in reproductive organs and sexual behavior under the regulation of reproductive hormones, including external manifestations and internal changes. External manifestations are directly observable behavioral changes and changes in external reproductive organs; while internal changes refer to changes in reproductive organs, mainly the changes in the development of follicles on the ovary, which are the essential changes in estrus. Therefore, in estrus identification, it is not only necessary to observe the external changes of the female animal, but more importantly, to judge the development of the follicles on the female animal's ovary.

1.1.1 General diagnosis technology

There are many methods for estrus identification, but in practical application, the principle of combining key methods and general ones should be adopted according to the estrus characteristics of various animals. However, no matter what method is used, before estrus identification, the animal reproduction history, estrus process and corresponding reproduction records should be understood.

1.1.1.1 External observation

This method is the most commonly used for estrus identification of various animals. The external performance and mental state of animals are mainly observed to judge their estrus status. For example, animals in estrus often show mental restlessness, singing, loss of appetite, frequent urination, sensitivity to the surrounding environment and male animals, and vulva congestion, swelling, moisture, and mucus outflow. Different animals often have special manifestations, such as cows crawling across, sows making troubles, mares neighing, labia constantly everting out to expose the clitoris, female donkeys "beeping" and so on. The above-mentioned characteristics change from weak to strong, and then from strong to weak with the estrus process, and disappear after the estrus is over. Therefore, in the estrus identification, it is necessary to judge according to the estrus characteristics of different animals.

1.1.1.2 Male exposure

This method is to judge whether the female animal is in estrus and the degree of estrus according to the response of the female animal to the male animal in terms of sexual desire and sexual behavior. When female animals are in heat, they usually show willingness to approach the male, bow their waists and raise their tails, open their hind limbs, urinate frequently, and seek copulation. If they are not in estrus or the estrus is over, they will stay away from the male. When they are forcibly drawn to approach, they will often avoid the target and even show resisting behaviors such as kicking and biting. The back-pressure method is commonly used for sows, and the degree of estrus is judged by the degree of its "immobility response"(Fig. 1-2A). Expectations are that at least half of the treated gilts will show estrus within 7-10 d of initial male exposure. Gilts can be mated at the first estrus but, since most responders will subsequently cycle regularly, service can be delayed for three or six weeks. If substantially more animals than the required number of replacements are exposed, the responding gilts can be retained for breeding and the remainder can be marketed before they are overweight.

Generally, non-breeding male animals with strong physique, strong sexual desire and no vicious habit are used to test the female animal.

1.1.1.3 Vaginal examination

This method is mainly to check the vagina including the appearance of the vagina for redness and mucus discharge (Fig. 1-2B, C, D), or insert a vaginal dilator or vaginal dilation tube into the animal's vagina and expand it, and use a light source to observe the color of the vaginal mucosa, the degree of hyperemia, the state of cervix relaxation, the color of the external opening, the degree of hyperemia and swelling, the size of the opening, the color of the secretion, the viscosity and amount, the presence or absence of mucus outflow, etc. During the inspection, the vaginal dilator or dilation tube should be cleaned and disinfected to prevent infection; care should be taken when inserting it so as not to damage the vaginal mucosa and the external urethra orifice. This method usually applies to large animals, such as cattle, horses, donkeys, etc. Since this method can not accurately determine the ovulation time of animals, it is currently only used as an auxiliary inspection method.

1.1.1.4 Rectal examination

The method of rectal examination is to put the arm into the rectum of the female animal, and touch the ovary and follicle development with fingers across the rectal wall, feeling for the size, shape and texture of the ovary, the location, size, elasticity and thickness of the follicle wall, and determining whether the follicle has ruptured, whether there is a corpus luteum, etc. The inspection should be carried out according to the steps, and in order to avoid squeezing the follicle, do not squeeze it hard. Through rectal examination combined with external estrus manifestations, the degree of follicle development and ovulation time can be accurately judged, so as to accurately determine the suitable mating period. Therefore, this method is mainly used in large livestock such as cattle and horses, and is widely used in production because of its direct reliability. The disadvantage of this method is that the judgment result depends on the experience of the surgeon. Therefore, the operator needs to go through many times of practice and accumulate rich experience before he can operate effectively.

1.1.1.5 Reproductive hormone detection

When female animals are in estrus, the level of progesterone decreases and the level of

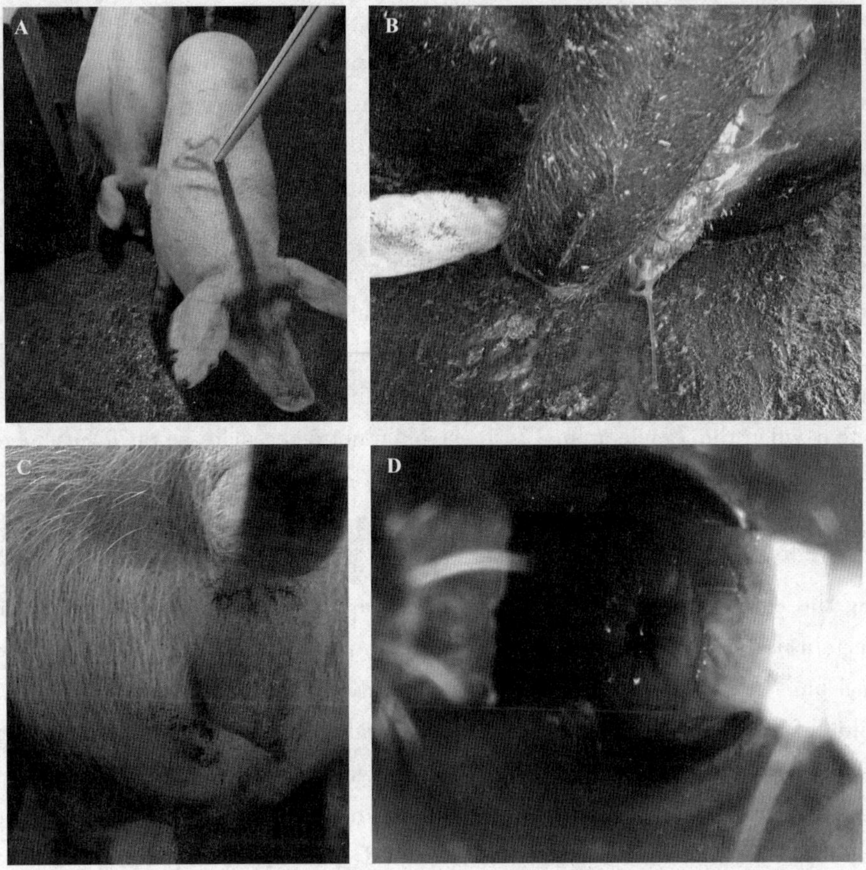

A. Immobility response B. Clear slime mucus discharge C. Flushed vulva
D. Vaginal dilator observation for cow cervix

Fig. 1-2 The back-pressure method and vaginal examination for estrus diagnosis

estrogen increases. Estrus can be judged by using hormone measurement techniques, such as enzyme-linked immunoassay (enzyme immunoassay, EIA) or radioimmunoassay (RIA), through the determination of reproductive hormones (FSH, LH, estrogen, progesterone, etc.) in body fluids (plasma, serum, milk, urine, etc.), and according to the change level of reproductive hormones in the estrous cycle. At present, there are a variety of hormone ELISA test kits for estrus identification abroad (see the Chapter 3 for details).

1.1.1.6 Bionics

Through simulating the male animal's voice (playing tapes) and smell (natural or synthetic aerosol preparations) by bionics to stimulate the female animal's auditory and olfactory organs, the arousal, if any, of the female animal is observed after being stimulated, and whether the female animal is in heat is determined. In production practice, there are many estrus identification tests for pigs using this method. The results show that only 48% of the sows in estrus show the static reflex when only the back-pressure test is used; 90% of sows in heat exhibit the standing reflex when the boar is not present—they can hear the call and smell the boar. The use of natural or synthetic boar sex hormones, sprayed in the sow herd, can stimulate the standing reflex in sows in estrus. Therefore, it has been suggested that the use of boar odor and sound, combined with the appearance of a simulated boar, can optimize the effect of sow estrus identification.

1.1.1.7 Electrical measurement

The resistance value of vaginal mucus of the female animal is measured by resistance meter for estrus identification, so as to determine the suitable insemination time. This method has a certain reference value for determining the suitable insemination time. The rationale of this method is that when the female animal is in estrus, due to the increase of mucus secretion, the ion concentration in the reproductive tract reduces the resistance value.

1.1.1.8 Determination of pH value of reproductive tract mucus

During the estrus cycle of female animals, the pH value of reproductive tract mucus shows a certain pattern of change, generally neutral or alkaline in the estrus peak, and acidic in the luteal phase. The pH value of the cervix fluid of cows is between 6.0 and 7.8. When the pH value of multiparous cows is between 6.7 and 6.8, the insemination and fertilization rate is the highest. The pH value of vaginal mucus was greater than 7.3 on the day when the three breeds of Landrace, Large White and Hampshire started estrus, 7.2 at the peak of estrus, and less than 7.2 during the gestation period, and the insemination was performed at pH 7.2-7.3. The conception rates in estrus were 93.8%, 96.7% and 92.3%, respectively.

1.1.1.9 Ultrasonography

Ultrasonography is a physical examination method established by using the physical characteristics of ultrasound and the acoustic characteristics of animal tissue structure, which can accurately detect different tissues and organs of the body. Ultrasound diagnosis is an important imaging technique used in reproductive research and diagnosis. This technology can be used for pregnancy diagnosis, fetal growth and development monitoring, recording changes in ovarian follicle structure, monitoring abnormal fetal development, and diagnosing double lambing in horses and cattle. Ultrasound diagnostic technology has the advantages of minimal trauma to the body and no need for surgery. Ultrasonic wave is a kind of sound wave whose vibration frequency is above 20 000 Hz/s and exceeds the range of human hearing threshold, referred to as ultrasound. Under the action of high-frequency alternating electric field applied by the host machine, the special chip in the probe produces strong compression and tensile vibration of ultrasonic waves in a certain direction, which generates the desired ultrasonic waves. In addition to transmitting ultrasonic waves, the probe can also be used to receive ultrasonic waves and convert electroacoustic signals, so it is also known as the transducer. Probe type ultrasonic measuring instrument adopts the form of ultrasonic device equipped with a certain power probe. When the probe touches the corpus luteum or follicle on the ovary through the vaginal wall, the probe is subjected to different reflective shock waves, and an image of the structure of the corpus luteum or follicle is displayed on the display screen. The estrus stage is determined by the size of the follicle diameter.

In the mid-1960s, ultrasonic diagnostic technology was applied in the field of veterinary medicine. The A-type and D-type (doppler-type) diagnostic instruments were initially used, and the M-type and B-type were developed in the mid-1970s. At present, the B-type (brightness mode) ultrasonic diagnostic technology is the main ultrasonic technology for monitoring the structure of the corpus luteum or follicle. The B-mode ultrasonic diagnostic instrument belongs to the brightness modulation (brightness modality) type, which causes the piezoelectric crystal of the ultrasonic probe to vibrate through the pulse current. At the same time, multiple beams of ultrasonic waves are emitted to detect on one section, and the reflection of the sound wave is converted into a pulse current signal by the probe, and light spots with different brightness are formed on the display screen to display a section tomographic image of the probed part. Due to the different acoustic resistance values of various tissues of the body, the intensity difference of sound wave reflection is shown. When the non-reflection (liquid) and strong reflection (dense tissue) parts are detected, the anechoic black and strong reflection are displayed, respectively. The white color of the echo waves may vary depending on how the scan is displayed. Therefore, B-mode ultrasonography is also called ultrasonotomography. The imaging speed of this method can be fast or slow, and the fast one can immediately display the activity state of the inspected organ, which is called real-time ultrasound diagnosis. The B-mode real-time ultrasonic tomograph can be equipped with a line scan probe and a sector scan probe, which are called real-time line scanner and real-time sector scanner, respectively. In addition, there are also some B-mode ultrasonic tomographs that can be used with line array and sector probes. Line scan images are rectangular, while sector scan images are fan-shaped. Resorting to instrument identification, this method is accurate and reliable, but the operation is complicated and the cost is high.

1.1.2 Modern estrus monitoring technology

Estrous detector enables the breeder to determine the exact time during the ovarian cycle when ovulation occurs. This method is mainly used for cattle and horses, and sometimes also used for sheep and pigs. There are several types of heat detectors.

1.1.2.1 Chin-ball marker

The device consists of a ball-shaped dye box with a steel ball piston valve fixed on a solid leather bridle, and the dye box contains a colored dye. When in use, the device is tied under the jaws of the male animal, when it gets on the back of the female animal in estrus, the steel ball of the movable valve touches the back of the female animal, so the dye in the dye box flows out and is printed on the back of the female animal, accordingly, you can know that the female animal thus marked is in heat.

1.1.2.2 Kamar Heatmount Detector

Kamar Heatmount Detectors are a valuable heat detection aid that assist in identifying cows ready for artificial breeding (Fig. 1-3). The device consists of a plastic capsule filled with white dye. The device is firmly adhered to the tail root of the cow when in use. When the female animal is in estrus, the test male animal will climb on it and apply pressure to the capsule, so that the dye in the capsule changes from white to red, so according to the degree of color change, it can be inferred that the female animal has accepted the safety level of mounting. Over several decades, the Kamar Heatmount Detector has become the accepted standard device for economical, accurate bovine heat detection worldwide.

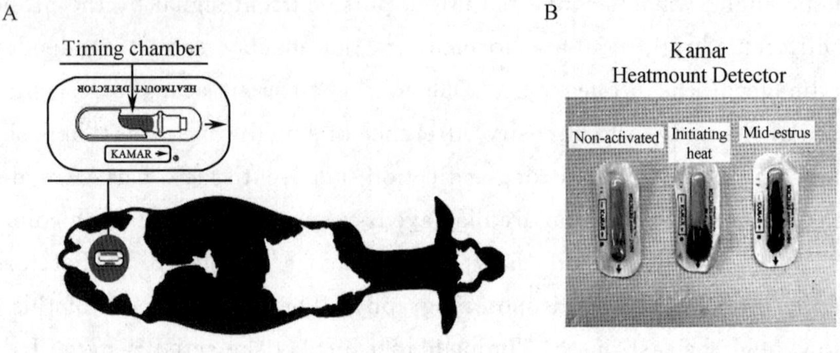

A. Glued onto the sacrum (tail head) B. Detector activation status

Fig. 1-3 Kamar Heatmount Detector usage diagram (from https://kamarinc.com/)

1.1.2.3 Tail paint chalk

The application of this method not only improves the efficiency of estrus identification, but also reduces the labor intensity of estrus identification. During operation, comb the loose hair at the base of the cow's tail neatly with a brush or comb, and then rub it with a special crayon on the tail of the cow, that is, apply a 2.5-5 cm wide mark from the bulge on the back of the cow to the upper part of the tail. Crayons should be reapplied daily for best viewing. After the first rubbing, you only need to reapply once a day (the crayon should not be too heavy, otherwise it will be difficult to identify whether it has been rubbed off, especially for cows that are difficult to determine the estrus period and are less active). The crayon on the tail being rubbed off is a sign that the cow may be in estrus, but it is necessary to observe other signs of estrus to confirm (Fig. 1-4), especially for cows whose crayon is only partially rubbed off, you should pay more attention to the following conditions: ①changes in genital mucus, the degree of disarray of tail hair, whether there are stains on the back and sides of the cow; ②whether the vulva is swollen, and whether there is mucus or blood on the tail of the cow; ③whether the mucous membrane of the vulva is red, swollen and slippery; ④records to confirm whether the cow is in a normal estrus cycle; ⑤palpate the cow's uterus to accurately determine whether the cow is in estrus; ⑥the color of the crayons applied to non-pregnant cows is different from that applied to pregnant cows, so that it can be observed whether pregnant cows have miscarried or not.

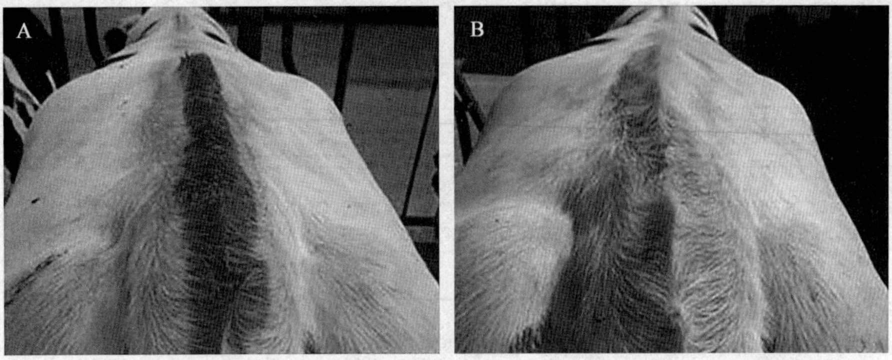

Fig. 1-4 Tail chalk after it is freshly applied (A) and after estrus (B)

(Skenandore and Cardosoa, 2017)

1.1.2.4 Step counting

This method can be used for auxiliary estrus identification of cows in large and medium-sized dairy farms with high level of automation management. The theoretical basis for this is that cows exercise more when they are in heat. The specific steps are: fix a pedometer with a signal transmitting device on the cow's leg, the receiver in the machine room can receive the pedometer signal of each cow, and then transmit it to the computer terminal, and use special software to analyze it, so as to record the average daily exercise of the cow. The pedometer records the daily exercise of each cow and draws a graph. If a cow's exercise continues to increase during a certain period of time, in addition to other external factors such as diseases, it means that it may start estrus, thus prompting the breeder. If combined with other detection methods, such as rectal examination, the effect is better.

1.1.2.5 Cervical mucus resistance value measurement

During the estrus cycle of animals, the cervical mucus resistance value shows a certain regularity of change and is intrinsically related to the ovulation of the female animal. In the non-estrus state, the cervical mucus resistance value is the highest. With the appearance of estrus, the cervical mucus resistance value gradually decreases, and the resistance value drops to the lowest point during ovulation. After ovulation, the mucus cervical resistance value gradually increases to the non-estrus value state at the time. Therefore, the estrus state and stage can be judged according to the cervical mucus resistance value.

1.1.2.6 Rumen-reticular temperature (RRT) monitoring

During the estrous cycle of cows, there are changes in rumen-reticular temperature (RRT) at which the intensity of estrus expression can be determined by a rumen-reticular bolus thermometer. Fig. 1-5 represents the schematic of estrus and ovulations alerts based on rumen-reticular temperature monitoring method.

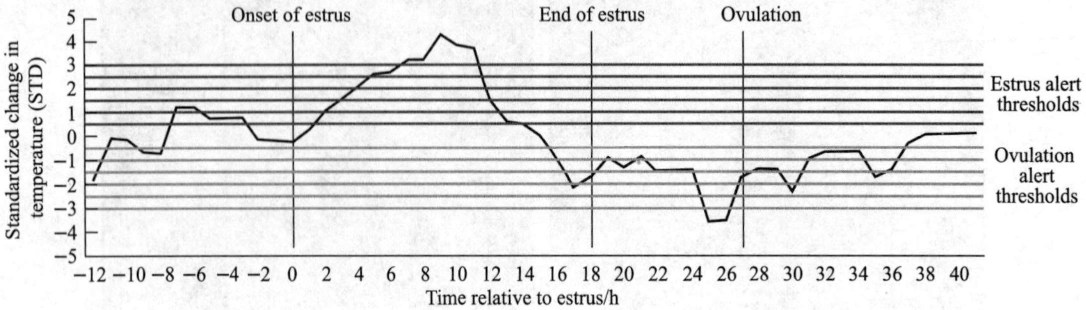

Fig. 1-5 Schematic of estrus and ovulations alerts based on standardized change in rumen-reticular temperature relative to baseline (Burnett et al., 2020)

1.1.2.7 Audio-based estrus detection

Early detection of abnormalities is an important issue in group livestock management. Failure to detect estrus in a timely and accurate manner can be a limiting factor in achieving effective reproductive performance during reproductive management on dairy or pig farms. The audio-based estrus detection technology has been applied in dairy cattle breeding management, taking a system reported in the literature as an example (Fig. 1-6). The real-time detection system of dairy cow estrus consists of four modules: feature extraction and dairy cow estrus detection module, two online processing modules, attribute subset selection and support vector data description (SVDD) training modules, and two offline processing modules. In the feature extraction module, the Mel frequency cepstrum coefficient (MFCC) algorithm is used to simulate feature extraction from audio signals through sound sensors or closed circuit television (CCTV) cameras. By recording and characterizing the vocal characteristics of cows in heat, they are divided into two spectral samples: heat sound and normal sound. Therefore, the above-mentioned system is used to automatically detect the estrus state of the cow.

1.1.2.8 Video-based estrus detection

Computer vision and machine learning have been deeply delved into in animal behavior; and it also has many advantages in animal estrus monitoring. In the case of cows, in order to detect estrous stage with certainty, it is necessary to pay particular attention to the clear signs of heat that will allow you to determine which stage of the estrus cycle the cow is in.

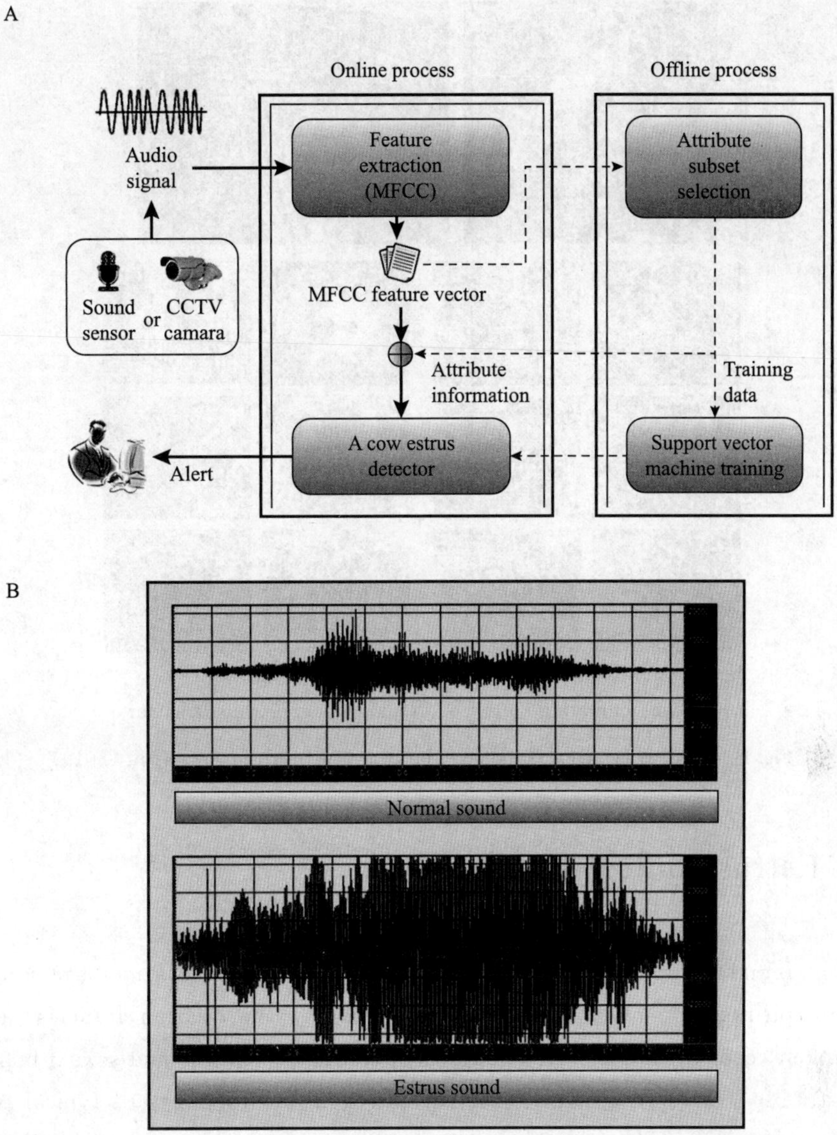

A. The overall structure of the system B. Spectrums of normal sound and estrus sound of cows

Fig. 1-6 Audio-based estrus detection system(Chung et al. , 2013)

During the day, keepers can make observations to determine the animals' signs of heat. However, night is a blank period. There is one major advantage in using an auto-detected camera: they fill the surveillance gap while respecting the herd's circadian rhythm. Each cow is equipped with a certain number of sensors, combined with camera image data, and analyzed by artificial intelligence deep learning computers to finally determine the animal's estrus stage (Fig. 1-7).

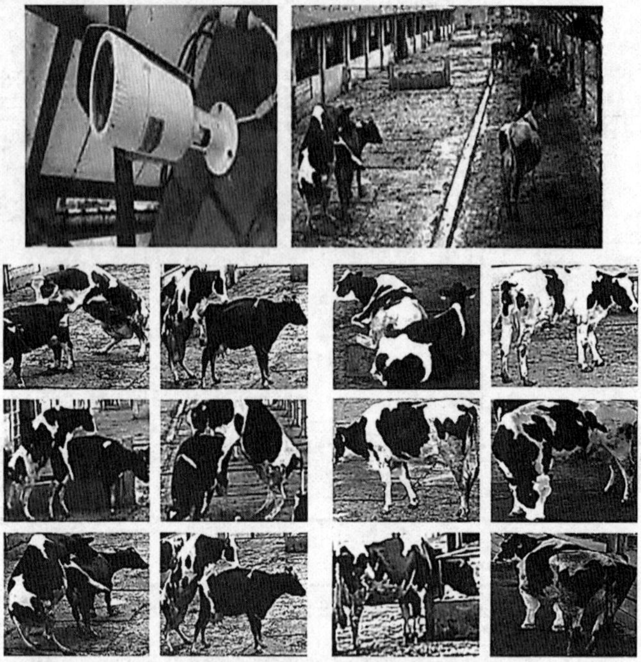

Poitive: mounting behavior Negative: non-estrous behavior

Fig. 1-7 Estrus behavior cows based on neural network in sows (Liu, 2019)

1.2 Estrus control technology

After the female animal grows and develops to a certain age, under the control of the hypothalamus-pituitary-gonadal axis and gonadotropins, the ovarian follicles develop and secrete estrogen, causing a series of changes in reproductive organs and sexual behavior, and sexual desire (Fig. 1-8 represents the level of estrogen and intensity of typical preliminary signs of estrus of sows). The use of exogenous hormones, drugs or some animal husbandry management measures to artificially control individual or group estrus and ovulation technology is called estrus control technology, including estrus induction and simultaneous estrus technologies.

1.2.1 Estrus induction technology

Estrus induction refers to the technology of artificially inducing estrus and ovulation of female animals. It is mainly used for estrus and mating of estrus females, such as seasonal estrus sheep, lactating sows and postpartum dairy cows that are not in estrus for a long time. The use of estrus induction technology can shorten the farrowing interval and increase the number of litter and parity; the farrowing season can be adjusted to make the dairy animals produce milk evenly within one year; for seasonal estrus animals, it can be used in

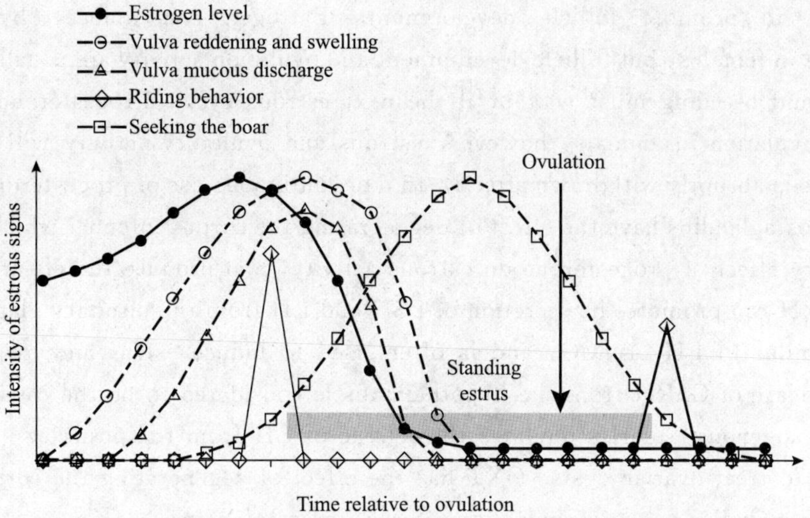

Fig. 1-8 Level of estrogen and intensity of typical preliminary signs of estrus relative to standing estrus and ovulation in sows (Brad et al., 2007)

any season of the year. Estrus induction can enable seasonal estrus animals to be in estrus in all seasons; it can reduce the reproductive loss caused by pathological aesthesia such as ovarian cysts and persistent corpus luteum, thereby improving the fecundity of livestock.

1.2.1.1 Principles of estrus induction

Female anestrus can be divided into two types: physiological and pathological. Physiological anestrus is characterized by lack of follicle development or lack of corpus luteum present on the ovary, and the ovary is in a static state, such as female animals before puberty, and seasonal estrus animals during non-estrus seasons. Pathological anestrus is mainly caused by ovarian dysfunction, such as ovarian cysts, persistent corpus luteum and other reasons, so that female animals can not show a normal sexual cycle.

Estrus induction is to use reproductive hormones or animal husbandry management measures to change the ovaries from a relatively static state to an active cycle state based on the regulation of reproductive hormones on the estrus, promote follicle growth and development, maturation and ovulation, induce female animals to estrus and be ready for breeding.

1.2.1.2 Methods of estrus induction

The hormones involved in inducing estrus mainly include follicle-stimulating hormone (FSH), luteinizing hormone (LH), pregnant mare serum gonadotropin (PMSG), human chorionic gonadotropin (hCG), gonadotropin-releasing hormone (GnRH), oxytocin (OXT), estrogen (E_2) and its analogs, progesterone (P_4) and its analogs, prostaglandin (PG) and its analogs, and sexual pheromones.

FSH, LH, PMSG and hCG can directly promote the follicle development, maturation and ovulation of female animals. Among them, FSH and PMSG are the first hormone preparations to promote follicle development. Estrogen can induce obvious estrus performance in females, but follicle development and ovulation activity are usually lacking on the ovary, and breeding must wait until the next estrous cycle. Progesterone can inhibit estrus and ovulation in animals, however, estrous and ovulatory activity will occur when progesterone is abruptly withdrawn after 12-16 d of continuous use of progesterone in estrous females. Prostaglandins have the effect of degenerating the corpus luteum, which can relieve the inhibitory effect of progesterone on estrous activity, and produce the effect of inducing estrus. GnRH can promote the secretion of FSH and LH from the pituitary gland, so it has an effect similar to LH or hCG, and is often used to induce estrus and ovulation. The abnormal release of GnRH from the hypothalamus is considered to be the cause of ovarian cysts, and exogenous GnRH stimulates the release of LH from the pituitary gland, which can be used to treat ovarian cysts. OXT has the effect of degenerating the corpus luteum, and thus has a similar ovulation induction effect to prostaglandins.

Male pheromones are also reproductive hormones that induce estrus in females of the same species. The "boar taint" in pork is due to the concentration of boar pheromones in body fat. The α and β androstenone derived from the testes and the saliva are two pheromones secreted by the boar. The "ram effect" that induces estrus in ewes is also the result of the action of pheromones.

In addition, management factors are also important factors that affect the estrus of animals. This is because the nervous system and endocrine system of female animals are affected by management factors. Therefore, in addition to hormonal treatment, the nervous system also induces the production of estrus signs. For example, changing light conditions, giving male livestock stimulation or weaning stimulation (termination of lactation), etc., can be used to induce the animals to initiate estrus. For young females or house-raised livestock, it is possible to stimulate the livestock to come into estrus by strengthening exercise. This effect of neurostimulation is achieved through modulation of the hypothalamic-pituitary-ovarian axis.

1.2.2 Simultaneous estrus technology

The technique of making a group of females come into heat and ovulate in the same period is called synchronization of estrus, also called estrus synchronization. Simultaneous estrus is a superficial phenomenon, while synchronization of ovulation is the inner manifestation and essence of synchronous estrus. In animal production, inducing a batch of female animals to be in estrus at the same time in the same week or a few days is also called simultaneous estrus; in the embryo transfer process, when using frozen semen for breeding and fresh embryo transfer, it is generally required that the difference in estrus time should not exceed one day. Therefore, the effect and accuracy of synchronous estrus must be strictly controlled; if frozen embryo transfer is used, synchronous estrus is not necessarily required.

1.2.2.1 Significance of estrus synchronization

Simultaneous estrus technology has the following significance in modern animal husbandry.

1. Conducive to the promotion of artificial insemination technology

Conventional artificial insemination requires estrus identification for each female animal, which is time-consuming and labor-intensive for large-scale farms with large breeding stocks, which is not conducive to promotion. The use of simultaneous estrus technology combined with timed artificial insemination (TAI) technology can save the intermediate step of estrus identification, reduce missed mating caused by dark estrus, and improve animal production efficiency.

2. Facilitate management and improve efficiency of animal husbandry

Modern animal husbandry production involves a large-scale breeding. Simultaneous breeding, pregnancy, parturition, fattening and slaughtering can be achieved by using the simultaneous estrus technology. Production activities can be centrally managed to improve production efficiency and reduce production costs. At the same time, animals can be marketed in batches or crops to improve the uniformity of commodity groups and boost sales.

3. Improve the reproductive rate of low-reproductive herds

For low-reproductive herds, such as cattle and water buffalo in southern China, the reproductive rate is generally lower than 50%. Excessive labor and other measures often fail to restore the normal estrus cycle a long time after childbirth, so induction of estrus, breeding, and conception at the same time can improve the reproductive rate.

4. Simultaneous estrus is the basis of embryo transfer technology

When using fresh embryo transfer, a donor can obtain more than ten embryos, which requires a certain number of recipient dams that are in estrus at the same time as the donor dam. In addition, sometimes embryo production and transfer are not carried out in the same place, and it is also necessary to synchronize the estrus of the recipient and the donor in different places to ensure the successful embryo transfer.

1.2.2.2 Technical principles

In a female animal's estrous cycle, the luteal phase accounts for about 70% of the entire estrous cycle, for example, the luteal phase of cattle is about 15 d. Progesterone secreted by the corpus luteum has a strong inhibitory effect on follicle maturation. Only after the corpus luteum subsides and the progesterone level drops, can the follicles develop to the mature stage, and the female animal show estrus and ovulate. Therefore, controlling the ebb and flow of the female animal's luteal phase is the key to controlling the female animal's estrus during the same period.

In the natural state, the estrus of a single female is random, and in a large group of non-pregnant adult females, about 1/21 of the females are in estrus every day. The randomness of this estrus can be changed by artificially controlling the ebb and flow of the corpus luteum on the ovary, so that a group of domestic animals can enter estrus and ovulate according to people's wishes. Artificially prolonging the luteal phase or shortening the luteal phase are currently two technical approaches used for simultaneous estrus.

1. Synchronous estrus method to prolong the luteal phase

A herd of dams was treated with progestin simultaneously, during which period luteal regression on the ovaries of the dams was performed. Due to the action of exogenous hormones, follicle development is inhibited and can not mature. If the exogenous progestin is treated for too long, the corpus luteum of all dams will regress and no follicles will develop to maturity during the treatment. All female animals release the inhibition of progesterone at the same time, and they can go into estrus at the same time.

At present, progesterone treatment is usually used for 9-12 d for sheep. After treatment, the corpus luteum of all dams can not be regressed. Therefore, a certain dose of estrogen can be given at the beginning of progesterone treatment to accelerate luteolysis, shorten the luteal phase, and improve the follicle development.

2. Synchronous estrus method to shorten the luteal phase

The most effective way to eliminate the corpus luteum on the ovary of the female animal is to use prostaglandins and their analogs (PGs). After the female animals are treated with PGs, the corpus luteum regresses, the follicles mature, and they come into estrus.

The sensitivity of various livestock to PG is different. The corpus luteum of sheep must be sensitive to PG on the 4th day after the last ovulation, the corpus luteum of cattle must be sensitive to PG on the 5th day after the last ovulation, and the corpus luteum of pigs is only sensitive to PG after the 10th day after ovulation, so the theoretical estrus rates of sheep, goats, cattle and pigs after one PG treatment are 13/17, 13/21, 16/21 and 11/21, respectively.

With the use of PG twice treatment method, the disadvantage that some female animals can not be in estrus at the same time in one treatment can be overcome. Usually, the second treatment is done 9-12 d after the first treatment. It is used for simultaneous estrus of cattle and sheep. This method may induce higher simultaneous estrus and insemination rates.

1.2.2.3 Hormone application

1. Hormones that inhibit follicular development

Hormones that inhibit follicular development include progesterone, medroxyprogesterone, flugestone, clogestone, megestrol, and 18-methyl norethisterone. The effectiveness of these drugs can be divided into long-term (14-21 d) and short-term (8-12 d), generally not more than one normal estrous cycle.

2. Luteolytic hormones

Prostaglandin $F_{2\alpha}$ ($PGF_{2\alpha}$) and its analogs (such as cloprostenol) have significant luteolytic effects, and are only effective in female animals in the luteal phase when used for simultaneous estrus treatment.

3. Hormones that promote follicle development and ovulation

The use of estrus drugs at the same time, if combined with gonadotropins, can enhance estrus synchronization and increase the estrus rate, and promote better follicle maturation and ovulation. Commonly used drugs of this type are PMSG, hCG, FSH, LH, GnRH, and chlordiazepoxide.

1.2.2.4 Treatment procedure for animals

1. Cows

(1) Long-term (12-16 d) administration of progesterone is mainly conducted through vaginal embolization and subcutaneous implantation. Within 5 d of treatment, most female animals showed signs of estrus.

(2) PG dosing schedule Given once intramuscular injection of 0.2-0.4 mg of cloprostenol or 1-2 mg of $PGF_{2\alpha}$, or two injections of $PGF_{2\alpha}$ or its analogs at intervals of 9-12 d, most cows can ovulate within 2-5 d after drug treatment.

(3) Progesterone-PG program The females were first treated with progesterone for 7 d, and PG 0.4 mg intramuscularly or 0.2 mg intravaginally was injected 1 d before the withdrawal of the progestin. Cows can be in heat and ovulate within 2-3 d after treatment.

(4) PG-GnRH program First, give intramuscular injection of GnRH 500 μg, 6 d later, intramuscular injection of cloprostenol 0.4 mg.

2. Sheep

(1) Progestorone-eCG program Progesterone vaginal suppository was first used for 12-14 d, and then 1-2 d before the suppository was withdrawn, 350-800 IU eCG was injected depending on the body weight, and the ewes were in estrus within 2-3 d.

(2) $PGF_{2\alpha}$ program In the breeding season, one or two intramuscular injections of cloprostenol are given, and the dosage is generally 0.1 mg.

3. Pigs

At present, the simultaneous weaning method is usually adopted for the induction of estrus in lactating sows, and estrus generally occurs within 3-9 days after weaning. If combined with injection of cloprostenol 0.2-0.3 mg at weaning, the estrus rate can reach 85%-90% within 1 week, or injection of eCG 400-800 IU or injection of CSG600 or PG600 (eCG 400 IU + hCG 200 IU).

Progesterone drugs (such as Altrenogen) can be used for estrus treatment of sows at the same time. For example, if Altrenogen is taken orally through feed every day, the dose is 12.5-15 mg/(head · d), and the drug is continuously used for 14-18 d, the estrus rate within 3-8 d after the treatment is over 90%, but ovarian cysts incidence is higher.

How simultaneous estrus technique can improve reproductive estrus rate remains to be studied, to enable ovulation to be normal and ovulation more concentrated, so as to formulate a set of comprehensive technology, implementation plan and work schedule including drug administration and insemination, which is conducive to the implementation of timed artificial insemination.

1.3 Ovulation control technology

The process of ovulation is initiated soon after exposure of the follicle to the LH surge, and it quickly becomes independent of subsequent gonadotropin control (the relationship of reproductive hormones, estrus, and ovulation in pigs is shown in Fig. 1-9. Ovulation control includes two aspects: controlling the time of ovulation and controlling the number of ovulations. Although estrus control (inducing estrus and simultaneous estrus) also controls ovulation, the ovulation time control mentioned here refers to the use of exogenous ovulation-stimulating hormone to induce ovulation to replace the natural ovulation that occurs under the influence of gonadotropins in the body. Ovulation can also be said to trigger the premature rupture of mature follicles to release the egg.

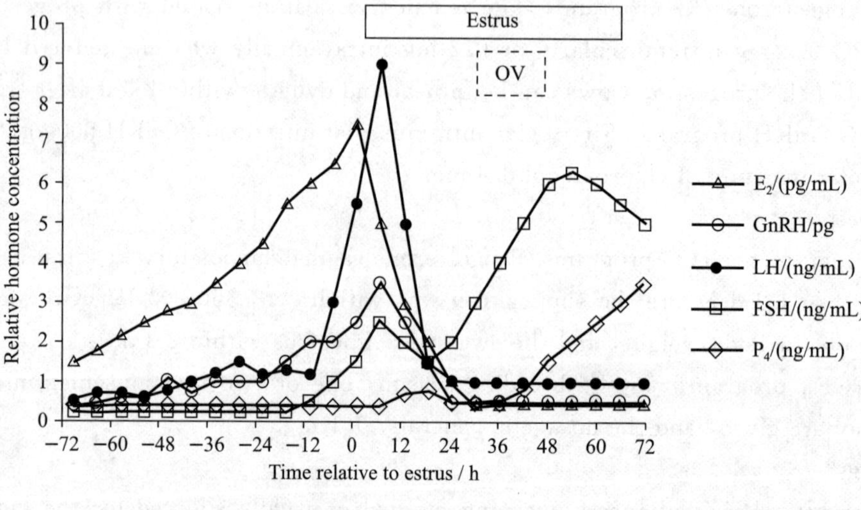

Fig. 1-9 The relationship of reproductive hormones, estrus, and ovulation in pigs(Knox, 2015)
(Representative hormonal profile of LH, FSH, estradiol, progesterone, and GnRH during proestrus in relation to the timing of estrus and OV in the mature female pig)

Ovulation number control refers to the use of exogenous hormones to increase the number of ovulations. There are two situations: first, when performing embryo transfer, superovulation is performed on the donor; second, the number of ovulations is appropriately increased in a restricted manner to achieve the purpose of producing multiple births, such as increasing the number of ewes from a single birth to a single birth, to produce twins, or to increase what would normally be a twin to a triple. Superovulation and induced multiples differ in the number of ovulations. There are more of the former than the latter, and their purposes are also different. Transplantation must be carried out after superovulation, and the induced multiple birth is a natural pregnancy.

1.3.1 Ovulation timing control and Ovsynch

1.3.1.1 Ovulation timing control

Control of ovulation is generally performed by hormonal treatment when mature follicles are present, prior to estrous behavior. It is important to control the time of ovulation treatment before the endogenous LH peak appears, otherwise the ovulation time will still be affected by the LH peak in the body and will not be advanced, resulting in: ①no estrus behavior; ②discharge of immature oocytes instead of mature eggs; ③no ovulation at all, or follicle luteinization.

The main hormones that induce ovulation are hCG, LH, GnRH and their analogs. Dosage of hCG: 250-500 IU for sheep, 500 IU for pigs, 500-2 000 IU for cattle, LH can also be used. The time from hCG injection to ovulation is similar to the time from the onset of natural estrus (reception of mating) to ovulation.

For pigs, exposing sows in heat to more sexual stimulation from the boar, or doing double mating can have the same effect.

1.3.1.2 Technical principle of Ovsynch

One of the hormonal protocols is Ovsynch, which enables artificial insemination (AI) to be performed at the precise optimum time without control of the ovaries and uterus. Ovsynch-timed artificial insemination (TAI) technology regulates the estrus cycle of animals such as cows by applying reproductive hormones to promote estrus, thereby improving the breeding efficiency of the animals.

The technology is based on the synchronization of estrous cycle and ovulation, the former being the basis of the latter. Taking dairy cows as an example, after weaning, prolactin (PRL) begins to decrease, and GnRH secretion begins to increase. This is the basis for restarting the growth and development of the follicle. Ovsynch technology uses this principle to inject GnRH on any day of the cow's estrous cycle to promote the synthesis and secretion of pituitary FSH and LH; next, the follicle development will be synchronized, the corpus luteum will be dissolved by injecting $PGF_{2\alpha}$, and the content of progesterone in the plasma will be reduced, so that the inhibitory effect of progesterone on follicular development will be released. At this time, the dominant follicle begins to develop, thereby achieving the synchronization of follicle development and ovulation. A typical Ovsynch proposal for cows is that injections of GnRH be administered 7 d before and 48 h after an injection of $PGF_{2\alpha}$, and cows are inseminated 16-24 h after the second injection of GnRH (Fig. 1-10).

In addition to the typical Ovsynch, there are other Ovsynch proposals for dairy cow and

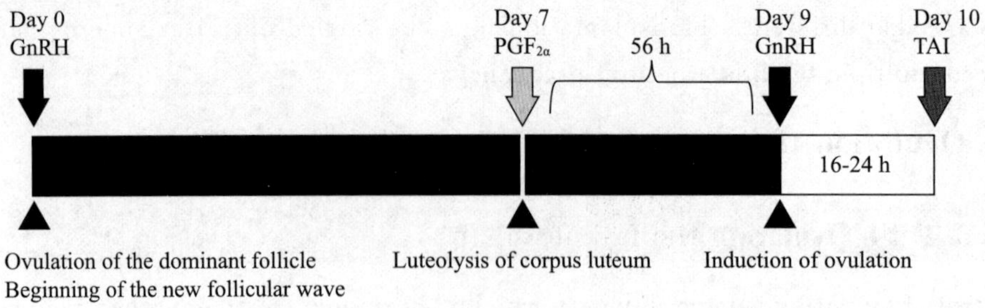

Fig. 1-10 One example of Ovsynch protocol(Nowicki et al, 2017)

beef, such as Ovsynch-48, 56, 72, Cosynch-CIDR and PRID-$PGF_{2\alpha}$-eCG (Feng, 2021), which can promote the control efficiency of ovulation time.

1.3.2 Ovulation count control

1.3.2.1 Hormone induction

In order to control the ovulation number of cows in the same estrus period, Yang et al. (1991) proved through three experiments that injecting an appropriate dose (1 800-2 000 IU) of PMSG can achieve the expected effect, which is much better than the use of large doses of PMSG reported in the literature. Compared with other methods, it can save cost and reduce the adverse effect of PMSG on ovary. Because PMSG remains in the body for a few days after the cow is injected with PMSG, it can stimulate the ovary for a long time, make it swollen, and secrete a large amount of gonadotropins and gonadal steroid hormones, which not only affects ovulation, but also interferes with the normal operation of the cow after pregnancy. In cows with PMSG induced ovulation, rabbit anti-PMSG serum can be injected at the beginning of estrus, which can neutralize the residual PMSG in the body. In this way, it can not only promote follicle ovulation, reduce the number of unovulated large follicles, but also accelerate the cystic ovary. The recovery of estrus is manifested in: ①shortening of estrus duration of cows; ②after breeding, the level of gonadal steroid hormones in the body quickly returns to normal, which is beneficial to maintaining a constant intrauterine environment of twin-pregnant cows.

1.3.2.2 Hormone immunization

Yang et al. (1990) used imported testosterone-3-carboxymethyl oxime and estrone-3-hemisuccinate to connect with bovine serum albumin by mixed acid anhydride method to obtain testosterone and estrone active immunization preparations. Two herds of Soviet Merino ewes were treated with these two immunization formulations on two consecutive occasions 6 and 3 weeks prior to the breeding season. The results showed that the percentage of ewes producing double lambs after active immunization with testosterone and estrone increased by 320.1% and 103.2%, respectively, and the lambing rate was increased by

42.7% and 35.4%, respectively.

Yang et al. (2002) used inhibin extracted from pig semen to actively immunize cattle, and the 2 mg inhibin immunization group induced a twinning rate of 26.3% (15/57, n= 247). Morris et al. (1995) reported that they treated a total of 25 Heads of Heyford with five synthetic fragments of bovine follistatin α-subunit (amino-terminal 8-20, 8-30, 63-72, 107-122, 153-167). Crossbreeding heifers resulted in 7 twins with a twin rate of 28% (7/15). In this study, the twin rate (26.3%, 15/57) of active immunization cows with semen inhibin was close to the effect of active immunization with follistatin.

1.3.2.3 Hormone gene immunization

It refers to the technology of transforming the expression vector with the target gene into livestock by a certain method, inducing the body to produce specific antibodies and causing corresponding immune responses, also known as nucleic acid immunity or DNA immunity. Such vaccines are called genetic vaccines or naked DNA vaccines. The specific method is to fuse inhibin α (1-32) with a carrier protein and insert it into a eukaryotic expression vector to immunize animals.

1.3.2.4 Moderate superovulation method

Superovulation is used to generate embryos for collection and transfer to living recipients or cryopreservation to store the desired strain. Induction of twins and induction of estrus are two different concepts and also different from superovulation. Induction of twins is mainly used in sheep and dairy cows of single-parity breeds. After the treatment, breeding is carried out to make ewes and cows pregnant normally. In sheep, PMSG (or FSH) is usually injected 4 d before the sheep is expected to come into estrus (the 12th to 13th day of the estrous cycle), or treated with progesterone for 14 d, and then injected with PMSG (300-700 IU) and hCG (200 IU). This can also be done in anestrous state, but the dosage of hormones needs to be appropriately increased. In the past, it was more difficult for cows to have twins. If there is one male and one female, the female is infertile and has no breeding value, which is called the Freemartin phenomenon. The use of sexed semen now solves this dilemma by achieving a female rate of over 90%. It can be seen that the technology of inducing twins needs to be used in combination with the technology of sex-controlled semen in order to achieve the expected purpose.

The amount of ovulation is not only directly related to the amount of gonadotropins, but also affected by many factors. In order to achieve 2-3 eggs from the ewe, it is difficult to ensure that the ewe has 2 eggs and no more and no less than 2 eggs. When using PMSG, it also needs to be matched with antiserum to neutralize excess PMSG in a timely manner.

In practice, it can be considered a success if more than half of the treated ewes can increase the number of lambs, such as a 50% increase in the lambing rate.

In general, the method of inducing ovulation is still blemished with the difficulty of prediction due to the accuracy and reliability of the results, so it has not yet been supported by conventional technical measures, and there are many problems that need to be studied and explored.

Assignments

1. What are the methods of estrus detection?
2. What is the significance of estrus and ovulation control?
3. What are the commonly used hormones during estrus control?
4. What are the methods of estrus control?
5. What are the principles and methods of estrus synchronization?
6. What is the ovulation control?
7. Please describe one example of Ovsynch protocol.

(Wei Quanwei)

Code 1-1 Excerpt from the Report to the 20th National Congress of the Communist Party of China

Chapter 2 Fertilization Technology

Summary

Fertilization technology includes artificial insemination (AI) and in vitro fertilization (IVF). AI is one of the important practical techniques in modern animal husbandry production, which can significantly improve the utilization rate of outstanding breeding stock and speed up the breed improvement. The basic technical steps of AI include semen collection, quality inspection, dilution, packaging, preservation and insemination. IVF is a technique of assisted reproduction in which the fertilization of an egg by sperm is achieved under suitable conditions outside the body through artificial operations. The basic technical steps of IVF include in vitro maturation (IVM) of egg, in vitro capacitation of sperm, in vitro fusion of egg and sperm, and in vitro culture (IVC) of the early embryo, etc. This chapter mainly introduces the technical steps of AI and IVF; in addition, the microfertilization technique is also introduced.

Fertilization is the process by which sperm and egg interact and combine to produce a new individual (zygote) either *in vivo* or *in vitro*. Fertilization techniques include artificial insemination (AI) which is *in vivo* fertilization and *in vitro* fertilization (IVF). AI is a method of mating in which semen is collected under artificial conditions, and then transmitted to the reproductive tract of estrus females after quality inspection, dilution and preservation. IVF is a technique that sperm processed specially fertilizes oocytes *in vitro*. Babies born after embryo transfer in IVF are also known as "test-tube babies". Compared with AI, IVF requires less sperm and improves the efficiency of semen utilization. The completion of IVF requires mature sperm and eggs and culture conditions conducive to the survival and metabolism of sperm and egg.

2.1 Artificial insemination

In 1780, the Italian physiologist Spallanzani carried out the first artificial insemination (AI) of dogs. From the 1940s to the 1960s, AI developed rapidly in developed countries such as the Soviet Union, Britain, Denmark, the Netherlands, the United States of America, Canada and Japan, and was used in a variety of livestock and poultry production. In 1949, British scientist Polge succeeded in freezing chicken semen. In 1950, British scientists Smith and Polge developed the cryopreservation technology of bovine semen. Frozen semen has been widely used in AI since 1960. In 1995, Pursley et al. invented the synchronous ovulation and timed artificial insemination (TAI) of dairy cows, and the classic processing program is GnRH-$PGF_{2\alpha}$-GnRH-AI, also known as Ovsynch/TAI. Since then, TAI schemes such as double-Ovsynch, Cosynch, Presynch-Ovsynch and PRID/CIDR-Ovsynch emerged, which solved the problems of low participation rate and conception rate of dairy cows caused by adverse estrus monitoring and reproductive diseases.

2.1.1 Advantages and disadvantages of artificial insemination

1. Maximize the utilization rate of excellent breeding male animals (poultry)

With the use of AI technology, once they ejaculate, male animals (fowls) can mate with females (fowls) dozens, if not hundreds, of times as many as they would naturally mate with.

2. Accelerate breed improvement of livestock and poultry

AI technology, especially the use of frozen semen, greatly improves the breeding ability of male animals (poultry), thus rapidly expanding the genetic genes of excellent breeding male animals (poultry), and rapidly improving the production performance of their offspring, thereby accelerating breed improvement of livestock and poultry.

3. Significantly reduce the number of breeding male animals (poultry)

After the use of AI technology, due to the great improvement of the utilization rate of breeding male animals (poultry), only a very small number of excellent individuals need to be retained to meet the reproductive needs, thereby saving a large number of male animals (poultry) feed and management costs.

4. Overcome the mating difficulties caused by the disparity in the size of male and female animals (poultry)

Breeding male animals (poultry) are generally larger, and there are many obstacles to mating with small local females (poultry), and the use of AI techniques can overcome this problem.

5. Control the spread of disease

AI prevents the spread of infectious and other diseases related to sexual intercourse by avoiding direct contact between male and female animals (poultry). Insemination is the last link of AI, and one of the most important techniques, and the timely and accurate delivery of semen to the appropriate part of the reproductive tract of the female animal is the key to ensuring fertility.

However, AI technology also has some shortcomings. The skilled technicians who strictly abide by the operation specifications are required, and their great advantages can be brought into play. If the operation technology is not standardized, it will not only affect the pregnancy rate, but also spread diseases. In addition, the semen freezing technology of some animals (such as pigs, dogs, etc.) (Parrilla et al., 2019; Hermansson et al., 2021), especially special economic animals and wild animals, has not yet passed the test, so further research is needed.

2.1.2 Brief introduction of development history for artificial insemination technology in livestock and poultry

AI revolutionized the livestock and poultry industry worldwide. AI has a long history in animal production, initially used in 1748, and AI of poultry began in 1930 (Ouyang et al., 2021). For several decades, the application of AI remained obscure due to high-cost and inefficiencies, but with advancements, AI revolutionized the animal breeding industry worldwide. Intra-uterine insemination (IUI) technology is used in humans to treat fertility problems. AI is the most common ART used in dairy cattle (80% in Europe and North America) and pigs (90% in Europe and North America). AI is used in other livestock species like horses, donkeys, buffalo, deer, and goats, poultry species including chicken, turkey, goose, guinea fowl, and chicken, and small domestic animals like cats and dogs.

Horse AI in China was first carried out on Jurong horse farm, Jiangsu Province, in 1935. In the 1960s, the AI of dairy cows was popularized. Since 1977, the frozen semen of cows has been popularized throughout the country (Ouyang et al., 2021). The number of crossbreeding cattle improved by AI technology is more than ten million every year (Froman and Feltmann, 1998). The AI technology of buffalo and yak is also popular. The scale of AI and the research level of frozen semen of sheep in China rank first in the world. AI has been popularized in Xinjiang, Inner Mongolia, and other northern provinces where sheep breeding is a developed endeavor (Manafi, 2011). AI of pigs began in Guangxi in the 1950s. At present, an academic conference on AI of pig is held every year. Both the level of AI technology and the number of inseminated sows are among the highest in the world. In addition to the above domestic animals, AI of camels, deer, dogs, rabbits, pandas, foxes and other special economic animals and some wild animals in China have also been successful (Manafi, 2011). It is worth mentioning that the Giant Panda breeding team led by researcher Hourong in Sichuan Province of China has taken the lead in the world in terms of Giant Panda breeding, especially in the technology of combined application of electric stimulation sperm collection and rectal massage of the ampulla of the vas deferens to improve the success rate and amount of sperm collection from Giant Pandas (Wang et al., 2020; Cai et al., 2018).

In recent years, China has issued and implemented a series of relevant standards and technical regulations on AI of livestock, such as the national standards for frozen semen of goat (GB 20557—2006), frozen semen of cattle (GB 4143—2022), technical regulations for AI of pigs (NY/T 636—2021), semen collection instrument for cattle and sheep (NY 530—2002), semen injectors for animals (NY/T 531—2002), AI straws (NY 1181—2006), technical regulations for the production of frozen semen of cattle (NY/T 1234—2006), technical regulations for AI of cattle (NY/T 1234—2006), bovine frozen sexed-semen (GB/T 31582—2015), technical specification of AI in dairy cattle with frozen sexed semen (NY/T 3646—2020), frozen semen of horses and donkeys (NY/T 3796—2020), technical specifications for AI of yaks (NY/T 3797—2020), etc. The implementation of the above standards and technical regulations has significantly improved the quality of frozen semen of livestock and poultry and the pregnancy rate of AI, and played a positive role in promoting the expansion and improvement of improved livestock and the healthy development of livestock and poultry industry in China.

2.1.3 Technical process of artificial insemination

2.1.3.1 Semen collection

1. Preparation before semen collection

(1) Preparation for semen collection site Semen collection should have a good and fixed place and environment to establish a solid conditioned reflex for male animals, ensure the safety of humans and animals and prevent semen pollution. For this reason, the semen collection site should be spacious, flat, quiet, clean, and fixed, and the semen collection rack (or breeding rack) or the false dummy animals for male animals to climb must be strong, and the placement should be convenient for the entry and exit of the male animals and the operation of the semen collectors. The floor should be flat and not too smooth, and it is best to lay a rubber mat to prevent slippage. The place should be cleaned before semen collection and equipped with a device to disinfect and sterilize by ultraviolet radiation.

The ideal semen collection site should be equipped with indoor and outdoor parts connected to the semen processing room. The size of the indoor semen collection site for large livestock is generally 10 m × 10 m and 4 m × 5 m for pigs and sheep.

(2) Preparation for the dummy for semen collection The dummy for semen collection can be either a live animal, i.e. "real dummy animal", or a shelf made equipment, i.e. "fake dummy animal". Female animals healthy, strong, moderate in size and gentle in temperament should be chosen as real dummy animals, of which estrus females are the best. Male and castrated animals of the same species can also be used as dummy animals. According to the size of female animals, the false dummy animals are made of steel pipes or materials, and then wrapped with a layer of animal skin, gunny bags, artificial leather and other flexible and soft items on their surface. It is a convenient, clean, safe and effective method to collect semen from animals on artificial platform, which is suitable for all kinds of livestock.

(3) Preparation for the male animals Before collecting semen from the male animal, the method of seduction must be used to entice the male animal to have sufficient sexual excitement and sexual desire, especially for the male animal with slow sexual arousal, it is

necessary to change the dummy, change the position and watch other domestic animals mounting and straddling. For the use of fake dummy animal to collect semen, the male animal must be trained to establish conditioned reflexes.

Male animals must be trained to climb over the dummy platform (Fig. 2-1). During training, vaginal mucus or urine of estrus female animals can be smeared on the hindquarters of the dummy, or an estrus female animal can be placed next to the dummy animals to stimulate the male animals' sexual desire and induce mounting, but mating is not allowed. Once the male animal mounts onto the female animal, it should be pulled down, which should be repeated for several times. When the male animal's sexual desire reaches a climax, quickly take away the female animal, and then let it climb across the dummy. At this time, if the surface of the dummy animal is coated with vaginal secretions of estrus females, the effect of semen collection is more ideal.

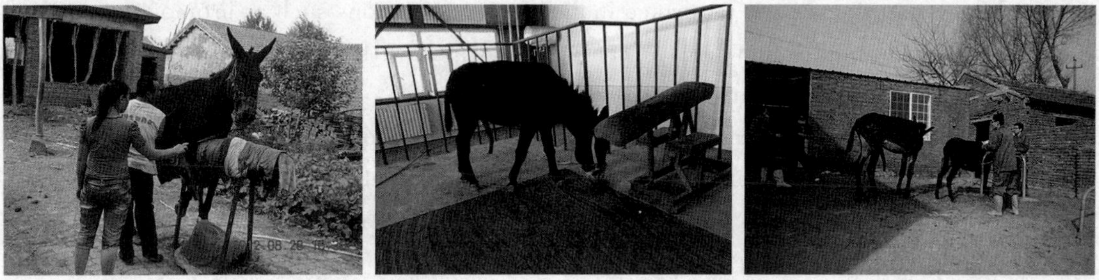

Fig. 2-1 Training of Dezhou donkey breed to climb over the dummy before collecting semen

In addition, when training boars, you can use another boar to climb over the dummy, or apply the urine of other sperm collecting boars to the fake platform pig to induce the tuned boar to contact and cause crawling reflexes. In the process of training, operators should be skilled, patient, and not apply undesirable stimuli such as persecution, whipping, and intimidation so as not to cause difficulties in training. After the first successful semen collecting, it is necessary to go through several iterations, and pay attention to the regular collection of semen in the non-breeding season, so as to consolidate the conditioned reflex established by the male animals. Attention should be paid to the cleanliness and hygiene of the external genitalia of the male animals during training; The best choice is to train in the morning when animals are full of energy and sexual desire. The time and place of training should be fixed, and the time of each training should not be extended. Boar training may begin at 7 months of age, the duration is about 4-6 weeks, up to once a day, 3-4 times a week, once lasting 15-20 min. The general desire of good reserve pigs 1-2 times can be successfully trained, not too strong 7-8 times. After successful training, it is necessary to collect semen for another 2 consecutive days.

(4) Cleaning and disinfection of equipment All equipment used for semen collection should be clean and sterile and disinfected strictly before use and washed after each use. The traditional detergent is 2%-3% sodium bicarbonate solution or 1%-1.5% sodium carbonate

solution. Soap or washing powder is often used instead in grass-roots units, but the safety is not as good as the former. After the equipment is washed with detergent, it must be rinsed with water several times without leaving a trace and then strictly disinfected before use. The disinfection method varies depending on the material of each vessel.

① Glass equipment is disinfected using an electric blast drying box for high temperature drying and disinfection, the temperature is required to be 130-150 ℃ and maintained for 20-30 min, until the temperature drops below 60 ℃, it can be taken out of the box to use. Autoclaving can also be sterilized for 20 min.

② Rubber products are generally disinfected with 75% alcohol cotton balls, and it is best to wipe with 95% alcohol cotton balls once to accelerate the volatilization of residual moisture and alcohol odor on the rubber, and then rinse with normal saline. For pigs, horses and donkeys, soak in boiling water for 3-5 min, and then rinse with normal saline.

③ Metal instruments can be soaked in disinfectant solutions such as bromogeramine and then rinsed with normal saline, etc. It can also be wiped with a 75% alcohol cotton ball or disinfected with an alcohol lamp flame.

④ Solutions such as lubricants and normal saline can be boiled in water for 20-30 min, or disinfected with autoclave steam. The container cap should be removed or inserted with a large injection needle, and the bottle port is bandaged with gauze to avoid bursting.

⑤ Other supplies, such as cotton, gauze, tampons, towels, cork, etc., can be disinfected by steaming or autoclaving.

(5) Preparation for the artificial vagina A pseudo-vagina is an artificial vagina designed to mimic the environmental conditions in the vagina of female animals. Although the artificial vaginas used by various domestic animals are different in shape, size, and type, the design principle and basic structure are the same, that is, it is composed of basic components such as the outer cylinder (also known as the shell), the inner tube, the semen collection cup (bottle, tube), the nozzle and the fixed rubber ring (Fig. 2-2). In addition, cattle should also have a set of fine cup protective sleeves or set of fine funnels; pigs must have a set funnel and, at the same time, have a double balloon. There are two types of bovine pseudo-vaginas, European and Soviet. The Soviet-style collecting cup is a double-layer design of insemination, which is suitable for use in low temperature weather and cold areas (Fig. 2-3). The artificial vaginal outer tube of the male rabbit can be made by a bamboo tube with a length of 4-5 cm and an inner diameter of 3.5-4 cm, and the inner tube can be modified with a condom or rubber finger sleeve.

Before the installation of the artificial vagina, the outer barrel and inner tube should be checked for broken cracks, trachoma, aging and stickiness, otherwise there will be water leakage and air leakage and affect the semen collection. The artificial vagina must be washed, tube installed, disinfected, rinsed, watered, lubricant applied, and temperature and pressure adjusted before use. The temperature of the artificial vaginal inner tube is generally controlled at 38-40 ℃, which can be maintained by injecting warm water equivalent to 2/3 of

Fig. 2-2 Appearance and installation of the artificial vagina for horse/donkey semen collection

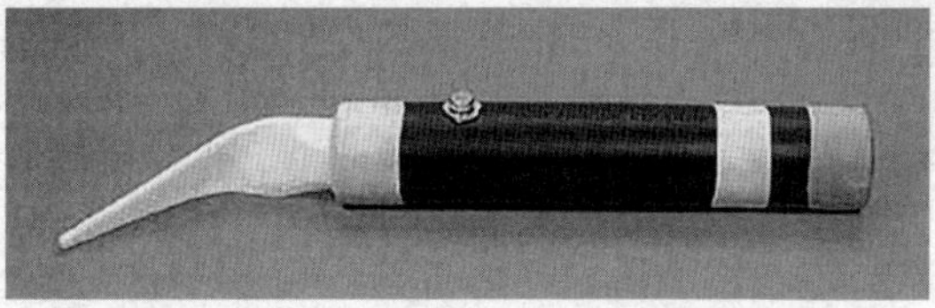

Fig. 2-3 Physical appearance of the bovine artificial vagina

the volume of the inner outer shell of the artificial vagina. The semen collection cup should also be maintained at 34-35 ℃. Pressure in the fake vagina is regulated with the help of water injection and air. Insufficient pressure can not stimulate ejaculation, excessive pressure makes the penis difficult to insert and can not ejaculate after insertion. Generally, the pseudo-vaginal inner tube entrance with an X-shape or Y-shape is appropriate. Then lubricate the inner surface of the artificial vagina with a disinfectant lubricant, and the application site is 1/2-2/3 of the total length of the artificial vagina. However, the lubricant should not be applied too much, so as not to mix with semen.

2. Semen collection operations

(1) Collecting bull's semen using the artificial vaginal method(Fig. 2-4)

① Lead the breeding bull to the semen collection room, and let the breeding bull stay in the semen collection room for a while so that the bull can prepare before ejaculation, which is conducive to improving the quality of the semen collected.

② Lead the breeding bull to the side of the dummy animal and the semen collector holds an artificial vagina facing the dummy cattle and standing on the right posterior side of the dummy cattle.

Fig. 2-4 Collection of bull's semen by the artificial vaginal method

③ When the bull's penis erects and the animal mounts the cow or the dummy cattle, the semen collector holds the artificial vagina, quickly takes a step forward, tilts the artificial vaginal barrel down to the direction of the bull's penis, close to the right side of the cow's hip, with the left hand behind the foreskin opening, the palm of the hand supports the foreskin upwards (do not use the hand to grasp the penis, otherwise it will make the penis retract). Induce the penis to the right and into the artificial vagina. When the bull rushes forward with force, it means that the ejaculation is complete. After the bull ejaculates, the semen collector simultaneously tilts one end of the semen collection cup of the artificial vagina downwards so that the semen flows into the semen collection cup.

④ When the bull jumps down, the artificial vagina should move back with the penis and not withdraw. When the penis is protruded by the pseudo-vagina, the pseudo-vagina is immediately erect, the cylinder mouth is up, and the semen bottle is down and immediately sent to the semen processing room. After deflation, remove the semen cup and cover the lid.

(2) Collecting boar semen by hand Take out a clean semen collecting cup from the incubator (37 ℃). If the refining column is not close to the laboratory, place the refining container in the incubator.

The semen collector shall wear latex gloves on the one hand, a food glove on the other hand, a semen collecting cup (preferably a thermos cup in winter), and tie a double-layer sterile gauze at the port of the semen collecting cup to filter the semen and remove the colloid.

When the boar mounts, the semen collector squats on the right side of the semen collection platform, cuts off the long hair of the boar foreskin, squeezes the foreskin to drain the accumulated urine after the boar climbs, cleans it with clean water, and then wipes the foreskin with a dry towel or paper and discards the outer food gloves. When the penis is stretched out, hold the spiral penis head with a gloved hand to prevent it from rotating. At the same time, the fingers continue to press rhythmically, which can cause the boar to ejaculate. When boars ejaculate, their fingers stop pressurizing. When ejaculation stops, their fingers continue pressurizing until all semen is collected. Do not let go during ejaculation, do not collect the first part of clear liquid (about 25 mL), and keep the boar to complete the whole ejaculation process (3-5 min).

After the boar gets off the fake platform pig, throw away the gauze and gel used for filtration to ensure that the semen is not overturned. Send the semen to the semen processing room for semen quality inspection as soon as possible, and dilute and preserve it.

(3) Semen collection by the artificial vagina in rams Before semen collection, cut off the excess long hair around the ram penis, wash the foreskin with 0.1% potassium permanganate solution, squeeze out urine and other residues in the foreskin cavity, and wipe them dry. After pulling the ram to the semen collection site, do not let it climb across the platform immediately, but control it for a few minutes to continuously strengthen the ram's sexual excitement. When the penis is fully erect and stretched out, let the ram mount the platform, so as to improve the semen quality. During semen collection, the semen collector holds the back end of the artificial vagina with his right hand, fixes the semen collecting cup (bottle), squats down on the right rear side of the dummy with the air nozzle piston facing

down, and makes the artificial vagina close to the hip of the ewe. When the ram crosses the back of the dummy, and the penis has not touched the dummy, quickly guide the ram's penis into the artificial vagina with the left hand, pressing into the skin of the penis or massaging the penis. The ram's hindquarters rushing forward quickly means that ejaculation is over. At this time, move the artificial vagina backward with the ram action and quickly erect the artificial vagina. One end of the semen collecting cup is downward, then open the air nozzle on the piston to release air. Remove the semen collecting cup, add the bottle cap, and send it to the semen examination and treatment room for examination.

(4) Collection of horse (donkey) semen by the artificial vaginal method When collecting horse(donkey) semen, the sperm collector holds the handle of the sperm collection tube in his right hand, and leans the sperm collection tube against his right elbow. When the breeding male horse (donkey) climbs, he immediately stands at the right rear side of the horse (donkey), fixes the sperm collection tube at the rear end of the horse at an angle of 30-35 ℃, and quickly guides the penis into the sperm collection device. In general, after the penis is inserted into the sperm collector, it can ejaculate after several times of twitching. To judge whether a male horse ejaculates or not, we mainly observe the rhythmic contraction of tail muscles and the incitement of tail root and tail hair. After ejaculation, withdraw the penis from the semen collector, slowly tilt the end of the semen collecting cup of the semen collector downward, open the air release valve for air release, and quickly send the semen to the semen processing room(Fig. 2-5).

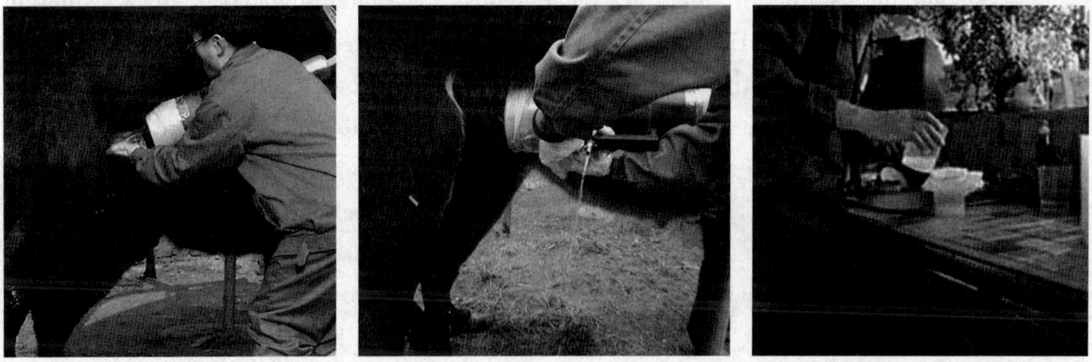

Fig. 2-5 Collection and filtration of donkey semen by the artificial vaginal method

(5) Semen collection of poultry Cut off the feathers around the anus of the selected and adjusted breeding roosters and disinfect them with 75% alcohol for standby.

Massage semen collection method is a commonly used method in China at present. It is also a semen collection method that can be used for chickens, ducks and geese. The massage method is divided into a single semen collection method and double semen collection method.

① Single person operation method: After doing a good job of self-disinfection protection, the semen collector sits on a stool and takes a breeding rooster and puts it between his two thighs. The rooster is required to face the semen collector with the tail slightly higher than the head. The right hand of the semen collector is placed on the abdomen of the breeding rooster. The middle finger and ring finger clamp the handle of the semen collecting cup. The mouth of the cup is downward to prevent pollution. At the same time, the left hand is naturally extended to massage the chicken by sliding from the back of the chicken to the tail of the chicken, so as to reduce the stress of the breeding rooster and cause sexual pleasure (Fig. 2-6).

The strength of the left-hand sliding massage should be appropriate, depending on the size of the breeding rooster. However, when the left hand slides to the tail, the index finger and thumb should gently pinch and press the cloaca from above, while the right thumb and index finger pinch and press the cloaca from below, alternating with the left and right hands. About 1 min later, the breeding rooster begins to have the ejaculation reflex (the ejaculation reflex begins in 8-10 s when the breeding rooster is sexually aroused). At this time, the right hand quickly turns the semen collecting cup, and the cup mouth faces the cloacal cavity to collect semen. If there is less semen, repeat the above operations, but prevent too much transparent liquid or even feces from entering the semen collecting cup to avoid polluting semen. The semen collection volume of cocks is generally 0.3-0.8 mL, up to 1-1.2 mL.

② Two persons operation method: The team is composed of a semen collector and an assistant. The assistant holds the keel with one hand and gently holds the chicken legs with the other hand. The semen collector massages the semen (the method is the same as the single semen collector method). As this method wastes labor and takes a long time, it will not be introduced in detail.

③ Female bird seduction method: When collecting semen by estrus fowl seduction method, the feathers around the anus of the breeding rooster are cut off and disinfected, and then operator A immobilizes the rooster. At this point, operator B holds the hen by the wing roots and belly and faces operator A with both hands. Operator B is required to place the thumb of both hands on the hen's wing roots at the same time and the other fingers on the hen's chest and belly so that the tail is slightly higher than the head. Operator A manually fixes the rooster about 10 cm from the ground. At this point, the hen's back is facing the breeding rooster. Finally, operator A releases the breeding rooster. The breeding cock will move closer to the hen when he sees the hen. When the breeding cock steps on the back of the hen for ejaculation, operator A can quickly put the prepared semen collecting cup around the hen's anus and wait for the opportunity to receive the semen from the breeding cock.

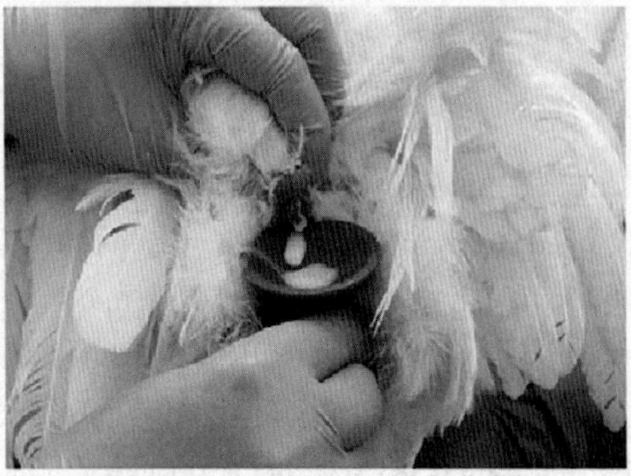

Fig. 2-6 Semen collection of cock

④ Notes: The water supply and feeding shall be stopped about 5 h before semen collection to prevent the feces from polluting the semen during semen collection. The semen quality of roosters who have not collected semen for a long time is not high, and they should be ejaculated first.

Because continuous semen collection consumes a lot of breeding cocks, it is necessary to rest for 1-2 d to recover the body condition during continuous semen collection for 3 d, so as to prevent the decline of semen quality.

The semen collector shall be fixed. It is forbidden to treat the breeding rooster roughly so as to avoid being scared and unwilling to expel sperm.

3. Semen collection frequency

Semen collection frequency refers to the number of semen collection times of male animals in a week. In order to maintain the normal sexual physiological function of breeding males, maintain healthy body condition, and maximize the ejaculation volume and semen quality, it is very important to reasonably arrange and utilize the semen collection frequency of breeding males. In production practice, adult bulls generally collect semen twice each day, with an interval of more than 0. 5 h each time. They can also collect semen three times a week and every other day. The sperm production of young bulls is 1/3-1/2 less than that of adult bulls, and the number of semen collections should be reduced. Rams about 1. 5 years old can be used to collect semen for the first time, usually 1-2 times/d; 2. 5 years old and above, they can be harvested 2-4 times/d, or the second time after the first time of semen collection at an interval of 5-10 min. The semen collection frequency of male horses should be done once every other day. If the production requires semen collection once a day, the semen collection should be stopped for 1-2 d after a few days of continuous semen collection within a week. The frequency for adult boar is twice a week or one five days, and for young reserve boar, once a week. Most boars are used for the given semen collection and breeding frequency. Once the habit is formed, increasing the semen collection frequency will not increase sperm production.

4. Semen quality evaluation

(1) Objective and principle of semen quality evaluation The purpose of semen quality inspection is to identify the quality of sperm and to serve as the basis for semen treatment. At the same time, it is also the basis for evaluating the breeding and management level and breeding value of male animals. In order to objectively and accurately evaluate sperm quality, it is necessary to avoid light, keep constant temperature, finish detection quickly and take representative samples. In order to ensure the objective and accurate inspection results, the following basic operating principles should be paid attention to during sperm quality inspection.

① Semen collected at a constant temperature and away from light should be quickly placed in a constant temperature water bath or thermos at about 30-37 ℃ to prevent a sudden drop in temperature, which may cause low-temperature damage to sperm. According to the specified requirements, pay attention to maintain the appropriate temperature around the working chamber (20-30 ℃) and the microscope or its stage (37-38 ℃). At the same time, try to avoid light exposure to reduce the impact of light on sperm.

② For quick inspection, all inspection preparations should be made in advance. The quality inspection should be carried out immediately after the semen is collected. It is required to act quickly and shorten the inspection time as far as possible to prevent the sperm quality from declining during the inspection.

③ The inspection process and environment shall not damage the quality of semen. For example, the glass rod dipped in semen and other utensils shall be disinfected and sterilized, and the disinfectant and its smell shall not remain.

④ The sampling shall be representative. Samples should be taken from all the semen collected and gently shaken or stirred evenly to strive for objective and accurate results.

⑤ Combination of routine key inspection and regular comprehensive inspection. There are many items for semen quality inspection, which usually adopts the method of combining routine key inspection one by one and regular comprehensive inspection, which can not only save time, but also comprehensively grasp the dynamic changes in semen quality.

⑥ Comprehensive analysis by synthesis, accurate evaluation of semen quality grade. Generally, the evaluation of semen quality can not be concluded by one or two indicators. Some items shall be repeated 2-3 times when necessary, and the average value shall be taken as the result. The evaluation of semen quality and the value of a male breeding should not be based on a few inspection results, but on multiple evaluation records at a certain stage.

(2) Visual examination of semen

① Semen volume. After semen collection, the semen in the collecting bottle is filled in a graduated test tube or semen bottle in an isothermal environment, and the volume of semen collection of male animals can be measured. Pig and horse (donkey) semen shall be filtered with 4-6 layers of sterile gauze or centrifuged to remove colloidal substances, and then the volume of semen collection shall be measured according to the above methods.

② Semen color. The semen of domestic animals is generally opaque milky white, sometimes pale yellow, which may be related to the higher content of β-carotene in the feed of male animals. The semen of cattle and sheep can be seen in the form of cloud and fog when they are just collected. The semen of pigs, rabbits, donkeys, and horses contains colloidal substances, which are the secretion of seminal vesicle glands. Colloidal substances may form uterine plugs during natural mating, which can prevent semen backflow and inhibit bacterial growth.

③ Semen smell. Normal semen has no odor and a slightly fishy smell. During the examination, gently stir the air above the semen container with your hands, and smell the semen with your nose.

④ Semen cloudiness. The semen of cattle and sheep has a high density, so when observed in a glass container, the semen rolls up and down like a cloud, which is called cloudiness. This is a sign of sperm motility. "+++" can be used to indicate the cloudiness significantly higher; obviously higher, it is indicated by "++"; if it is not clearly higher, it is indicated by "+".

⑤ Semen pH value. Immerse the pH test paper in semen for a moment, then take it out and compare it with the standard chromatography to determine the pH of semen. Table 2-1 shows pH value ranges of fresh semen from livestock and poultry.

Table 2-1 pH value ranges of fresh semen from livestock and poultry

Cattle & Sheep	Horse & Donkey	Swine	Chicken	Goose
6. 5-7. 1	7. 4-7. 5	7. 4-7. 5	7. 0-7. 6	6. 9-7. 4

(3) Microscopic examination

① Sperm motility assessment: Motility refers to the percentage of sperm that move forward in a straight line in the total number of sperm. Generally, it is evaluated by the ten grades scoring method of 0-1. For example, 80% of the sperm in semen move in a straight line, and the sperm motility is 0. 8; 70% of the sperm in semen move in a straight line, and the sperm motility is 0. 7, and so on. During the examination, several visual fields shall be observed and the average value shall be taken. Generally, sperm motility should be checked at each stage of semen treatment in order to understand the effect of semen treatment in time. The original semen of cattle and sheep and the concentrated semen of pigs have high density. For the convenience of observation, it can be diluted with normal saline before the examination. The temperature has a great impact on sperm motility. When checking sperm motility, put the microscope in the incubator and evaluate sperm motility at 38-40 ℃.

The methods of evaluating sperm motility include the plate pressing method, dead and alive staining method, etc.

Plate pressing method Take a drop of semen on a glass slide, cover it with a cover glass, place it on the constant temperature stage of the microscope at 38-40 ℃, magnify it by 200-400 times, and subjectively observe the semen sample. If the video microscope is used to reflect the sperm movement image from the screen, several people can observe and evaluate at the same time, which is conducive to improving the accuracy of the results.

Dead and alive sperms staining This staining is to distinguish dead and living spermatozoa and calculate the proportion of dead and living spermatozoa by taking advantage of the fact that living spermatozoa are not colored by the dyes (such as eosin and Congo red) and dead spermatozoa are easy to be colored due to the increased permeability of the surface facial mask.

Methods: eosin and aniline black were prepared into 5% and 1% solutions with normal saline, respectively; drop a drop of eosin and aniline black on the slide, respectively; drop a drop of semen into eosin solution and mix it well, then mix it with black aniline solution and make a smear for microscopic examination; the dead sperm is pink, but the living sperm is not colored; sperm motility is the proportion of live sperm in 500 spermatozoa. The temperature of the whole dyeing process shall be maintained at about 37 ℃. Generally, the vitality of fresh semen should be above 0. 7; the semen after liquid preservation should be more than 0.5; the frozen semen should be more than 0.35 after thawing before it can be used for AI.

② Sperm density determination: Sperm density refers to the number of sperm contained in semen per unit volume (1 mL). Inspection methods include estimation, counting, photoelectric colorimetry, etc.

Estimation method The estimation method is usually combined with sperm motility test. Place the original semen at low magnification (100×) under the microscope, according to the density of sperm in the field of vision, the semen density is roughly divided into three grades: dense, medium, and rare. The evaluation results are highly subjective and may have large errors, but the method is simple and commonly used in grass-roots AI stations.

Counting method The sperm density of male animals can be measured accurately by blood cell count. But it takes time. The operation steps are as follows:

Find the calculation room of the blood cell computer with a low power microscope under the microscope; then use a high-power mirror to find the middle square of the computing room.

Sperm can be killed by diluting semen with 3% NaCl solution. The original semen of cattle and sheep is diluted 100-200 times with a red blood cell dilution tube or micropipette, and the original semen of pigs and horses is diluted 10-20 times with white blood cell dilution tube or micropipette. After fully mixing, discard a few drops in front of the dilution tube, and then fill a small drop of semen into the calculation room.

Five representative middle squares are selected to count the sperm under the magnification of 400×. The sperm density is calculated according to the following formula: the number of sperm in 5 middle squares × 5 × 10 × 1 000 × dilution ratio = number of sperm per mL of semen.

Photoelectric colorimetry Photoelectric colorimetry is a method to determine the number of sperms according to the light transmittance of semen. The higher the transmittance, the less the number of sperm, and vice versa. Determine the light transmittance of the standard semen sample with a spectrophotometer and establish a standard curve. Then dilute the tested sample with 2.9% sodium citrate solution containing 5 mL 10% formaldehyde per liter for 1∶(80-100) times. Determine the light transmittance of the sample with a 550-nm wavelength spectrophotometer. Then compare it with the standard or check the sperm density table to determine the sperm density of the sample. This method is accurate and rapid. At present, sperm densitometer is used to automatically measure and read data. When determining the semen of horses, donkeys and pigs, the jelly in the semen should be filtered first to avoid affecting the light transmittance.

Sperm density of normal original semen: sheep, 2-3 billion/mL; cattle, 1-1.5 billion/mL; pigs and horses, 200-300 million/mL.

③ Sperm morphology examination

Abnormality rate of sperm All sperm with abnormal morphology and structure are abnormal sperm. Sperm abnormality rate is the percentage of abnormal sperm in total sperm. The sperm deformity rate of high-quality semen is no more than 18% in cattle and pigs, 14% in sheep, 12% in horses (donkeys) and 20% in dogs. If the abnormal sperm exceeds 20%, it can be regarded as poor semen quality and can not be used for insemination.

Sperm deformities can be divided into four categories according to the location of occurrence: A) head deformity, for example, the head is huge, thin, slender, damaged, double head, etc; B) neck deformity, such as neck fracture, double neck, tortuosity, swelling, etc; C) midpiece deformity, such as protoplast drops, swelling, bending, etc; D) main segment deformity, such as bending, convolution, zigzag, double tail, etc. The sperm abnormality rate can be checked simultaneously on the stained-glass slide for calculating the proportion of dead and living sperm, or a small amount of semen can be made into a smear, stained with common staining solution (such as methylene blue) or blue-black ink for 3 min, washed and dried, and then observed and counted 200-500 sperm under a 400× microscope to calculate the percentage of abnormal sperm.

Abnormality rate of sperm acrosome There are several kinds of abnormal acrosome, such as enlargement, defect, partial abscission and total abscission. Sperm acrosome abnormality rate refers to the percentage of sperm with abnormal acrosome in total sperm. Under normal conditions, the average acrosome abnormality rate of bovine sperm is 5.9% and that of pig is 2.3%. If the abnormal rate of bovine sperm acrosome is more than 14% and that of pig is more than 4.3%, it will directly affect the fertilization rate. The abnormal acrosome may be related to the abnormal sperm production and the secretion of accessory gonads. At the same time, semen stored *in vitro* for a long time may suffer from low temperature, especially with improper freezing methods. The common examination method is making semen into smear and fix it in the fixed solution for a while; after washing, the sperm are stained with Gemsa buffer for 1.5-2.0 h. After washing and drying, the sperm are encapsulated with resin and placed under a 1 000× microscope or phase contrast microscope to observe more than 200 sperm, and the abnormal rate of acrosome is calculated.

Sperm survival time and survival index Sperm survival time refers to the total survival time of sperm under certain external environmental conditions, and survival index refers to the average survival time of sperm, indicating the rate of sperm motility decline. Sperm survival time and survival index are closely related to conception rate. Determination of these two indexes is also one of the methods to identify the effect of semen dilution treatment.

Methods: The semen (which can be divided into several equal parts) after dilution is stored at a certain temperature, and samples are taken every certain time to check the vitality until there is no motile sperm. The interval from the start of dilution examination to the penultimate examination plus half of the interval between the last and penultimate examination is the total sperm survival time. The sum of the product of the average sperm motility and its time before and after each adjacent examination is the survival index. The longer the sperm survival time, the greater the survival index, indicating that the stronger the sperm viability, the better the semen quality.

2. 1. 3. 2 Semen dilution

Semen dilution refers to adding a diluent suitable for sperm survival and maintaining its fertilization ability into semen. The purpose of semen dilution is to reduce the energy consumption of sperm, supply proper nutrition and protective substances, and inhibit the activities of harmful microorganisms in semen to prolong the life of sperm; to enlarge the capacity of semen and increase the number of mated females for one ejaculation; at the same time, it is convenient for the preservation and transportation of semen. Therefore, semen dilution is an important technical link to fully reflect and give full play to the advantages of AI.

1. Main components and functions of semen diluent

The components of the diluent must be able to provide the energy materials needed for sperm survival. Increase semen volume, maintain appropriate pH, osmotic pressure and electrolyte balance, enhance sperm resistance to low temperature, and prevent the growth of bacteria.

(1) Diluent The function of diluent is to expand semen volume. Generally, it is an isotonic solution with seminal fluid. Commonly used diluents include 0. 9% sodium chloride, 5% glucose and 2. 9% sodium citrate (the most commonly used).

(2) Nutrient The role of nutrients is to provide energy for sperm movement and survival. The energy substances that can be used by sperm are mainly monosaccharides, such as glucose, fructose, lactose, sodium pyruvate, glutamine, milk, egg yolk, etc.

(3) Protective agent The role of protectants is to protect the survival of sperm *in vitro*.

① Buffer substance: Maintain the stability of pH, ion concentration and osmotic pressure in the diluent. Commonly used substances are phosphate buffer (PBS), trimethylol aminomethane (Tris), sodium citrate, sodium bicarbonate, etc.

② Non-electrolyte: Prevent the high ion concentration in the diluent from leading to sperm coagulation reaction and aging caused by strenuous exercise. Common substances include monosaccharide, sucrose, etc.

③ Antibiotics: It mainly inhibits the growth of bacteria. Semen is inevitably mixed with bacteria during collection and treatment. Seminal plasma is a good medium for bacteria multiplication. Although some bacteria are not pathogens, their metabolites are harmful to sperm survival. Pathogens can cause reproductive tract infection in female animals and abortion in female animals. Penicillin, streptomycin, gentamicin, sulfa and kanamycin are commonly used.

④ Cold shock resistant substance: The function of anticold stimulant substance is to prevent sperm from cold shock. Commonly used substances are egg yolk and milk. Cold shock refers to the phenomenon of sudden death or significant decline in vitality of domestic animal sperm when the temperature drops from 20 ℃ to 0 ℃. Causes of cold shock is that the melting point of acetal phospholipids in sperm is high, and it is easy to coagulate at low temperature, causing irreversible degeneration and death of sperm. Mechanism of cold shock prevention is that yolk and milk contain lecithin, which has a low melting point and can enter the sperm to replace acetal phospholipids to protect the sperm. In addition, lipoproteins or phospholipids have similar effects.

⑤ Antifreeze agent: The function of antifreeze is to prevent the freezing damage of sperm in ultra-low temperature environment. During the freezing process of sperm, the free water inside the sperm will form ice crystals, which will have a harmful impact on the sperm structure. Antifreeze can replace the water in the sperm and prevent the formation of ice crystals in the sperm, so as to protect the sperm. Commonly used antifreeze agents include glycerol, ethylene glycol, dimethyl sulfoxide (DMSO), etc.

⑥ Other additives: The substances that promote sperm motility include superoxide dismutase (SOD), β-amylase, oxytocin (OXT), prostaglandin E (PGE), vitamin B, caffeine, adrenaline, etc. Sperm protection substances include plant extracts (such as Chinese herbal medicine), ethylene diamine tetra acetic acid (EDTA), removing heavy gold layer ions, etc.

(4) Basic requirements for diluent It can supply the nutrition and energy for sperm. The osmotic pressure is similar to that of semen. The pH is similar to that of the original semen. It can inhibit the growth of bacteria and be of low cost and be simply prepared.

(5) Classification of diluents The diluent is classified into the following 4 categories according to the nature and purpose of the diluent.

① Prepare for immediate use: It is suitable for diluting semen immediately after semen collection for the purpose of simply expanding semen volume and increasing the number of inseminated female animals. This kind of diluent has a simple formula, and usually takes sugar or milk as the main body, so it is not suitable for preservation.

② Store diluent at normal temperature: It is suitable for short-term preservation of semen at room temperature. It is mainly composed of sugars and weak acid salts, and the pH is low (weak acid).

③ Store diluent at low temperature: It is suitable for cryopreservation of semen, with milk or yolk as the main body, and has the characteristics of anticold shock.

④ Freeze the diluent: It is suitable for semen cryopreservation, with yolk and glycerin as the main body. It has the characteristics of antifreeze. At the same time, there is a supporting thawing solution.

(6) Preparation principle and precautions of diluent There are many kinds of livestock semen diluents. The diluent used should be determined according to the different livestock species, the effect of diluent, the preservation method and whether the components are easy to obtain. Pay attention to the following principles during preparation.

① The drugs used to prepare diluent should be pure in composition, accurately weighed, fully dissolved, filtered, sealed and disinfected before use.

② Egg yolk should be taken from fresh eggs. After the shell is washed and disinfected, it should be extracted by breaking the shell. Try to avoid mixing egg white. Add it after the diluent is disinfected and cooled. Use it after it is fully mixed.

③ Distilled water or ionic water used to prepare diluent shall be fresh and neutral in pH. Ready to use and ready to prepare distilled water or ionic water may be better.

④ All utensils used for preparing diluent must be cleaned and disinfected strictly, and can only be used after being washed with diluent.

⑤ The diluent should be used and prepared immediately and kept fresh. If conditions permit, after disinfection and sealing, it can be stored in the refrigerator for 1 week, but egg yolk, antibiotics, enzymes, hormones and other components must be added when in use.

⑥ The milk shall be fresh (skimmed milk powder is preferred), especially the fresh milk, which shall be filtered and sterilized in a water bath for 10 min to inhibit the lactohydrocarbon that can kill sperm and remove the milk skin before use.

⑦ Antibiotics, milk, hormones, vitamins and other additives must be added accurately according to the dosage when the diluent is cooled to room temperature.

2. Semen dilution method

Semen dilution should be carried out as soon as possible after semen collection (generally no more than 30 min), and the contact with air and other vessels should be minimized. Therefore, preparations for semen quality inspection, dilution and preservation should be made before semen collection. Before dilution, sudden change of semen temperature should be prevented, especially low temperature stimulation below 10 ℃. The fresh semen should be quickly put into a 30 ℃ thermos, and then the semen and diluent should be adjusted to the same temperature in a 30 ℃ water bath pot, and diluted at the same temperature. When diluting, add the diluent slowly into the semen along the wall of the cup (bottle), and do not pour the semen quickly into the diluent. Then shake it gently or stir it with sterilized glass rod to make it mix evenly. Do not shake it violently. In case of high-fold dilution, it should be carried out in several times, first 3-5 times dilution, and then high-fold dilution, so as to prevent the sudden change of the environment where the sperm is located, resulting in dilution strike. After semen dilution, microscopic examination should be carried out immediately. If the viability decreases, it indicates dilution or operation improper.

3. Determination of semen dilution ratio

Too large dilution ratio of semen is harmful to sperm survival and seriously affects the conception rate; However, too small dilution ratio is hard to give full utilization of semen, so the dilution ratio of semen should be calculated accurately according to the number of effective sperm required for each fertilization, the type of diluent, and the number of mated females. The dilution multiple of semen and the amount of diluent to be added can be calculated according to the following formula:

Dilution multiple/head = [semen collection (mL)] × (sperm count/mL) × viability/(effective sperm/head) = total number of effective sperm/(effective sperm/head) = amount of diluent to be added = sperm collection (mL) × (dilution multiple/head portion)

2.1.3.3 Preservation of semen

1. Normal temperature preservation

(1) Principle The principle is that the metabolic activity of sperm is inhibited in the weak acid environment, the energy consumption is reduced, and its activity is inhibited. Once the pH returns to neutral, the sperm can be resuscitated. Therefore, weak acidic substances are added to semen diluent to adjust the acidic environment of sperm, so as to inhibit sperm activity and achieve the purpose of sperm preservation. Different acids have different inhibitory regions and protective effects on sperm production. It is generally believed that organic acids are better than inorganic acids. However, the preservation of semen at room temperature is also conducive to the growth and reproduction of microorganisms, so antibiotics must be added. In addition, the addition of necessary nutrients (such as monosaccharide) and isolation of air are conducive to the preservation of semen.

(2) Methods Normal temperature preservation is more suitable for the preservation of pig semen. Pig sperm at 17 ℃ is ideal for the preservation at room temperature. In order to reach this temperature, special thermostatic incubator, dry well, and water well can be used for storage, and special insulated transport box or wide port thermos bottle can be used for storage during transportation. Generally, the diluted semen is bottled and sealed, wrapped with gauze or towel after labeling, and stored at a temperature of 17 ℃ away from light. Although the temperature is allowed to change between 15 ℃ and 20 ℃ during storage, the temperature difference should be reduced and kept constant as far as possible. The effective time of pig semen preservation in extender at 17 ℃ can last as long as 7 d.

2. Preservation at low temperature

(1) Principle The preservation principle is that when the temperature slowly drops to 0-5 ℃, the sperm will be in a dormant state, and the sperm metabolic function and motility will be weakened. Therefore, low temperature is used to inhibit sperm activity and reduce the energy consumption of metabolism and movement. When the temperature rises, the sperm gradually recovers its normal metabolic function without losing its fertilizing ability. The key of semen cryopreservation is to control the cooling rate. When the cooling rate is too fast, it will cause irreversible cold shock reaction of sperm. The cause of sperm cold shock death is that the melting point of acetal phospholipids on the sperm cell membrane is high, which is easy to coagulate at low temperature, and is easy to cause the destruction of the cell membrane, which makes the K^+ and protein in the cell exude, seriously damaging the cell structure. It can also rapidly destroy ATP in cells, and can no longer make it synthesize, so as to seriously affect fermentation and respiration. In order to avoid sperm cold shock, cryoprotectants such as yolk, milk and other anticold substances can be added to

the cryopreservation diluent, and the method of slow cooling can be adopted to avoid sperm cold shock. Egg yolk and milk contain lecithin which is hydrophilic and difficult to assimilate at low temperature. It can replace some acetal phospholipids on sperm cell membrane and reduce the cold shock phenomenon during cooling.

(2) Methods The storage temperature at low temperature is 0-5 ℃. Generally, the diluted semen shall be placed in a refrigerator or a wide port thermos, and the temperature shall be kept constant during storage. During operation, the operating procedures of gradual cooling shall be strictly observed. In principle, after semen dilution, it shall be gradually cooled to 0-5 ℃ to avoid sperm cold shock. Therefore, it is generally necessary to reduce the temperature by 5 ℃ during 30 min, that is, by about 0. 5 ℃/min. However, in practice, in order to improve work efficiency, direct temperature reduction method is adopted. The test tubes (or vials) containing diluted semen shall be wrapped with several layers of gauze or cotton, then put into plastic bags, and then directly put into the refrigerator (0-5 ℃) or the wide port thermos bottle containing ice.

Semen stored at low temperature must be heated before insemination. Generally, the test tube or vial storing semen can be directly immersed in 30 ℃ warm water.

3. Cryopreservation

In 1950, British scientists Polge and Smith used egg yolk and glycerin as protective agents and dry ice as cold source to freeze sperm. After thawing, fertilization was successful, creating a new era of AI in the world. By the 1990s, all dairy cows and the vast majority of beef cattle in developed countries were inseminated with frozen sperm. Semen cryopreservation uses liquid nitrogen (−196 ℃) or dry ice (−79 ℃) as a cold source to dilute semen and freeze it for long-term preservation. It is the most effective method for semen preservation at present.

(1) Principle

① Change of substances during sperm cooling: Substances exist in gaseous, liquid and solid states. The solid state can be divided into crystalline state (regular arrangement of ice crystal molecules) and glassy state (irregular arrangement of ice crystal molecules) according to different molecular arrangements. At different temperatures, the two forms can be transformed into each other. It is proved by experiments that −60-0 ℃ is the temperature range for ice crystal formation, especially in the range of −25-−15 ℃. When the temperature drops gradually and passes through this temperature range, the crystalline state will be formed; if the solution rapidly cools down and passes through this temperature zone, the solution will form glassy freezing.

Semen cryopreservation is to add antifreeze and control the cooling rate to avoid the formation of crystals during freezing. Cryopreservation can make the sperm in a zero-metabolism state. After warming up, the sperm can be resuscitated without losing the ability to fertilize. However, ice crystals will be formed during sperm freezing, which will cause sperm damage. These damages include: the water outside the sperm will first form ice crystals, thus forming the osmotic pressure difference inside and outside the sperm, resulting in the extravasation of water inside sperm, increasing the osmotic pressure inside the sperm and the electrolyte concentration, causing acid and alkali imbalance, resulting in chemical damage to the sperm.

The formation of ice crystals produces mechanical pressure, which leads to the destruction of sperm cell membrane. Ice crystals penetrate into the sperm, destroy the sperm structure, and cause physical damage to the sperm. If the antifreeze such as glycerin is added to the semen solution, the antifreeze will penetrate into the sperm cells and reduce the electrolyte concentration. The hydrophilicity of glycerin can prevent the water in the sperm from suffering, avoid the formation of hypertonic sperm cells, and interfere with the lattice arrangement to form a "supercooled solution". At the same time, the cooling rate is controlled to achieve rapid cooling, so that water molecules lose energy and form a more "supercooled solution" before they have time to move and arrange. Rapid cooling makes the liquid only form tiny "microcrystals", i. e. vitrified state, so the sperm is not easy to be damaged and can recover its vitality after thawing. This is the vitrification theory.

② Changes during preservation and thawing: Ice crystals formed during semen freezing are very small, so the surface energy is very high. Microcrystals always try to combine into large ice crystals. Once a certain amount of energy is absorbed (such as heating up), the microcrystals will merge into large ice crystals, which will destroy the sperm. Therefore, in the process of preservation, even if the ice sperm temperature only rises to −60 ℃, it may cause sperm death. Thawing of frozen semen is the reverse process of freezing. Rapid temperature rise is also required during thawing to ensure that microcrystals are prevented from merging into large ice crystals, so that sperm can be resuscitated without losing the ability to fertilize. Although semen cryoprotectants can enhance the freezing resistance of sperm and avoid the formation of ice crystals, when the concentration is too high, they will do harm to sperm, such as damaging the acrosome and neck of sperm, bending the tail, destroying some enzymes, etc. , and affecting conception. Pig sperm is particularly sensitive to glycerol, so it is necessary to master the amount of glycerol. In addition to glycerol, there are also polyhydroxy compounds such as DMSO, Tris, sugars, etc. which have antifreeze effects.

(2) Frozen semen production process The production process of frozen semen includes semen collection and quality inspection, semen dilution, equilibrium cooling, semen dosage form and sub packaging, freezing and thawing.

① Semen collection and quality inspection: Semen freezing effect is closely related to semen quality. Semen collection preparation and operation must be done to strive for high-quality semen. In general, cattle and sheep adopt the false vagina method, and pigs adopt the hand-held method to collect semen. The frozen semen can be produced only when the microscopic activity of the original semen reaches more than 0. 6 and the density is above medium.

② Semen dilution methods: Semen dilution methods are different according to the types of frozen semen, sub packaging dosage forms and dilution times. At present, one or two dilution methods are often used in production.

One-step method The diluent containing glycerol and yolk was added to semen in a certain proportion, which was suitable for low dilution.

Two-step method In order to avoid the harm of glycerol to sperm caused by too long contact time, semen can be diluted by half of the final dilution ratio with the diluent without glycerol (liquid Ⅰ) at room temperature, and then semen and liquid Ⅱ are placed in the refrigerator to reduce the temperature to 0-5 ℃ (pig semen should be 5-8 ℃, about 1 h in the whole process), and then the second dilution can be made at this temperature.

③ Balance cooling: In the first dilution, in order to avoid the phenomenon of cold shock caused by too fast cooling, the semen container should be wrapped with cotton or gauze to control the cooling rate. Balance is to place the diluted semen in an environment of 0-5 ℃ for 2-4 h after cooling, so that glycerol can fully penetrate into the sperm and play an antifreeze role.

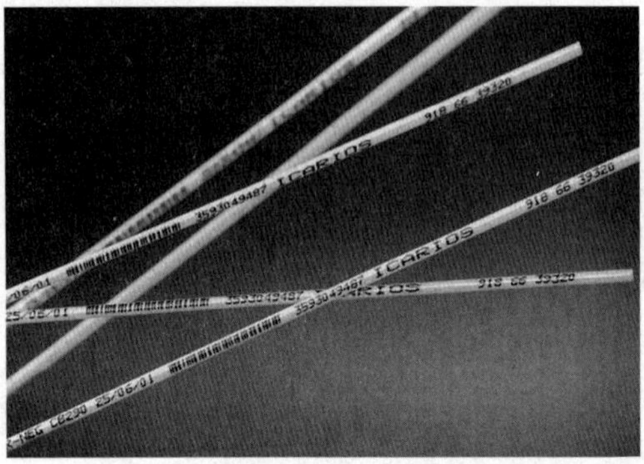

Fig. 2-7 Straw semen

④ Dosage form and sub packaging of frozen semen: Frozen semen can be prepared in ampoules, granules, bags and straws(Fig. 2-7). Straws are the most commonly used formulations due to the ease of industrialized production.

There are two dosage forms of 0. 25 mL and 0. 5 mL of straw frozen semen. Currently, 0. 25 mL dosage form is mostly used. Before freezing, the thin straw frozen semen shall be marked first, and then sub packed and sealed at low temperature. The advantages of thin straw frozen semen are closed packaging, no contact with the outside world during preservation and sperm delivery, and health and safety, and easy to mark. It can be directly clamped with tweezers and directly put into a warm water bath to remove ice. After thawing, insemination is carried out with a sperm inseminator gun. It is convenient for thawing and insemination. It is suitable for mechanized production and is an ideal dosage form. As pig insemination requires a large number of effective sperm, a 5-mL dosage form of large tube frozen semen specially used for pig semen freezing is being developed recently.

Liquid nitrogen is the cold source of frozen semen. During freezing, in order to control the cooling rate, attention should be paid to the distance between the freezing surface and the liquid nitrogen surface, and the balance time should be controlled.

Programmed freezing of fine straw frozen semen. First, lay the sub packed fine straw semen on the fine tube rack, and then cool and balance it for 2-4 h at 4 ℃. In mass production of frozen semen in thin straws, at present, the thin tube rack is usually put into a special freezer equipped with liquid nitrogen for program-controlled cooling and freezing.

⑤ Thawing of frozen semen in straws: The fine tube frozen semen can be directly thawed in 38-40 ℃ warm water, and can be taken out when the thin tube changes from milky white to transparent. Thawing of frozen semen is a necessary step to test the effect of semen freezing, and it is also a necessary preparation before insemination. After thawing of frozen semen, vitality test must be carried out, and insemination can be carried out only when it reaches more than 0. 35. The frozen semen of buffalo and horse can be used for AI if its viability is more than 0. 3 after thawing.

⑥ Preservation after thawing: Theoretically, frozen semen should be transfused within 15 min after thawing, but it is difficult to do so under the conditions of free-range breeding in rural areas, so there is a problem of preservation after thawing. The effect of cryopreservation after thawing is mainly determined by thawing solution and storage temperature. Due to the high concentration of antifreeze glycerin, the frozen semen in tubules can be stored well at 0-4 ℃ after thawing, and the absolute survival time can reach 161 h. Moreover, the pregnancy rate will not be affected within 10 h at 0-4 ℃ after thawing. After thawing, it can only survive for about 12 h at room temperature (24-27 ℃).

⑦ Storage and delivered of semen: At present, liquid nitrogen is widely used as cold source and liquid nitrogen tank as container for storage and transportation of frozen semen.

Liquid nitrogen and its characteristics Liquid nitrogen is a colorless, odorless and nontoxic liquid formed by the separation and compression of nitrogen in the air, with a specific gravity of 0. 8 and a boiling point temperature of −196 ℃. Liquid nitrogen is highly volatile. When the temperature rises to 18 ℃, its volume can expand 680 times. In addition, liquid nitrogen is an inactive liquid with poor permeability and no sterilization ability. Based on the above characteristics of liquid nitrogen, attention should be paid to preventing frostbite, splashing, suffocation, etc. When using large amount of nitrogen, air circulation should be maintained.

Liquid nitrogen tank The liquid nitrogen tank is composed of tank wall, tank neck, tank plug and lifting cylinder. It is a high vacuum insulation container made of thermal insulation materials, the insulation principle is similar to that of thermos bottle. Handle with care during use to avoid impact and toppling. Pay special attention to protect the tank neck and vacuum nozzle. The consumption of liquid nitrogen shall be checked regularly. When the liquid nitrogen is reduced by 1/2, it shall be supplemented in time. When taking frozen semen, the frozen semen shall not leave the liquid nitrogen surface for too long to avoid temperature rise. Therefore, a lifting cylinder or small plastic pipe without leakage of liquid nitrogen can be used.

Storage semen in liquid nitrogen tank When storing semen, it must be quickly put into the precooled lifting cylinder and immersed below the liquid nitrogen level in the tank. When taking frozen semen or transferring frozen semen to another container, the operation shall be quick and quick. The lifting barrel shall be lifted to the base of the neck pipe, and shall not be lifted out of the liquid nitrogen tank mouth. Be careful not to rub the inner wall of the neck pipe, and do not bend the handle of the lifting barrel too much. After taking it, the frozen semen shall be quickly immersed in the liquid nitrogen again.

Storage sow semen in extender at 17 ℃ received from boar studs (commercial or internal) can be delivered via domestic carrier or by company vehicles.

2.1.3.4 Insemination

1. Preparation before insemination

Preparations should be made in all aspects before insemination to ensure the normal implementation of insemination.

(1) Preparation of female animals After estrus identification, the female animal is determined to have reached the time of insemination, and it is fixed, and the tail is pulled to the side to clean and disinfect the external vulva. Wash with clean water, then disinfect with disinfectant, rinse with normal saline, and finally dry with a sterilizing cloth or paper.

(2) Preparation of the apparatus The instruments used in insemination should be thoroughly washed and strictly sterilized, preferably after rinsing with diluent or sterilized saline. Each female animal should prepare an insemination utensil, but when the utensils are not enough, the outer wall can be disinfected by rubbing with a 75% alcohol cotton ball, and after the alcohol odor volatilizes, rinse the outer wall and lumen 2-3 times with dilution before use. Currently, disposable utensils are generally used to prevent cross-infection between female animals.

(3) Preparation of semen When fresh semen is transfused, sperm motility should not be less than 0.7; sperm motility preserved at room temperature is not less than 0.5; the low-temperature preserved semen is first slowly warmed to about 30 ℃, and the viability is not less than 0.5; the viability of frozen semen after thawing is not less than 0.35.

(4) Preparation of technicians Inseminators should wear overalls, cut their nails short and polished, wash and dry their hands and sterilize them with 75% alcohol cotton balls, and if the arm is extended into the vagina, they should wash and disinfect the arm and apply it with sterile dilution. When the bovine rectal is grasped for insemination, long-arm gloves should be worn and lubricated with a lubricant.

2. Basic requirements for insemination

(1) Time of insemination Whether the female animal is fertilized is closely related to the appropriate insemination time. The time of female sperm infusion should be comprehensively analyzed and judged according to the time of ovulation of the female animal.

① Cattle: The average estrus cycle of cows is 21 d, the estrus time is short, 6-18 h, and ovulation generally occurs 10-16 h after the end of estrus. Therefore, it is found that the first insemination can be performed 12-20 h after the cow is in estrus, and the second insemination is performed 8-12 h later. Cows estrus in the morning are generally inseminated in the afternoon or evening of the same day, and the second insemination in the next morning; if estrus occurs in the afternoon or evening, the insemination is usually performed in the next morning, and the next afternoon or evening is inseminated again, and the effect is better. The initial mating cow has a slightly longer duration of estrus, and the rate of pregnancy is not high, usually starting to insemination around 20 h after estrus. Before the second insemination, it is best to check the follicles, if the cow has ovulated, it generally does not need to be inseminated.

② Pigs: The external manifestations of sow estrus are particularly pronounced, the vulva is red and swollen for a long time, and it is difficult to determine the accurate time of insemination through external observation. In general, sows are inseminated twice within 20-30 h after estrus, with an interval of 8-12 h, and one AI can also be performed when the "pressure reflex" still occurs after the estrus period.

③ Sheep: The time of insemination of ewes should be determined according to the circumstances of the estrus detection. Ewes are inseminated twice within the same day after estrus, with an interval of 4 d when 1 time of the estrus detection each day, once a day, once on the day of estrus and once after 4 d. Ewes are inseminated within the same day after estrus, with an interval of 12 h when 2 times of the estrus detection each day, and then a second semination every other half a day.

④ Horses (donkeys): According to the follicle development of the mare (jenny). The follicle development of the mare (jenny) is divided into 6 periods, and the insemination time can generally be arranged in accordance with the principle of "insemination may be selected at the third stage of follicular development, AI must be performed at the fourth stage of follicular development, and flexible AI again after ovulation". The so-called "insemination may be selected" is based on the speed of follicle development, combined with the physical condition of the mare (jenny) and changes in the environment. After ovulation, if the corpus luteum has not yet been generated, the sperm still has a certain fertility rate.

⑤ Rabbits: Rabbits are induced ovulation species, and the appropriate time for insemination should be arranged 2-6 h after induction of stimulation. It can also be determined according to the performance of estrus, when the female rabbit raises her tail to meet and has the performance of receiving mating, it can be inseminated, and the color of the visible mucous membrane of the vulva should be redness. Insemination can begin 3-4 d after the mare (jenny) is in estrus, once a day or every other day, but the insemination does not exceed 3 times.

(2) The amount of semen insemination and the number of effective sperm input The semen dose and effective sperm count required for AI in different livestock are different. It should vary according to the species of female animal, age, parity, uterine size and other physiological conditions and semen type. The amount of semen insemination of pigs, horses, and donkeys is larger than that of cattle, sheep, and rabbits. For large, postpartum breeding and loose uterus of female animals, the amount of sperm insemination should be appropriately increased. The amount of sperm in liquid semen is generally greater than the amount of sperm in frozen semen. Females undergoing superovulation treatment should have

an increased insemination dose and effective sperm count compared to normal females. In order to ensure a high fertility rate, it is important to deliver a sufficient amount of effective sperm count.

3. Artificial insemination methods for various animals

(1) Cows and heifers

① Vaginal opener insemination method: With the help of the opener, the cow's vagina is enlarged, and the external opening of the cervix is found with the help of a certain light source (flashlight, frontal mirror lamp, forehead lamp, etc.), the cattle insemination gun is inserted into the cervix 1-2 cm, the semen is injected, and then the cattle insemination gun and the opener are removed. Although this method is simple and easy to master, the site of insemination is shallow and susceptible to infection, and the fertility rate is low. As a result, it is currently rarely used.

② Insemination by rectal grasping: Rectal-grasped cervical insemination is now commonly used in bovine insemination. One hand of the operator reaches into the rectum to grasp the cervix, the other holds the inseminator which first extends diagonally upward into the vagina for 5-10 cm, avoids entering into the urethra, and horizontally into the cervix, and then the two hands work together to induce the inseminator into the cervix or the uterine body, and slowly release semen. During the insemination process, the sperm inseminator should not be held too tightly, and should be flexibly extended with the swing of the cow. Pay attention to maintaining the horizontal state of the cervix so that the inseminator can be slightly extended forward.

(2) Sows and gilts The boundaries between the vaginal and cervical junctions of sows are not obvious and therefore it is easier to inseminate. During insemination, apply a small amount of lubricant to the inseminator hose, and first insert it into the vagina at an angle of 45° obliquely, avoid the urethral opening, and then move straight forward. When obstruction force is felt, rotate slowly clockwise while moving back and forth until the front of the inseminator tube is felt locked by the cervix (gently pulled back without moving), connect the inseminator bottle (bag), and slowly enter the semen. During the insemination process, the sow should be continuously touched on the ventral side, breast (vulva) or back press to stimulate the sow, causing the uterus to contract and produce negative pressure to absorb the semen. Do not be too fast when inseminating, it generally needs to inseminate in 3-5 min. When the insemination is complete, slowly withdraw the inseminator and press the sow's buttocks for a moment to prevent the semen from flowing back.

(3) Ewes and doe The opener insemination is used. Open the ewe's vagina with an opener, find the cervical opening with the help of a light source, insert the inseminator into the cervical opening for 0.5-1 cm, and slowly inject semen. For ease of work and efficiency, a pit can be set up behind the infusion frame, or a liftable insemination table can be installed. In some places, it is also convenient to use the method of grasping the hind limbs of the sheep by an assistant or hanging its hind feet on a horizontal log higher than the sheep's

body, so that it is more convenient to make it upside down to fix for insemination.

(4) Mares (jennies) The inseminator of the mare (jenny) consists of a rubber tube about 60 cm long and 2 mm in diameter and a syringe. First, install the sperm-sucking syringe on the inseminator on the operating table, hold the joint part of the syringe with the hose in the left hand to prevent it from falling off, and hold the tip of the inseminator with the right hand so that the tip of the hose is hidden in the palm of the hand. Reach into the mare (jenny) vagina, find the vagina of the cervix, use the index and middle fingers to prop open the cervix, introduce the inseminator into the cervix 10-15 cm, lift the syringe, so that the semen flows naturally, or gently push in, and then pull the syringe from the hose, and then pump 5-10 mL of air to reload on the hose and push in, so that all the semen in the inseminator is drained. After the insemination is complete, gently withdraw the hose and press the cervix to prevent the sperm from flowing back (Fig. 2-8).

Fig. 2-8 Artificial insemination of donkeys with fresh and frozen semen

(5) Rabbits Female rabbits do not have special inseminator and one can use human catheters as rabbit inseminator. First, fix the female rabbit (inverted clamp, crawling, supine), wipe the vulva with an alcohol cotton ball, and then the sterilized inseminator before it is lubricated, and 0. 5-1 mL of semen is inhaled with a syringe. The inseminator parts the lips with his left hand, holds the inseminator tube in his right hand, inserts the vagina about 15 cm along the dorsal side till it reaches the bottom of the vagina, then twitch it back and forth several times, and injects semen with a syringe. When the inseminator is withdrawn, the buttocks of the female rabbit can be gently patted to contract the muscles and prevent the outflow of semen.

(6) Hens Collected and stored semen can be inseminated in hens by holding the hen upright and applying pressure on the abdomen around the vent area. Insemination is usually performed after 16:00. The assistant uses the thumb and index finger of the left hand, the little finger and the ring finger to pinch the legs of the hen, the palm is close to the chest of the chicken, and then the hand is upright, so that the hen's back is close to her chest, the chicken head is down, and then the right thumb and the rest of the fingers are separated in a

"figure eight" shape, spanning the soft parts on both sides of the cloaca, gently pressing down, and at the same time pushing up with the left hand supporting the hen's breast, make the fallopian tube that is open in the cloaca out of the cloaca (located on the left side of the chicken body). It causes the cloaca to open and the oviduct to protrude out. At this point, a syringe or plastic straw sucking 0.025-0.05 mL of semen can be inserted (2.5 cm) into the oviduct to deliver the appropriate amount of semen. After insemination, pressure around the vent area should be released; it aids the hen in holding sperms in the vagina or oviduct. Hen's squatting behavior can help us to identify the time of the first insemination. For caged chickens the inseminator can hold the legs of the hen with left hand and squeeze the left abdomen of the hen forward with right hand to turn the fallopian tube opening out for insemination. For optimum fertility, AI should be started before oviposition in turkeys, while it's not necessary for chickens.

2.2 *In vitro* fertilization technique

In vitro fertilization (IVF) is a technique of assisted reproduction in which the fertilization of an egg by sperm is achieved under suitable conditions outside the body through artificial operations, which is of great significance to mammalian reproductive physiology research, animal production, reproductive medicine, and endangered animals' conservation. More specifically, the mechanisms of mammalian gametogenesis, fertilization and early embryonic development can be better clarified in rodents by IVF. In the improvement of livestock breeds, IVF provides a cheap and efficient method for the embryo production and places a high value on making full use of superior germplasm resources, shortening the reproductive cycle, and speeding up the breed improvement.

Couples seeking IVF treatments may be experiencing either male or female infertility. The treatments are frequently utilized when the female produces normal eggs (oocytes), but the eggs can not travel to the uterus due to damaged fallopian tubes. In addition, IVF also provides both rich experimental materials and integral research methods for the embryo biotechnology (mammalian embryo transfer, cloning, transgenic, sex control, etc.).

2.2.1 A brief history of *in vitro* fertilization

It has been more than 140 years since the development of IVF. In 1878, Schenk began putting mature eggs extracted from rabbit and guinea pig with their epididymal sperm into the uterine fluid for *in vitro* maturation (IVM), and then firstly observed the phenomenon of second body discharge from the eggs and the cleavage of zygote. Subsequently, Pock and Menkin conducted IVF experiments on human eggs in 1944. Zhang Mingjue, a Chinese American, successfully conducted IVF experiments on rabbits in 1945, and discovered the sperm capacitation in mammals with Austin, an Australian scientist, in 1951. Until 1959,

Zhang Mingjue obtained the world's first batch of IVF rabbits, this technology has been truly recognized and accepted (Yang, 2019).

From 1960s to 1980s, rabbits, mice and rats serve as objects, scientists made a great progress on IVF, owing to their plenty of explorations on the mechanisms of sperm capacitation *in vivo* and the methods *in vitro*. At the beginning, a successful sperm capacitation was only allowed to incubate in the female reproductive tract, which was from either homogeneous or heterogeneous mammalian. Nowadays, we can achieve capacitation by IVC methods using the uterine fluid, the follicular fluid, the endometrial extract or serum, and finally complete *in vitro* capacitation with a well-defined chemical solution. Simultaneously, the comparative study on the *in vitro* capacitation effects of ejaculated spermatozoa and epididymal spermatozoa, found that mammalian semen contained decapacitative factors, and the key to capacitation was to remove decapacitative factors on the surface of spermatozoa. The IVF animals from about 30 species have been born after that by the above-mentioned theoretical and methodological achievements (Table 2-2). With the in-depth studies of IVF cattle, this technology becomes a major reproductive technology to expand the herd of elite cows for farmers in developed countries in Europe, America, and Oceania, when combined with the ovum pick up (OPU) technology.

Table 2-2 Brief description of the first *in vitro* fertilization study in mammals*

Species	Success time/year	Main finisher	Species	Success time/year	Main finisher
Mice	1968	Whitttngham et al.	Cattle	1982	Brackett et al.
Rat	1974	Toyoda et al.	Horse	1990	Zhang et al.
Hamster	1963	Yanagimachi et al.	Tiger	1990	Donoghue et al.
Cricetulus griseus	1969	Pickworth et al.	Lemur	1973	Could et al.
Guinea pig	1972	Yanagimachi et al.	Rhesus monkey	1984	Bavster et al.
Rabbit	1959	Chang et al.	Macaque	1984	Bavister et al.
Cat	1970	Hammer et al.	Baboon	1984	Clayton et al.
Dog	1976	Mahi et al.	Buffalo	1991	Suzuki
Goat	1985	Hanada et al.	Hungarian Grey Bull	1992	Solti et al.
American red sheep	1994	Conrod et al.	Hungarian Grey Bull	1984	Balmaceda et al.
Sheep	1985	Hanada et al.	*Callithrix jacchus*	1988	Lopata et al.
Pig	1986	Cheng et al.			

* From Guo (1998)

Since the 1980s, IVF has been widely used as a main method of treating infertility. In the United States of America, an estimated more than 48 000 test-tube babies were born in 2003, and the total number of test-tube babies in the world has exceeded one million, and this number is still rising year by year.

In China, although the work on IVF started late, it has progressed rapidly. Since Chen et al. from the Chinese Academy of Sciences successfully obtained IVF mouse in 1986, it has been born one after another, ranging from humans to rabbits, IVF sheep, cattle, goats, pigs and buffalo (Table 2-3). Especially in human and cattle, it is estimated now that nearly 1 000 test-tube babies are born every year in China, and hundreds of test-tube cows are born, which has reached the international advanced level (Zhu, 2015).

Table 2-3 First records for IVF humans and animals in China*

Species	The first creator	Work unit	Time/year	Oocyte source
Mice	Chen et al.	Institute of Genetics and Developmental Biology	1986	Ovarian oocytes
Human	Zhang et al.	Peking University Health Science Center	1988	Ovarian oocytes
Rabbit	Xu et al.	Northwest A&F University	1989	Fallopian oocytes
Sheep	Xu et al.	Inner Mongolia University	1989	Ovarian oocytes
Cattle	Xu et al.	Inner Mongolia University	1989	Ovarian oocytes
Goat	Qian et al.	Northwest A&F University	1990	Fallopian oocytes
Pig	Fan et al.	Jiangsu Academy of Agricultural Sciences	1990	Ovarian oocytes

* From Li (2005)

2.2.2 *In vitro* maturation of oocytes

2.2.2.1 Collection of eggs

1. Multiple ovulation

Mature eggs washed from the oviduct can fertilize directly with capacitated sperms after treatment with gonadotropins. Researchers apply these hormones in experiment animals (such as mouse, rat, and rabbit) and small multi-breed livestock (goat, sheep, and pig), but rarely use in large livestock due to the complex operating procedures and high cost. The key to the superovulation is to control the time that eggs maintain the fertilization ability when the eggs enter the oviduct or just release from ovary. Generally, it is required to be flushed before the egg has strong ability of fertilization.

2. Ovum pick up (OPU)

Oocytes could be directly obtained from the ovaries of living mammals with the ultrasonic detectors or laparoscopy. Oocytes, over 2 cm in diameter aspirated from follicles by laparoscopy are mostly mature, which can be fertilized with sperm after appropriate IVM.

Different animals have distinct methods for OPU. In small livestock, such as sheep and pigs, oocytes are often collected using laparoscopy while ultrasonic detectors are more suitable for assisting egg collection in the large livestock (cattle and horses). The latter firstly involves grasping the ovary by hand and inserting an egg suction needle through the vaginal wall, then absorbing oocytes from large follicles by using a B-mode ultrasonic image system. So far, a healthy cow can obtain 5-10 eggs per week by laparoscopy. The oocytes collected from living livestock generally undergo IVC before they are fertilized with capacitive sperms. OPU is of great significance for the high-quality female reproduction, which has been approved for commercial production in some countries.

3. Isolated oocytes from ovaries in the slaughtered dams

After washing and incubating at 30-37 ℃, the removed ovaries from the slaughtered dams are quickly transported to the laboratory. Then, using a syringe or a vacuum pump, the oocytes are sucked from the follicles of a certain diameter on the surface of the ovary (the diameter of the follicles in cattle is required to be 3-10 mm, that in sheep is 3-5 mm, and that in pigs is 3-6 mm at the sterile conditions). In addition, the ovaries can be sectioned to obtain oocytes. Most of the oocytes collected by this method are in the stage of germinal vesicle (GV), which needs to be experienced IVM before being fertilized with sperm. The key to harvesting oocytes from isolated ovaries is to be concerned with a sterile incubator for the ovary. Thus, the ovaries are washed several times with saline or phosphate buffered saline, and antibiotics are required to add into all solutions. This method is characterized by abundant ovaries and low cost, but it is hard to determine the production performance and pedigree of the female animals.

2.2.2.2 Selection of oocytes

The majority of collected oocytes generally form the cumulus oocyte complex (COC) with cumulus cells. The COC is required to be regular in shape, uniform in cytoplasm, neither black nor translucent, and surrounded by multiple layers of cumulus cells no matter how it is collected. In livestock, immature oocytes are often divided into four grades: A, B, C, and D before implementing IVF. Grade A oocytes are required to be closely surrounded by more than 3 layers of cumulus cells with uniform cytoplasm; grade B requires uniform cytoplasm, and the cumulus cell layer is less than 3 layers or partially surrounded by oocytes; grade C is naked oocytes without cumulus cells enveloped; grade D displays an oocyte attached with arachnoid cumulus cells. Only grade A and B oocytes are available for executing IVF.

2.2.2.3 *In vitro* mature culture of oocytes

Oocytes collected from the oviduct by superovulation or preovulatory follicles have been thought to be mature *in vivo*, which can be directly fertilized with sperm without IVC, otherwise, it requires to culture outside to arrive at maturity. Before culturing, the collected oocytes are selected and washed under a stereomicroscope, and then cultured in a medium for promoting oocyte maturation. At present, the medium for livestock oocytes maturation generally uses TCM199 supplemented with fetal bovine serum, gonadotropins, estrogens, antibiotics, etc. A droplet culture method is applied for IVC study, and 50-200 μL droplet ordinarily contains 10-40 oocytes (5 μL/egg). Then the droplet containing oocytes is cultured in a CO_2 incubator at 39 ℃ with 100% humidity and 5% CO_2 in air. In IVC, the incubation time of oocytes is 20-24 h for cattle, sheep, and goat, and 40-44 h for pig. During the first mature division, oocytes releases the first polar body into the perivitelline space along with the cumulus cells arranged in a radial coronal (called as the radial coronal). After that, the nuclear phase is observed under a microscope with DNA-specific dyes, suggesting that the oocytes are in the metaphase of the second mature division.

2.2.2.4 Markers of oocyte maturation

The secretion peak of FSH and LH promotes ovulation and oocyte meiosis, while the oocytes present in antral follicles become mature. The IVM of eggs is mainly related to the recovery and completion of the oocyte meiosis, changes in its cytoplasm, cell membrane, zona pellucida and cumulus cell. In mice, after a few hours of IVC, the oocytes arrives at the stage of germinal vesicle rupture, and the IVC time of germinal vesicle rupture in oocytes of various animals is inconsistent, ranging from 2-4 h for rats, 8 h for cattle, and 18-24 h for pigs, 8 h for goats.

During the first meiosis, the chromosomes of oocytes are arrested in metaphase Ⅱ after the first polar body is ejected. Chromatin becomes heterochromatin containing many dense granules, thereby exacerbating chromosomal condensation. Simultaneously, the centromeres of oocytes associate with microtubules and microtubule centers to achieve the attachment of the monomers of each pair of homologous chromosomes. After IVC for 10-13 h, mouse oocytes protrude bulges in the spindle region in anaphase and telophase, and finally forms the first polar body, which exhibits a smooth plasma membrane without microvilli, whereas the surface of the oocytes is filled with microvilli. As the oocyte matures, the sperm receptors on the zona pellucida (ZP, such as ZP1, ZP2 and ZP3) of the oocytes are rearranged. The outer layer of the ZP is reticular, and the inner layer is attached with broken microvilli and the cytoplasmic bulges of cumulus cells meanwhile the zona pellucida softens. In general, cumulus cell expansion is used as a characteristic of oocyte maturation *in vitro*.

The criteria for IVF of bovine oocytes were formulated by Hunter and Moor. According to this criteria, there are three types of mature oocytes in IVM: grade 1, mature oocytes are "completely expanded cumulus cells", and the expanded cumulus cell mass is 3-fold diameter of naked oocytes (300 μm); grade 2, mature oocytes are moderately expanded cumulus cells, and the expanded cumulus cell mass is 2-fold diameter of naked oocytes (200 μm); grade 3, mature oocytes are slightly expanded cumulus cells, and the cumulus cells are tightly adhered on the ZP (Hunter and Moor, 1987).

2.2.3 *In vitro* capacitation of sperm

Before fertilization, sperm should be characterized by the following capabilities: first, an ability to reach the oocyte and pass through the ZP; second, an ability to induce acrosome reaction; third, timely exposure of the protein binding site obtained during sperm capacitation to combine with the ZP of oocytes; fourth, an ability to successfully fuse with the cytoplasm of oocytes and stimulate the oocyte to complete the second mature division (Nagy et al., 2012). Most of the above-mentioned abilities have been acquired when sperm is transported in the epididymis. However, the final maturation of sperm depends on the microenvironment of female reproductive tract or IVC medium, which is physiologically called sperm capacitation.

2.2.3.1 Methods of sperm isolation

1. Suspension method

This is the most effective method for screening high motility sperm from semen. According to the upward movement features of sperm, the operator can separate high motility sperm from dead sperm and other seminal plasma components by suspension method. The method can increase the fertilization division rate of fertilized eggs from 46% to 59%, which is very suitable for poor semen with low sperm motility but needs careful consideration for valuable semen.

2. Glass cotton filtration method

In 1950s, the column layer filtration made up of small glass bead is used to filter out dead sperm in the diluted bovine semen in many laboratories, and this method can effectively filter out the dead sperm from poor semen to improve sperm fertilization rate in humans.

3. BSA/Percoll density gradient centrifugation

Since the density of normal sperm is higher than that of abnormal sperm, using Percoll, the normal live sperm are gathered in the high-density of Percoll after equilibrium centrifugation. The separated normal sperms have high fertilization ability. Generally, the density gradients at 30% and 45% are used in Percoll with equilibrium centrifugation for 10 min. Gradient centrifugation can obtain higher quality sperm to increase the sperm availability, and the sperm quality is more susceptible to the quality of Pecroll.

4. Centrifugal washing

This method is preferable to detect the semen with better sperm motility. Firstly, 0.1-0.2 mL of semen is placed into a centrifuge tube, then added 5-10 mL of capacitive fluid or semen, and centrifuged for 5 min at 300*g* to separate sperm from seminal fluid. Although this method is simple to operate and has high efficiency for sperm, it has no sorting effect on sperm, therefore, it is hardly suitable for semen with poor sperm motility and low fertilization rate.

2.2.3.2 Methods of sperm capacitation

1. Serum albumin method

Serum albumin, a macromolecular substance in serum, has a strong ability to bind cholesterol and zinc ions. Long-term incubation of sperm with serum albumin solution can remove part of the cholesterol and zinc ions from the plasma membrane of sperm, and reduce its proportion of cholesterol, thereby change the stability of the plasma membrane, and finally complete sperm capacitation.

2. Hypertonic salt method

When the sperm is treated with a high ionic solution, the proteins will peel off the surface of the cytoplasm of sperm, resulting in sperm capacitation. However, the sperm is generally damaged by a high ionic solution, then with a low sperm motility it will fail to complete fertilization. For this reason, now this method has been deprecated.

3. Follicular fluid incubation method

Follicular fluid contains substances from serum that induce sperm capacitation and acrosome reaction. Thus, the sperm capacitation can be induced by a certain concentration of follicular fluid. However, the follicular fluid also contains inhibitory factors, which is bad to the sperm motility. In follicular fluid, studies showed that the capacitation-promoting factors come from the endometrial cells of the follicle, while the inhibitory factors stem from serum. Therefore, follicular fluid can be used for sperm capacitation than IVF.

4. Calcium ionophore method

Ca^{2+} can enter sperm by the ionophore A23187, resulting in a higher concentration of intracellular Ca^{2+}, and finally trigger the sperm capacitation. Noticeably, the concentration and time of A23187 treatment should be controlled properly during the operation, otherwise, it will lead to a low sperm motility and dead sperm.

5. Heparin method

Heparin is a compound with a high-concentration sulfated aminopoly-saccharide, which can induce Ca^{2+} to pass through the cytoplasm of sperm, and activate sperm capacitation.

2. 2. 3. 3 Detection of sperm capacitation

1. Observation of sperm morphology and movement

The enlarged heads of capacitated sperm can be observed under a microscope, while they display an enhanced vitality, and hyperactivated movement.

2. Detection of acrosome reaction

To further judge whether sperm are capable of capacitation, it is required to induce their acrosome reaction and detect their acrosome integrity. Specifically, capacitated spermatozoa are incubated with 100 μg/mL lysolecithin for 15 min, and the integrity of the acrosome is checked by the double staining and transmission electron microscope methods.

2. 2. 4 Procedure of *in vitro* fertilization

The completion of IVF requires a suitable culture system that is conducive to the survival of sperms and oocytes (Fig. 2-9). The calcium ionophore is essential to the co-culture system of capacitated sperms and mature oocytes, while the capacitation fluid is an indispensable solution for the fertilization of sperms with eggs. The culture time of fertilization is related to the capacitation fluids. In general, the culture time is 6-8 h in BO solution and 18-24 h in TALP or SOF solution. Currently, there are two types of culture systems used for IVF: the droplet and the four-well culture plate methods.

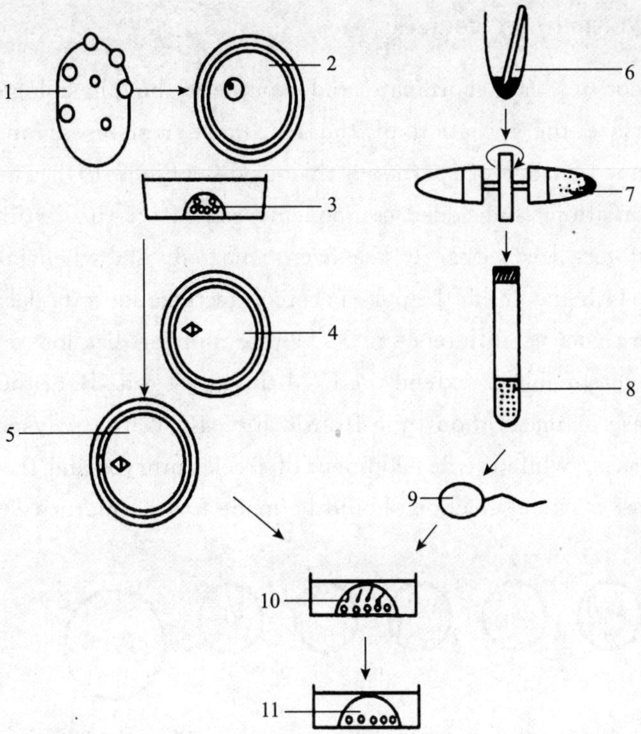

1. Ovary 2. Oocytes in the stage of GV 3. Oocyte maturation culture 4. Oocytes in the stage of MⅠ 5. Oocytes in the stage of MⅡ 6. Semen thawing 7. Sperm centrifugal washing 8. Sperm acquisition processing 9. Capacitated Sperm 10. *In vitro* fertilization 11. Embryo culture

Fig. 2-9 Mammalian *in vitro* fertilization (Li, 2005)

2. 2. 4. 1 Microdroplet system

It is the most widely used culture system. 20-40 μL microdroplets with fertilization fluid is placed on a plastic petri dish, coated with paraffin oil, and then the 10-20 IVM oocytes and the capacitated sperm [$(1.0\text{-}1.5)\times10^6$/mL] is put into each droplet to incubate for 6-24 h. The advantage of this method is less consumptions of fertilization fluid and semen, and better IVF performance. However, the limitations of this method are costly, time-consuming, and labor-intensive, and it is suspectable to the quality of paraffin oil.

2. 2. 4. 2 Four-well culture plate system

In a four-well culture plate system, each well contains 500 mL fertilized fluid and 100-150 IVM oocytes and $(1.0\text{-}1.5)\times 10^6$/mL capacitated sperm, and then incubates for 6-24 h.

This method is easy to operate beyond the quality effect of paraffin oil, but semen-consuming and unstable as that of the microdroplet system.

2.2.5 Main factors affecting *in vitro* fertilization

2.2.5.1 Maturity of oocytes

Only mature oocytes can coordinate and complete the physiological changes during fertilization, to ensure the initiation of the resistance responses for polyspermism and stimulate the continued change of sperms in the oocytes (Fig. 2-10). The oocyte maturation medium, culture conditions and added components can affect the fertilization rate and the development rate of fertilized eggs. It has been reported that when porcine oocytes were undergone IVM for 44 h and 48 h, the polyspermism fertilization rate decreased from 58% to 34%, although there was no difference in IVM rate and fertilization rate. To prevent the polyspermism, it's reasonable to extend the IVM time by 4-6 h. But other research suggests that with the increase of maturation time from 8 h to 48 h, the polyspermism fertilization rate of bovine increases, while the development of 4-cell embryos and 16-cell embryos is also inhibited. Therefore, a strict selection should be made for the maturity time.

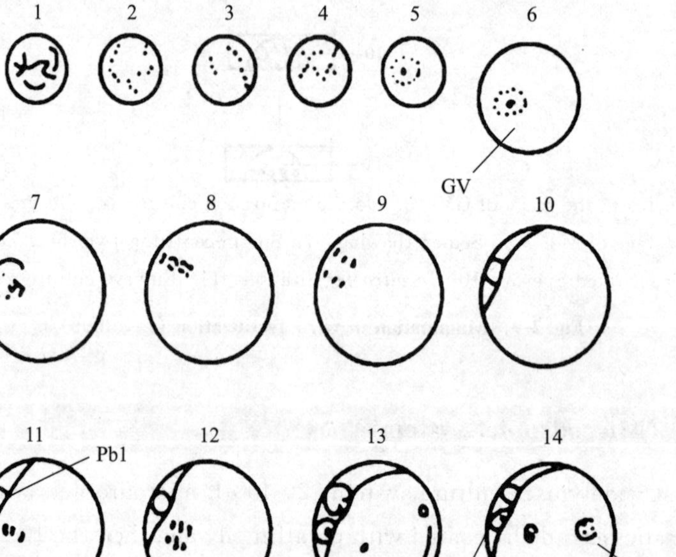

1. Leptonema stage 2. Zygotene stage 3. Pachytene stage 4. Diplotene stage 5, 6. Dictyate stage 7. Diakinesis stage 8. Metaphase-Ⅰ stage 9. Late-Ⅰ stage 10. Closed-Ⅱ stage 11. Metaphase-Ⅱ stage 12. Late-Ⅱ stage 13. Closed stage 14. Pronuclear GV: Germinal vesicle Pb1: First polar body Pb2: Second polar bodies SP: Sperm F: Sperm nuclei M: Male pronucleus

Fig. 2-10 Maturation process of the oocyte (Li, 2005)

2.2.5.2 Semen quality

There are both individual and batch differences for semen in the results of IVF. Therefore, the screening of semen should be conducted before mammalian IVF.

2.2.5.3 Fertilization system

The fertilization system mainly contains the osmotic pressure, the ratio of ion concentration, and pH in basic medium, as well as the capacitation fluid, etc., which directly influence the capacitation effect of sperm. Studies have shown that the polyspermism fertilization rate tends to increase when the pH is higher or lower than 7.4. For example, when the pH of culture medium reached 7.4 or 7.8, the polyspermia fertilization rate altered by 48% and 79%, respectively, although there was no change in the fertilization rate. Besides, the concentration of Ca^{2+} affected the IVF rate, the polyspermia fertilization rate and the cortical response. With observation of *in vitro* fertilized eggs by electron microscopy, it was shown that the cortical response induced by a higher concentration of Ca^{2+} (4.72 mmol/mL) was incomplete in the culture medium, which is a vital reason for the high polyspermia fertilization rate in IVF.

2.2.5.4 Encounter time of sperm with egg

The capacitation of sperm is prepared for the acrosome reaction during fertilization. If the sperm has undergone a complete acrosome reaction or the acrosome enzyme is early released before meeting the egg, the sperm loses the ability to pass through the egg. Neither immature nor aging oocytes can complete the normal fertilization process.

Only oocytes are fully mature and synchronized with the sperm capacitation, and they fertilize to get a higher fertilization rate. The question of how to control the time of sperm-egg encounter is worthy of discussion.

2.2.5.5 The ratio of sperm and egg

A higher sperm density will lead to polyspermia while a lower concentration will cause lower rate of fertilization. In general, the sperm-egg ratio is 2 000 : 1 for pigs, 7 000 : 1 for horses, (3 000-4 000) : 1 for cows, 5 000 : 1 for goats, (100-800) : 1 for mice, and 6 000 : 1 for humans.

2.2.5.6 Co-incubation time of sperm with egg

During fertilization, the co-cultivation time of sperm with egg in BO and TALP solution is 6-8 h and 18-24 h, respectively. The shorter time, the lower fertilization rate, and the longer time, the higher fertilization rate (Huang, 2011).

2.2.5.7 Fertilization solution

The TALP fluid developed by Bavister and Yanagimachi includes m-Tyrode's fluid,

BSA, sodium lactate and sodium pyruvate (Bavister and Yanagimachi, 1977). Using a BSA as a protein source, it was found that bovine oocytes matured *in vitro* could be fertilized in a BSA-free medium, and the probability of polyfertilization occurrence was low; a study clarified that protein in the fertilization fluid is critical for naked oocytes, but not for the cumulus-oocyte complex.

2.2.5.8 Heparin

In IVF, bovine spermatozoa are generally incubated with heparin for 15 min to 4 h to obtain the capacitation ability. Domestic researchers found that the egg cleavage rate and cleavage index could be significantly improved after the addition of various concentrations of heparin into bovine semen. The IVF effect of sperm mainly depends on the concentration of heparin in the semen and is independent of the capacitation treated by heparin before fertilization. After the heparin-capacitated sperm was washed and centrifugated to remove the heparin, the IVF effect was the same as that of the sperm without heparin treatment, and the cleavage rate was close to 70%. Therefore, heparin only promoted sperm capacitation in the condition of the presence of oocytes and cumulus cells because the oocytes and cumulus cells themselves truly played a key role in the capacitation. Further study showed that the effect of heparin concentration on bovine oocytes was independent of individual bulls during IVF. Indeed, an increasing concentration of heparin in the fertilization fluid improved the IVF effect of sperm but did not eliminate the individual differences between bulls. In addition, only if the concentration of heparin was above 18 mg/L in the fertilization fluid, it could improve the blastocyst hatching rate. A low concentration of heparin could only improve the fertilization rate, but not the fertilization quality, while a high concentration of heparin could promote sperm capacitation and increase the fertilization rate. The occurrence of poly-fertilization may rather depend on the maturation quality of oocytes, than the sperm concentration and heparin concentration.

2.2.5.9 Water

Water is the main component of all kinds of culture mediums. When preparing semen, except for fully washing and sterilizing the relevant glass containers, the quality of water must be emphasized in the fertilization solution. In hamster, the sperm motility assay system is used to detect the toxicity of glassware, filter systems, egg aspiration needles and other devices that contacts with sperm and oocytes or embryos, which is more convenient for detection.

2.2.5.10 Temperature

Temperature is essential to IVF. It was reported that if the penetrating temperature of bovine sperm-egg was lowered by 2 ℃, the fertilization rate would drop by more than 15%. In addition, it was found that the incubation temperature was the main factor affecting porcine sperm penetration in IVF. It was investigated that the fertilization rate was less than 1% at 37 ℃, and up to 89% at 39 ℃. For sheep, the sperm acrosome reaction was also much related to the change of temperature.

2.2.5.11 Light

It was found that sperm and early embryos were tolerant to normal light, although *in vivo* fertilization was carried out in the dark.

2.2.5.12 Gas

Gametes and embryos should be placed in an incubator containing either 5% CO_2 or 5% CO_2 +5% O_2 +90% N_2, as well as at 100% humidity to avoid evaporation of the culture medium. When the culture medium is only exposed to air outside the incubator, it should be shortened outside time as possible.

2.2.5.13 Somatic cytokines

Adding bovine follicular fluid to the fertilization fluid is proved to reduce the cleavage rate and blastocyst development rate of the oocytes. Instead, the addition of endometrial fluid significantly increases the blastocyst development rate of the oocytes.

2.2.6 Application, problems, and prospects of *in vitro* fertilization

2.2.6.1 Application

IVF is a reproductive technology with a rich prospect, which can extremely exert the reproductive potential of male livestock, but shorten the breeding time, reduce the breeding cost, and select the gender of offspring. If combining with other biological technologies (such as cloning technology and micro-insemination technology), the application of IVF can save endangered animals, maximize the use of germplasm resources of male livestock, and have great significance in improving economic traits of livestock and poultry, as well as the conservation of germplasm resources.

2.2.6.2 Problems

1. A low development rate of blastocyst with a few blastocyst cells

During IVC process, IVF eggs generally encounter a problem of developmental blockage, which is a degenerate phenomenon along with embryo growth at a certain stage. Bovine embryonic blockage occurs at the 8-cell to 16-cell stage, which results in a far lower development rate of blastocysts for IVF than that for *in vivo*. In addition, the total number of cells and the number of inner cell mass are significantly reduced in IVF blastocysts compared with blastocysts fertilized *in vivo*.

2. A low calving rate and a high birth weight of the fetus

The calving rate of IVF embryos, especially cattle, is 15%-20% lower than that of *in vivo* fertilized embryos, and the birth weight of the fetus is 3-4 g higher than that of *in vivo* fertilized offspring, and the latter leads to a high rate of dystocia in recipient dams.

2.2.6.3 Prospects

IVF is used to study the molecular mechanism of the maturation of oocyte and the development of early embryo, to invent an ideal culture system. The culture system contributes to a stable and orderly expression of embryonic genomes. It is beneficial to strengthening the research of preantral follicle culture if the use of the excellent female animals is maximized by IVF.

Secondly, to increase the success rate of IVF, the technology of live egg retrieval and the *in vitro* maturation technology of preantral follicles and small follicles will be improved in the future.

Also, the combination of IVF and other biotechnology for comprehensive development needs to be strengthened. The IVF is an indispensable method for rapid production and preservation of a larger number of high-quality and cheap oocytes and embryos under a complete in IVC conditions, with the extraordinary development of commercial embryo transfer technology. It is beneficial to fully tapping reproductive potentials of excellent female animals and accelerating the reproduction and breeding process of high-quality animals.

Furthermore, the IVF will provide suitable research materials for embryo engineering to facilitate the factory production of mammalian embryo. In addition, the technology is also of great significance in saving endangered animals and the development and utilization of rare animals, etc.

2.3 Microinsemination

Microinsemination (MI), also known as microfertilization (MF), developed in 1980s, is a new technology of IVF, which can save sperm. Using a micromanipulator, operators can directly inject sperms into either the cytoplasm of oocytes to achieve fertilization, which called as intracytoplasmic sperm injection (ICSI), or the perivitelline space to complete fertilization, which is called sub-zonal insemination (SUZI). Since this technology can ignore the setback of ZP and the cytoplasm membrane of oocytes when sperms fertilize with eggs, there is no strict requirements for the viability and integrity of sperm.

The study of MI began in 1962, a Japanese scientist, Hashimoto injected homologous or heterologous sperms into sea urchin oocytes to form male and female pronuclei, and the injection could activate oocytes to complete the second meiotic division. In mammals, the earliest MI study was on hamster by Uehara and Yanagimachi. With injection of the hamster into hamster oocytes, they found that sperm nuclei could depolymerize and develop into male pronuclei, which suggested that the plasma membrane fusion of male and female gametes was probably not required for the sperm nuclear disaggregation and prokaryotic development. Egg age exerted an impact on the nuclear depolymerization of sperm and the formation of pronucleus during MI. Only fully mature eggs were activated by sperm injection to induce the sperm nuclear depolymerization and the formation of pronucleus in the cytoplasm. Even though sperm can also depolymerize when they are injected into immature eggs, they can not form the pronucleus. In rats and mice, the study on the heterozygous fertilization found that the nuclear depolymerization of sperm in the egg plasma membrane and the formation of pronucleus were not species-specific, however, the ZP of egg was species-specific.

In 1988, Mann successfully obtained the first viable offspring of mice by injecting motile sperm into the perivitelline space. In 1993, Hosoi et al. injected immotile sperm into the cytoplasm of rabbit eggs by ICSI and obtained normal offspring. In livestock, the bovine sperm morphologically damaged by repeated freezing was injected into the cytoplasm of oocyte, and IVC was applied to obtain blastocysts, and the offsprings were successfully born after the embryo transplantation (Zhu, 2015).

In 1995, with the piezo-driven operating system, Kimura and Yanagimachi made a great improvement in the success rate of ICSI in mice, the surviving rate of injected eggs was 80%, 70% of which developed to blastocysts *in vitro*, and the litter rate reached 30%. Paleremo et al. decided to embark on ICSI study with human eggs, and finally the world's first ICSI baby was born (Zhu, 2010). Because ICSI possesses the advantages of high fertilization rate with single sperm, low polyspermy rate and is immune to the consequences of sperm concentration, morphology, and motility, it is widely applicable all over the world. At present, the fertilization rate of ICSI for human sperm can arrive at 65%-70%, and the implantation rate of embryos can reach more than 30%, it is contribution to assisting the men with severely defective quality of sperm or no sperm to obtain their own offspring.

In China, the similar studies progressed rapidly despite initiating relatively late. The fertilization rate of frozen sperm injection was 19.9% for cattle, and the cleavage rate was 9.4%; the fertilization rate of fresh and frozen epididymal sperm in ICSI was 25.2% for pigs, and the cleavage rate was 21.6% in 1992. The human MI technology was first successful in China by 1996. The Chinese Ministry of Health announced that 8 scientific institutes were qualified to apply ICSI technology in the field of artificial reproduction in 2001 (Zhu, 2010).

2.3.1 Operating procedure

MI mainly includes egg and sperm treatment, sperm injection, activation of injected eggs, and *in vitro* development of injected eggs, etc. The technical route of MI is shown in Fig. 2-11.

1. Egg collection and processing

The eggs currently used for MI can be obtained by superovulation and IVM eggs. The

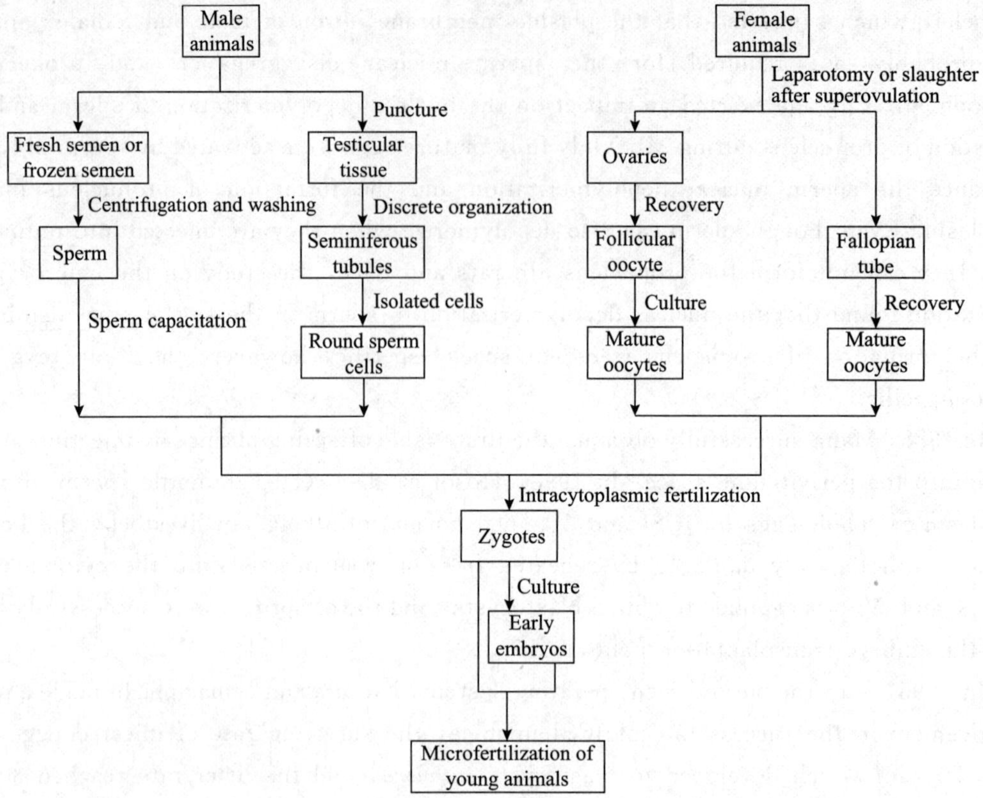

Fig. 2-11 The technical route of microinsemination (microfertilization) (Li, 2005)

cumulus cells of oocytes are removed with 0.1% hyaluronidase prior to sperm injection and then placed in paraffin oil-coated microdroplets. For bovine and porcine oocytes, due to their high fat content, they should be centrifuged (1 000g) to make the fat droplets gathered on one side in the oocyte, then the cytoplasm of oocytes becomes transparent for easy injection.

2. Preparation of sperm or sperm nuclei

Fresh or frozen-thawed sperm can be used for injection before accomplishing capacitation. If injection is only into the sperm heads, it is required to separate the sperm heads by a sucrose density gradient solution. The prepared sperm can be placed in droplets (manipulation solution, PBS or 5 mmol/L $NaHCO_3$-TCM199) which contains 8%-10% PVP or methylcellulose. In this way, the sperm can lose its motility, and hardly stick to the wall of injection tubes, which is convenient for capturing during injection. In addition, 5 mmol/L dithiotherie (DTT) can be added to the operation solution for sperm injection, to facilitate sperm nuclear depolymerization in the cytoplasm of oocytes, which is a common method to improve the formation rate of male pronucleus.

3. Microinjection of sperm or sperm nucleus

Currently, the methods used for sperm microinjection include conventional method and piezo-driven method, and the two methods are mainly different in the injection tube and driving power. The former requires an oblique needle with a spike, while the latter uses a flat needle. The inner diameter of the microinjection tube (6-8 μm) can only be slightly larger than the diameter of the sperm head. During micromanipulation, the sperm is immobilized in the operating droplets with an injection needle, and then a single sperm is drawn into the open end of the injection tube from the sperm tail to head. After that, the sperm is transferred to the operating droplet containing the oocyte, and the single sperm with a little operating solution is injected into the cytoplasm of the oocyte. To protect the integrity of oocyte spindle, the injection needle should be placed on the vertical line connecting the polar body and the egg center.

4. Activation of eggs

In addition to eggs for human, rabbit and mouse, which can be directly activated by the microinjection tube, the activation rate of other mammals is extremely low without other activation treatment. At present, the commonly used methods for oocyte activation are as follows: calcium ionophore A23187 treatment, electric shock, alcohol treatment, ionomycin + 6-dimethylaminopurine (6-DMAP), ionomycin + cycloheximide (CHX), etc. A more effective method is to activate the eggs with 5 μmol/L ionomycin for 5 min, and then culture them in a medium containing 2 mmol/L 6-DMAP for 3 h.

2.3.2 Significance of MI

1. MI is a supplement to IVF

Sperm microinjection technology is a supplement to IVF, which can solve a series of infertility problems caused by sperm itself, thereby this technology nowadays is widely used in male infertility treatment. In addition, the MI has an extremely positive effect on the basic theory research of developmental biology such as animal production, wild animal protection and fertilization mechanism, etc.

2. Improve the utilization rate of sperm of valuable breeds

The amount of sperm required for AI is at least 5 million/female animal, for IVF isat least 500 sperm/egg, and for MI is only one sperm/egg. Epididymal spermatozoa, testicular spermatozoa, round spermatids, and elongating spermatids during metamorphosis can also be fertilized by ICSI. In addition, only if the nuclear substance is intact, dead and morphologically abnormal sperm can fertilize eggs using ICSI. Therefore, MI can greatly improve the utilization efficiency of sperm collected from precious varieties, which is of great significance.

3. Promote the popularization of sperm separation technology

Using flow cytometry to separate X and Y sperm for sex control of animals is the most effective method at present. However, due to the limitation of sperm separation efficiency and high cost of the sperm separation technology, this technology has limited application in livestock production. Once combined with MI, application of this sex-control technology in livestock production can become a reality.

4. Improve the fertilization rate of IVM oocytes

Due to hardening of the oocyte ZP, the fertilization rate frequently decreases in IVM. The application of MI can improve the fertilization rate of such oocytes. In addition, the poor quality of oocytes collected from animals after IVM such as pigs and horses resulted in a high polysperm fertilization rate, and the use of single fertilization technology can solve this problem.

5. Solve the problem of males with oligospermia and poor motility

For males with oligospermia and poor motility of sperm, instead of conventional fertilization methods, MI can reproduce offspring. For infertile males, spermatids can be isolated from the seminiferous tubules, and then combined with MI technique to form a normal zygote. All these techniques are of particular significance for the conservation of endangered animals and the treatment of male infertility.

6. Provide methods to simplify transgenics

Combined with sperm-mediated transgenic technology, the target gene and sperm are incubated together in the sperm manipulation solution, and then injected into eggs to improve the efficiency of transgenic. Thus, the MI paves a simple and feasible technical road for the research of transgenic animals. By this way, transgenic mice and pigs have been obtained currently, which have confirmed the effectiveness of this strategy.

7. Study the interaction between sperms and eggs and the mechanism of fertilization

ICSI provides a convenient and rapid method for study on the fertilization mechanism and embryonic development. It can also be used to clarify the fertilization mechanism of homologous and xenogeneic animals, and to figure out the interaction details between sperms and eggs, ranging from the exact time and location of sperm-egg recognition, the mutual activation effect of sperm and egg, to the mechanism of interspecific sperm-egg interaction. These mechanisms have a great theoretical significance for understanding the fertilization mechanism in developmental biology and reproductive biology.

Assignments

1. What's the significance of AI?
2. What are the main steps of AI?
3. What's the semen collection methods for livestock and poultry?
4. How to measure the quality of semen?
5. What are the main components of semen dilutions?
6. What are the principles of different semen preservation?
7. What are the semen insemination methods for livestock and poultry?
8. What are the main factors affecting *in vitro* fertilization?
9. What are the main steps of *in vitro* fertilization?
10. What are the applications of *in vitro* fertilization?

(Wang Changfa, Gong Ting, Muhammad Faheem Akhtar)

Chapter 3 Pregnancy Diagnosis

Summary

Early pregnancy diagnosis is of great significance to improve reproduction performance. It can reduce non-pregnancy, shorten the birth interval, and reduce the feed cost due to ineffective feeding. The early pregnancy diagnosis techniques include general pregnancy diagnosis technique which consists of external examination, rectal examination, ultrasonic diagnosis, hormone assay, vaginal examination, etc. With rapid development of bioinformatics, smart pregnancy diagnosis technique has been established, which mainly contains omics technology and mid-infrared spectroscopy technology. This chapter introduces the principles and methods of pregnancy diagnosis.

Code 3-1

After animal mating or embryo transfer, it is necessary to know the status of pregnancy and embryo development in time, so as to strengthen the feeding management of pregnant animals, and find out the reason for non-pregnant animals for timely supplement or necessary treatment to shorten the interval between births and improve their fertility. The method of determining whether an animal is pregnant or not is called pregnancy diagnosis. With the development of intensive animal husbandry, early pregnancy diagnosis improves reproduction performance and the economic benefits. With timely and accurate of pregnancy diagnosis, animal breeders can find the failed pregnancy and arrange re-mating for none-pregnant animal early, thereby reducing feed cost due to ineffective feeding.

3.1 Basis for pregnancy diagnosis

During gestation, there are changes in reproductive organs, metabolism and endocrinology, and these changes have varied characteristics in different stages of gestation (Table 3-1). Based on these changes, we can make pregnancy diagnosis in animals.

Table 3-1 Characteristics in different stages of gestation

Change	Content
Body condition	Increased appetite, weight gain, and glossy hair. If the feeding level is not high, the weight loss of the female animal is significant in the middle and late gestation. Abdominal volume increases and edema appears in the lower abdomen and hind limbs. After pregnancy begins, the breasts become fuller due to progesterone and estrogen.
Behavior	No more periodic estrus; the breathing pattern changes from chest to belly breathing. In late gestation, the female animal's movements become steady, cautious and prone to fatigue. The frequency of defecation and urination increases, and the volume of feces and urine per time decreases.
Reproductive organs	The luteum becomes the pregnant luteum and persists throughout gestation. Uterine body and uterine horn were enlarged accordingly. Cervical plug is formed and cervical canal is closure; in early gestation, vaginal is tightening and dry while in late gestation, vaginal is edema and soft.
Hormone	FSH secretion does not change significantly, LH secretion increases in the beginning and then decreases. Estrogen level decreases and progesterone increases. Just before delivery, hormone secretion shows "three high and one low": the level of estrogen, oxytocin, and prostaglandin is high while the level of progesterone is low.

3.2 Principles of pregnancy diagnosis

The method of pregnancy diagnosis should be early, accurate, simple and fast. "Early" refers to the early diagnosis of pregnancy, to detect non-pregnant female livestock as early as possible and to take measures to shorten the pregnant period, which is particularly important for seasonal breeding animals. "Accurate" refers to accurate diagnosis. If the diagnosis is not accurate, the non-pregnant animal will be mistaken for pregnant, it will prolong the pregnant period and reduce fecundity. If a pregnant animal is diagnosed as non-pregnant and taken improper management measures, it will lead to miscarriage. "Simple" refers to the simple method and easy operation, less necessary instruments or reagents, especially suitable for actual production or field operations. "Fast" refers to the time to get diagnosis results. The faster the diagnosis speed, the more practical the value.

3.3 Methods of pregnancy diagnosis

The methods of animal pregnancy diagnosis include general pregnancy diagnosis technique and smart pregnancy diagnosis technique. The general pregnancy diagnosis techniques include external examination, rectal examination, ultrasonic diagnosis, hormone assay, vaginal examination and other diagnostic methods. The smart pregnancy diagnosis techniques include omics technology and mid-infrared spectroscopy (MIRS) technology.

3.3.1 Routine techniques for pregnancy diagnosis

3.3.1.1 External examination

After mating, the animal's posture and behavior change with the progress of gestation. External examination is to observe these changes to make a preliminary judgment of pregnancy status.

1. Consultation

Learn about the general physiological and reproductive status of the animals from the breeding staff, such as age, number of fetuses, date of last delivery and postpartum situation, estrus cycle and estrus behavior, mating mode and number of mating, date of last mating and whether estrus returns after mating, recent drinking water, appetite, behavioral changes and medical history, etc. Based on the understanding of these conditions, we can make a preliminary diagnosis of the physical condition of the subject and the possibility of pregnancy.

2. Visual diagnosis

Visual examination is to observe the body shape, behavior and the changes of some systems and organs on the basis of consultation.

(1) Physical observation After mating, the nutritional status of the animal is improved, the body shape becomes plump and the hair color is smooth due to the enhancement of anabolism. In the late stage of gestation, abdomen is asymmetry, and pregnant side is protruded. The abdomen of pig, dog, cat, rabbit and other multiparous animals is sagging. At the same time, the pregnant animals become gentle and cautious in their actions. At the end of gestation, some pregnant animals have edema in the lower abdomen or two hind limbs.

(2) Fetal movement observation At the end of gestation, especially near delivery, irregular abdominal wall vibration can be observed in the most prominent part of the abdomen. Auscultation is to listen to fetal heart sounds. The auscultation site is on the inside of the right knee fold in cattle and sheep, and on both sides of pigs and dogs. The auscultation time is late stage of gestation, such as cattle after 6 months of gestation. The method to distinguish fetal heart sound from their mother is based on the signs that fetal heart rate is fast, generally more than 100 times/min, and maternal heart sound is not heard in the lower abdomen. When hearing fetal heart sounds, pregnancy can be determined. Though, none-pregnancy or fetal death can not be determined if fetal heart sound is absent.

(3) External palpation The examiner palpates the fetus through the maternal abdominal wall. Those with the fetus touched can be diagnosed as pregnant, but those without the fetus touched can not be diagnosed as to whether they are pregnant. This

method can only be used in the middle and later stages of pregnancy. For small animals such as pig, sheep, dog, cat, and rabbit, this method can not be used because those animals can not undertake rectum examination.

① Pig: Try to make animal quiet and lie down first. Use left and right hands to hold the abdominal wall above the last two pairs of breasts and slide it up, down, left and right. On the 2 months of gestation, it is easy to touch the fetus, and the boundaries between the fetuses are obvious. But for the overweight pigs, it is not easy to touch.

② Sheep: After 3 months of gestation, the examiner clamps the neck of the sheep with both legs, covers the lower abdomen with both hands, and slides the right lower abdominal wall back and forth with his left hand. If there is a hard object in the abdomen, it can be judged as pregnancy.

③ Rabbit: 7-10 d after mating, the examiner can hold the rabbit ear and neck skin with one hand, fix the rabbit on the table, head toward the examiner, and gently touch the rabbit from front to back along the back of the abdominal wall. If the abdomen is soft as cotton, it is not pregnant. If the small smooth meatball is felt on both sides of the back of the abdomen, pregnancy can be diagnosed. From 7-10 d of gestation, the fetus is as big as peanut, 15 d as big thumbling, 20 d as big walnut, after 22 d, no sense of ball, but one can touch the head of the fetus. During palpation at 7-10 d of gestation, the fetus must be distinguished from the fecal ball. The fecal ball is oblate and inelastic with rough surface. The position of the fetus is relatively fixed, spherical, and most evenly distributed on both sides of the back of the abdomen, smooth and elastic when pressed.

④ Dog: 20-25 d after mating, keep the animal in a standing position and stroke her under the jaw to keep her stable. The position of the fetus is between the lumbar spine and the lower abdominal wall between the umbilical orifice and the fourth nipple. Place the left palm against the lower abdomen of the female, thumb on the right abdominal wall, and middle finger on the left abdominal wall. When the dog has low expiratory pressure, compress with both hands pointing to the abdomen and gently slide up, down, left and right to determine fetal position. Compared with other organs, the fetus is a grape-like lump, elastic and easy to distinguish. The fecal masses in the rectum are arranged in strings under the spine and are hard and inelastic. The position of the bladder is between the fourth and fifth pairs of nipples, the middle of the lumbar spine and the lower abdomen. When there is more urine retention in the bladder, it appears soft and fluctuating. The position of bladder is after the fetus, so attention should be paid to the difference in the examination. For small dog, the characteristics of different gestation age are shown in Table 3-2.

Table 3-2 Breast changes and the fetal size and shape in different gestation age of dogs

Gestation age/d	Fetal size/cm	Fetal shape	Breast changes
10	0.4	Circular	—
15	0.6	Circular	The fifth pair of nipples are slightly thicker

Table 3-2 continued

Gestation age/d	Fetal size/cm	Fetal shape	Breast changes
17	0.8	Circular	The fifth pair of nipples are slightly thicker
20	1.0	Circular	The fourth and fifth pairs of nipples are slightly thicker
25	1.5	Circular	The hard core is discovered in nipples
27	2.0	Circular	The hard core enlarges
30	2.5×3	Oblong and forward moving	The hard core enlarges

From Yang (2010)

⑤ Cat: Cats are prone to palpation because their abdominal wall is usually flabby. The best palpation time is between 20 and 30 d of gestation because in this period the separation between fetuses occurs. After this stage, the spacing between fetuses becomes indistinct and can not be palpated. In late gestation, the fetus can be felt outside the abdominal wall.

3.3.1.2 Rectal examination

In current animal husbandry production, rectal examination of cattle, horse and other large livestock is the most reliable method to check pregnancy and embryo development. The advantage of rectal examination is that the preliminary diagnosis can be made after 20 d gestation and confirmed diagnosis can be made after 40 d gestation.

1. Cattle

First touch the cervix, slide the middle finger forward to find the groove between the uterine horns, then move hand forward, down and back, try to hold the uterine horns in hand and touch them respectively. The uterine horn of a pregnant cow falls into the abdominal cavity and is not easy to touch. At this time, first hold the cervix and pull the uterine horn backward, then slide forward quickly with hand and intestinal tube, and hold the uterine horn. In this way, gradually move forward to find out the whole uterine horn. After touching the uterine horn, look for the ovary on the outside or below the tip, which can be palpated with one hand (Fig. 3-1).

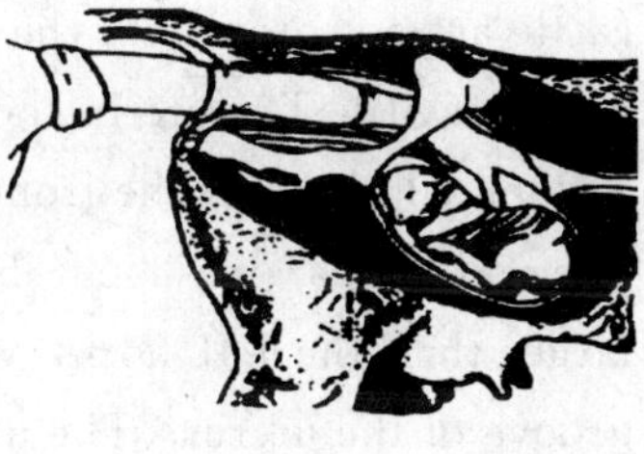

Fig. 3-1 Diagram of rectal examination (Yang, 2009)

(1) Method of finding uterine artery Slide palm forward against the top of pelvis. After crossing the headland, you can clearly touch the last pair of thick branches of the abdominal aorta, namely the left and right internal iliac arteries. The uterine artery and umbilical artery originate from the starting point of the internal iliac artery. The uterine artery enters the broad ligament not far from the internal iliac artery, so it feels free during tracking. The method of palpating the uterine branch of the vaginal artery is to extend the fingertip to the end of the sacrum, stick it near the superior sciatic spine of the lateral wall of

the pelvis, and slide the fingers back and forth. The vaginal artery is a relatively free artery in the pelvis, from up to down, and is very short, so it is easy to identify (Fig. 3-2).

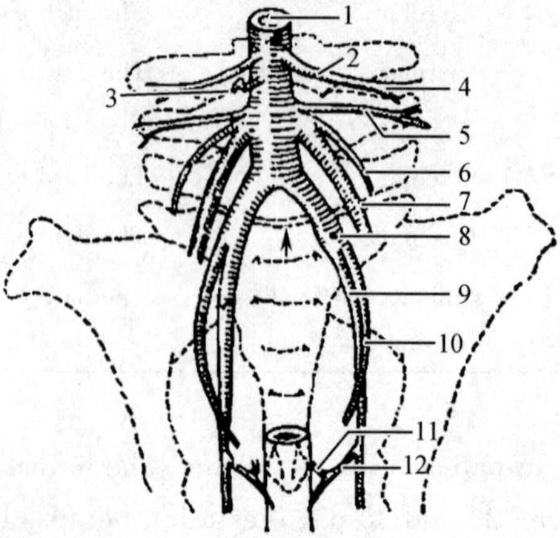

1. Abdominal aorta 2. Ovarian artery 3. Posterior mesenteric artery 4. The tributary of uterine in ovarian artery 5. Arteria circumflex iliac profunda 6. Uterine artery 7. Arteria iliac externa 8. Medial frontobasal artery 9. Umbilical artery 10. Vaginal artery 11. The tributary of uterine in vaginal artery 12. The tributary of vaginal in vaginal artery

Fig. 3-2 Schematic diagram of the various arteries in cattle (Yang, 2010)
(From the bottom to the top, promontory is located in where the arrow points)

The cattle rectal mucosa is easy to exude blood. When operating in the rectum, you can only touch it with finger belly when the intestine is loose, the mucosa should not be touched. In case of rectal contraction, the hand should retreat slightly with the natural contraction wave of the intestine, and should not be strongly extended forward. Due to the short mesentery, the hand should not be forced to extend forward with the intestine. When cattle are not pregnant, the cervix, uterine body, uterine horn and ovary are located in the pelvic cavity. The uterine horn of cattle that has given births for many times can fall into the abdominal cavity at the front edge of the pelvic entrance. The two uterine horns are equal in size and similar in shape, bending like sheep horns. Sometimes the right horn is slightly larger than the left horn, which is loose and plump, and can clearly touch the interhorn groove of the uterus. The uterine horn shrinks upon touching and becomes elastic and hard. The uterus can be held in the hand with obvious interangular groove.

(2) Morphological changes of ovary and uterus On the 20th-25th day of gestation, there is corpuslutea protruding from the surface of ovary, which is much larger than none-pregnant ovary. The thickness of uterine horn is unchanged, but the uterine wall is thick and elastic.

In the 1st month of gestation, the groove between uterine horns is still clear, the pregnant horn and uterine body are thick and soft, and the angular curvature of horn is not obvious. During palpation, the gestational horn generally does not contract, and sometimes it feels elastic with liquid fluctuations, like a soft-shell egg. The empty horn often shrinks, feeling elastic and bending obviously. Gently hold the pregnant horn with fingers and slide it from one end to the other. You can feel the fetal membrane sac sliding between fingers, or gently pinch the uterine horn with your thumb and index finger, then relax slightly, and you can feel a layer of film sliding away in the uterine wall. The thickness of uterine horn depends on the number of births, and those with more births are slightly thicker than those with less births (Fig. 3-3).

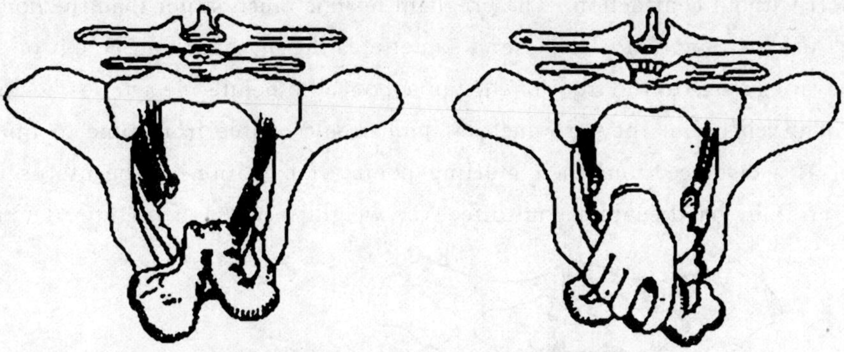

Fig. 3-3 The location of uterus at 1st month of gestation (Zhao et al. , 2002)

In the 2nd month of gestation, the groove between uterine horns is not clear, but the bifurcation between the two horns is still obvious. The uterine horn enters the abdominal cavity. In cattle and calves with fewer births, part of the uterine horn also enters the abdominal cavity. The size of the two uterine horns is obviously different. The pregnant horn is about twice thicker and longer than the none-pregnant horn. The wall of gestational horn is soft and thin with fluid fluctuation. If you can't touch the uterine horn clearly in front of the cervix and only touch a pile of soft things, this condition indicates that the cow may be pregnant. Conducting palpation carefully, you can touch the two uterine horns clearly (Fig. 3-4).

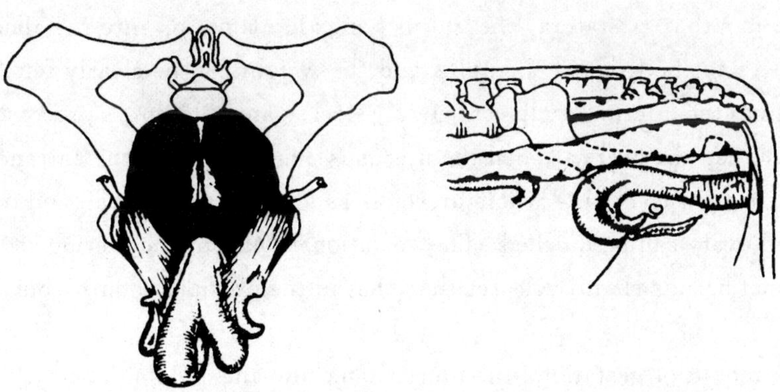

Fig. 3-4 The location of uterus at 2nd month of gestation in front view and left side view
(Zhao et al. , 2002)

In the 3rd month of gestation, the cornual sulcus disappears. The cervix moves to the anterior edge of the pubis. The enlarged uterus can be touched forward from the cervix, sagging from the pelvic cavity to the abdominal cavity, and the width between the two corners is more than one palm. When there are many contents in intestines, the uterus is squeezed into the entrance of the pelvis, and when the uterine wall shrinks, the scope of the whole uterus can be touched. Occasionally, you can touch the fetus floating in the amniotic fluid, and sometimes feel the fetal movement like shrimp jumping. The uterine wall is generally soft without contraction. The pregnant horn is much larger than the none-pregnant horn. The fluid wave motion is clear, and sometimes the placental can be felt on the uterine wall. The ovaries move to the front of the pubic bone. Some uterine arteries begin to display slight pregnant veins, but they are unclear, hidden and visible from time to time, and are easy to feel at a distance from their starting point. If palpation of the uterus is unclear, lifting the cervix by hand enables you to feel the weight increase of the uterus (Fig. 3-5).

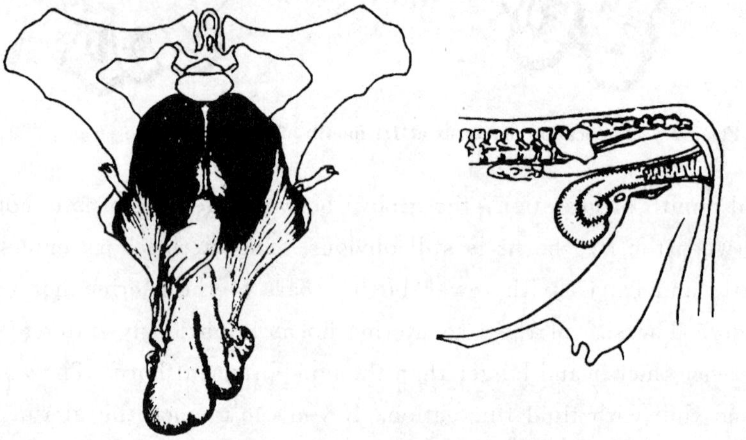

Fig. 3-5　The location of uterus at 3rd month of gestation in front view and left side view
(Zhao et al. , 2002)

In the 4th month of gestation, the uterus hangs like a pocket into the abdominal cavity. The cervix moves to the front of the pubis and the weight can be clearly felt by holding the cervix. Touching the uterine wall can clearly feel many placental processes, which are slightly smaller than the ovary. When the uterus is pushed back to the entrance of the pelvis by the gastrointestinal tract, the whole uterus is as large as a volleyball and occasionally the fetus, horn and ovary can be touched. The gestational pulse of the uterine artery on the side of the gestational horn is slightly clearer than that in the previous month, but it is still slight (Fig. 3-6).

In the 5th month of gestation, the uterus sank into the abdominal cavity. The cervix is palpable just below the front edge of the pubis. The placental process is larger and the fetus can be touched while both ovaries can not be touched. The pregnancy pulse of uterine artery on the pregnant side has been obvious.

In the 6th month of gestation, the uterus sinks to the bottom of the abdomen Due to the short mesocolon, the fetus can only be touched when the stomach and intestines are full and

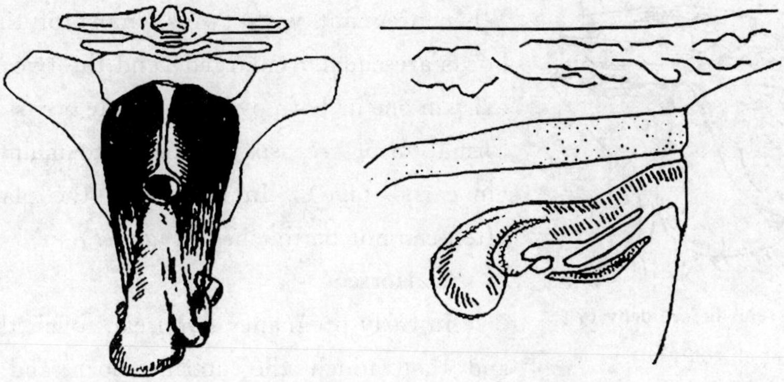

Fig. 3-6 The location of uterus at 4th month of gestation in front view and left side view

(Zhao et al. , 2002)

the uterus moves backward and rises. The placental process is as large as pigeon egg and is easy to touch on both sides of the gestational horn. The uterine artery on the side of pregnancy side is thick, and the pregnancy pulse is also obvious. Weak pregnancy pulse appears in the uterine artery on the none-pregnant side.

In the 7th month of gestation, the fetus is larger and is easier to touch. The placental process is also larger. Both uterine arteries have obvious pregnant veins, but the none-pregnant side is weak. Individual cattle have no obvious pregnancy pulse even before delivery. The uterine branch of the vaginal artery on the side of the pregnant horn began to appear pregnancy pulse.

In the 8th month of gestation, the cervix returns to the leading edge of the pelvis or pelvic cavity, where it is easy to touch the fetus. The placental process is as large as a duck egg. The pregnant veins of both uterine arteries are obvious. The pregnant veins of the uterine branch of the vaginal artery on the side of the pregnant horn are clear.

In the 9th month of gestation, the anterior part of the fetus enters the pelvic entrance. All uterine arteries have significant pregnant veins. When hand reaches into the anus, do not need to touch the vaginal branch of uterine artery. As long as you stick it to the side wall of pelvis, you can feel the vibration of pregnant veins.

(3) Precautions Within 2 months of gestation, the uterine body and horn expand together, and the wall is soft. Unlike horses with clear elastic round bubbles, a pile of soft tissue with wave motion of cattle can be touched. In 3-4 months of gestation, the uterus is enlarged and perpendicular to the abdominal cavity. It is difficult to find out the uterine wall, and the diagnosis may be difficult. At this time, you can hold the cervix and feel the weight of the uterus, which can be judged as pregnancy. In addition, it is easy to touch the placental process at 4th month while the ovary in the pregnant side is not easy to touch. After 5th month of gestation, it is easy to make a diagnosis according to the cervical moving into the abdominal cavity, the uterus expanding and sagging to the abdominal cavity, the inability to touch the ovary, the presence of placental process, fetus and pregnancy pulse.

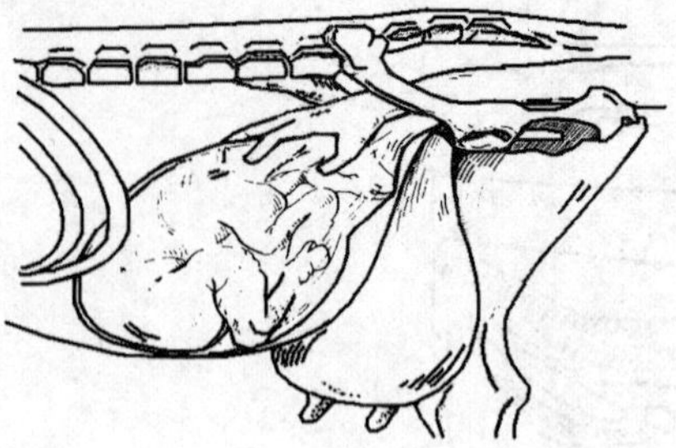

Fig. 3-7 The fetus before delivery

(Zhao et al. , 2017)

When pregnant with twins, most of the two uterine horns are equally enlarged, and the two corpus luteum exist in one or both ovaries. Cattle occasionally have the formation of accessory placenta to maintain pregnancy (in early stage). In this case, the placental process often can not be touched (Fig. 3-7).

2. Horse

In early pregnancy, you can touch the ovary first, and then touch the uterine horn and uterine body backward. You can also find the uterine body and uterine horn directly according to the hook bottom method. Beginners can change hands when touching the uterine horn, that is, right hand touching the left corner and left hand touching the right corner. Skilled people can also use one hand to check the uterine horns on both sides. When looking for the uterine horn, first pinch the broad ligament of the uterus with fingers under the back of the ovary, and then slide fingers towards the midline to feel that the uterine horn is like the edge of a woolen sweater. After touching the uterine horn, you should touch the uterine body and pay attention to whether there is any abnormality. When touching the uterine body, the hook bottom method can be used, that is, put the palm in the enlarged part of the rectum, put the finger in the narrow part, bend the finger downward and pull it back, and then hook the uterine body in the hand for palpation. More than 3 months after mating, the ovary can still be touched first, and then the uterine horn and uterine body in turn. If horse is pregnant, the ovaries are falling or can not be touched. At this time, it is necessary to palpate the uterus, fetus and uterine artery. The way to find the uterine artery is to slide the palm forward against the top of the pelvis, cross the bifurcation of the promontory and internal iliac artery, and then move forward not far, the thick external iliac artery can be touched. At the leading edge of the starting point of the external iliac artery, the uterine artery can be touched. There is deep circumflex iliac artery at the starting point of external iliac artery which is not easily confused with uterine artery because it is located above the peritoneum. The method of finding the uterine branch of the vaginal artery is the same as cattle. If horse is not pregnant, both ovaries are usually close to the spine. The two uterine horns are equal in size, flat and soft. After being stimulated by touch, uterine often contract and become round. After a short time, they relax and become banded again (Fig. 3-8).

On the 16th-18th day of gestation, the degree of uterine horn sclerosis increases, the tip is not flat when pinched gently, the inside is hard and the outside is soft. At the base of uterine horn and 2-3 cm in the middle of uterine fundus, the downward protruding blastocyst feels obvious, whose size is as pigeon egg. The none-pregnant horn is more curved and longer while the pregnant horn is more straight and shorter. Some pregnant horses have concave grooves at the two corners of the uterine fundus. There is no alternation of shape,

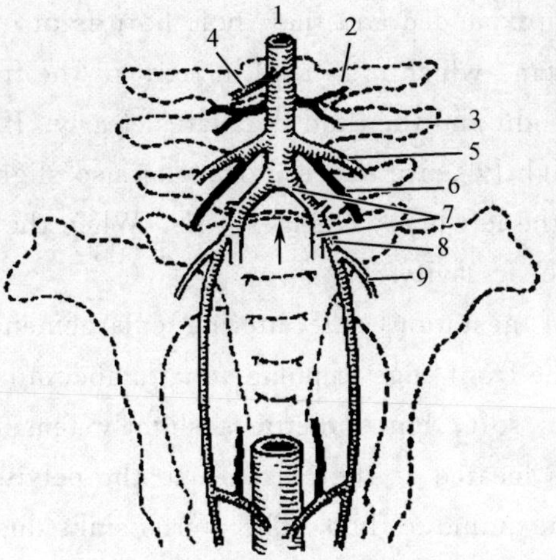

1. Abdominal aorta 2. Ovarian artery 3. Anterior uterine artery 4. Posterior mesenteric artery 5. Arteria iliac externa 6. Uterine middle artery 7. Internal iliac artery 8. Internal pudendal artery 9. Umbilical artery

Fig. 3-8 Schematic diagram of the various arteries in horses (Zhao et al., 2017)

hardness, straightness and softness when palpating the uterine horn.

On the 20th-25th day of gestation, both uterine horns are cylindrical and pointed at the front. The uterine wall is thick, hard and elastic, but the none-pregnant horn is soft and thin. The alternation of uterine muscle contraction and relaxation disappeared. A swelling which size is like a table tennis appears at the junction between the base of the horn of pregnancy and the uterine body. The gestational horn is short, and a bend is often formed on the none-pregnant horn. Sometimes a bend is also formed before the enlargement of the gestational horn.

On the 30th-35th day of gestation, the pregnant horn is thick and short while the none-pregnant horn is soft. The blastocyst is located on the transverse oval swelling which is below the junction of the pregnant horn and the uterine body. Sometimes, the blastocyst is located at the base or middle region of the gestational horn. The diameter of the expanded swelling is 4-5 cm, which is the same size as an egg. Generally, the swelling is larger if horse has experienced pregnancy more times, otherwise it is smaller. The touch feeling is hard and elastic, sometimes is soft and fluctuating. On the 35th day of gestation, the lower wall of the expanded part is more prominent than the upper wall, and a concave groove is formed between the lower wall of the expanded part and the none-pregnant horn.

On the 50th-60th day of gestation, the diameter of the expanded part is 10-12 cm, which is as large as softball or even slightly larger. The wall feels tense and hard when touching, but sometimes it feels soft and thin. The liquid in the swelling part fluctuates clearly. The thickness of the two uterine horns is significantly different, and the top half of the pregnant horn is thicker than the none-pregnant horn. The uterine body hangs to the pelvic floor.

On the 75th day of gestation, in addition to the uterine body and horn, the base of the none-pregnant horn is also expanded and the whole horn is of oval shape. The diameter of the enlarged part is 14-16 cm, which is located in front of the front edge of the pubis. The blastocyst wall is not too tight and the fluid fluctuates clearly. If right horn is pregnant, the right broad ligament is slightly tense and the ovary is also slightly sagging; if left horn is pregnant, the position of the left ovary changes little. When the intestine is full, the uterus can be squeezed into the pelvic cavity.

In the 3rd month of gestation, the uterine enlargement is slightly smaller than volleyball, sinking from the front edge of pubic bone to abdominal cavity and slightly to the right. The uterine wall is soft, but sometimes slightly tense, and the liquid fluctuates obviously. If the uterus is located at the entrance to the pelvis, the fetus can be touched occasionally and float in the amniotic fluid. The uterus sinks due to the increase of weight, and the broad ligament of the uterus and themesangium of the ovary (especially the anterior edge of the mesangium) become tense. The ovaries on the side of the pregnant horn decrease to the level above or below the middle of the pelvic cavity, and the distance between the two ovaries decrease.

In the 4th month of gestation, the uterine wall sinks forward in a bag shape at the front of the pubic bone (Fig. 3-9). The wall is soft and the liquid fluctuates clearly. The whole uterus is as big as a volleyball. The cervix is located at the anterior edge of the pubis. The uterus can sometimes be squeezed by the filled intestine to the front edge of the pelvic entrance, touching the fetus floating in the amniotic fluid. Two ovaries are closer to the midline of the body. The ovary on the side of the pregnant horn drops to the level of the front of the pubis or lower, depending on the intestine filling. The uterine artery on the pregnant horn is thicker than that on the none-pregnant horn. There is a slight pregnancy pulse, which can be felt only by slight palpation.

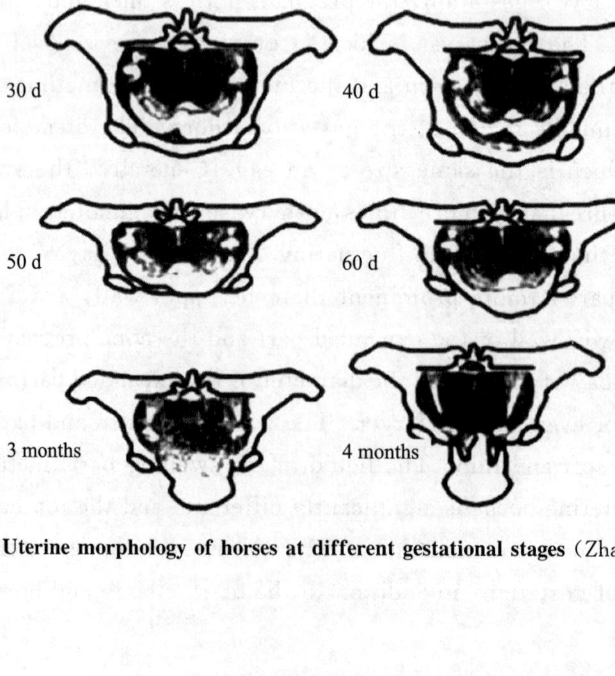

Fig. 3-9 Uterine morphology of horses at different gestational stages (Zhao et al. ,2017)

In the 5th month of gestation, the phenomenon of pregnancy is basically the same as that at the 4th month. The uterus is larger and sinks deeper into the abdominal cavity, so it is not easy to feel fluctuations. The ovaries are also lower and close, and their position is not fixed, so they are often inaccessible later. The pregnancy pulse of uterine artery on the side of pregnancy horn is still slight, but some individuals are stronger.

In the 6th month of gestation, the uterus sinks into the abdominal floor and is covered by the intestinal tract, which is not easy to touch, but the downward uterine wall can be touched at the front edge of the pubic bone, and the fetus can also be touched. The broad ligament of the uterus is tense. The uterine artery on the side of gestational horn is obvious.

In the 7th-8th month of gestation, the fetus is already large, so it is easy to touch. The pregnant veins of the uterine artery on the side of the pregnant horn are obvious and weak on the side of the none-pregnant horn. The uterine branch of vaginal artery on the side of pregnancy horn also begin to appear pregnant pulse.

In the 9th month of gestation, the cervix enters the pelvic cavity. The uterine arteries on both sides are thick, the pregnant veins on the side of pregnant horn are obvious, and the pregnant veins on the side of none-pregnant horn are also clear. The pregnant veins of the uterine branch of the vaginal artery on the side of the pregnant horn also begins to be clear.

In the 10th month of gestation, the thickness of uterine arteries on both sides is almost the same, and the gestational veins on the side of gestational horn are more obvious. The pregnant veins of the uterine branch of the vaginal artery on the side of pregnant horn have been obvious, and the pregnant veins of the vaginal artery on the side of none-pregnant horn also have begun to appear.

In the 11th month of gestation, the pregnant veins of all uterine arteries are significant. Once hand reaches into the rectum, the tremor of the pulse can be felt. The anterior part of the fetus sometimes enters the pelvic cavity.

It should be noted that after 5 months of gestation, it is easy to determine whether pregnant by rectal examination, but early diagnosis is required in production. In rectal examination within 1 month of pregnancy, you should pay attention to the changes of uterine horn. During gestation, the uterine horn is flat ribbon, and after gestation, it becomes cylindrical sharp. The none-pregnant horn is long and curved while the pregnant horn is thick, short and straight, and there is vesicular swelling at the base. A concave groove also appears in the fundus of the uterus during gestation. The diagnosis of pregnancy should be based on the size, shape, position and texture of uterine horn; you especially pay attention to the vesicular swelling.

When rectal examination is used to diagnose pregnancy in horses and cattle, special attention should be paid to the following problems.

(1) Difference from false pregnancy False pregnancy often occurs within 40 d after horse (donkey) mating. There is no significant difference between the changes of uterus and real pregnancy. The uterine horn is thin, round and hard, there are concave grooves at the

bottom of uterus, the vagina is dry and pale, and there is cervical mucus plug. The difference is that the blastocyst can not be touched until 30-40 d or more. Most of the ovaries have no follicles, occasionally have follicular development, and do not ovulate.

(2) Difference from bladder filled with urine Uterus on the 60th to 90th day of gestation may be confused with bladder filled with urine. Especially the uterus of bovine gestation for 2 months becomes longitudinally oval when contracting, the transverse diameter is about one palm wide, and the wall is tight, which is very like a urinary bladder. But the bladder contour is very clear, the surface has a sense of reticulation, and there is no involvement on both sides. There are uterine horns and broad ligaments in front and on the side of the enlargement of the horse's uterus. There are bifurcations at the front two corners of the cow's uterus and cervix at the back, which can be clearly felt. Therefore, it is not difficult to distinguish them from the bladder.

(3) Difference from uterine diseases The uterus at 90th-120th day of gestation is easy to be confused with uterine pus and effusion. When the uterus accumulates pus or effusion, one side of the uterine horn and uterine body expand and increase in weight, which makes the uterus sink and the position of the ovary decrease accordingly. In case of effusion, the liquid can flow from one corner to another. When pus accumulates, as part of the water is absorbed by the uterine wall, the thick fluid thickens and feels like dough. Whether it is pus or effusion, the uterus has no pregnant signs, no pregnant pulse, and the cotyledons can not be touched in cattle.

(4) Difference between pregnant pulse and some quivering arterial beats The uterine artery of none-pregnant cattle that has been delivered for many times also has similar quivering of pregnant pulse, which should be paid attention to. When horses suffer from parasitic aneurysms, the posterior mesenteric artery also has fibrillation pulse. Do not confuse it with the pregnant pulse of ovarian artery and uterine artery. There is only one posterior mesenteric artery, which originates below the aorta at the junction of the 4th-5th lumbar vertebrae. In addition, the abdominal segments of the bilateral mammary arteries of cattle also have fibrillation arterial beats during lactation. However, they are located on both sides of the front edge of the pubis, which is very thick and easy to identify.

3. Sheep

When using rectal examination for sheep, the instrument used in this method is a plastic rod or wooden rod with length of 45-50 cm, diameter of 1-1. 5 cm and smooth surface. The tested sheep will go on a hunger strike for one night. Before examination, enema with soapy water to discharge the feces in the rectum. During the examination, keep the sheep supine and keep its hind legs upward to relax the abdominal muscles. Apply lubricant to the probe rod, gently insert it into the rectum, and move it back and forth, left and right until it reaches 30-35 cm. After the probe rod is inserted to the required depth, gently press down the outer end of the rod to lift the inner end of the rectum slightly to hold up the fetal (Fig. 3-10 A and B).

When touching the pregnant uterus, you can usually feel the rotating direction of the probe rod. If you feel an entity like a uterus, it can be determined as pregnancy. If sheep is not pregnant, you can not feel the fetus from the left to the right at the back of the abdominal cavity. At the same time, you can clearly touch the plastic rod from the outside of the abdominal wall. Special attention should be paid to preventing the probe rod from sliding between the uterus and abdominal wall to avoid misdiagnosis (Fig. 3-10 C).

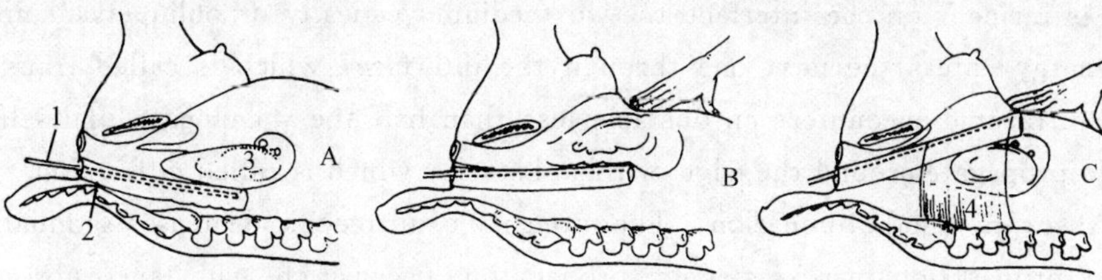

1. Probe 2. Rectum 3. Fetus 4. Probe slides

A and B. Correct method (press the probe to the rear of the abdominal cavity where the fetus is located)

C. False method (probe slides between uterus and abdominal wall)

Fig. 3-10 Probe rectal examination in sheep (Zhao, 2017)

3.3.1.3 Ultrasonic diagnosis

In the mid-1960s, ultrasonic diagnosis was applied in the field of animal medicine. Type A and type D diagnostic instruments were used, and type M and type B were found in the mid-1970s. According to the published data, ultrasonic diagnosis can be used in almost all kinds of mammals. In the United States, Japan, Australia and European countries, B-mode ultrasonic diagnostic instrument has entered horse sheds, sheep farms and pig pens, and has been popularized to grass-roots veterinary stations, especially in small animal diagnosis. With the continuous improvement of ultrasonic technology, in ultrasonic tomography, the imaging of ultrasonic detection has been achieved and become an effective method for the gross morphology of living tissues and organs (Fig. 3-11).

1. Physical characteristic of ultrasonoscope

(1) Beam feature or pointing The sound source transmitting ultrasonic is concentrated on one direction when working, similar to a flashlight beam because of its high frequency and short wavelength which is far smaller than the diameter of the transducer. The spatial range to which ultrasonic propagates from the source is called the ultrasonic field. The distribution of the field is narrow and cylindrical as well as the width is close to the size of the piezoelectric crystal of transducer, so that the

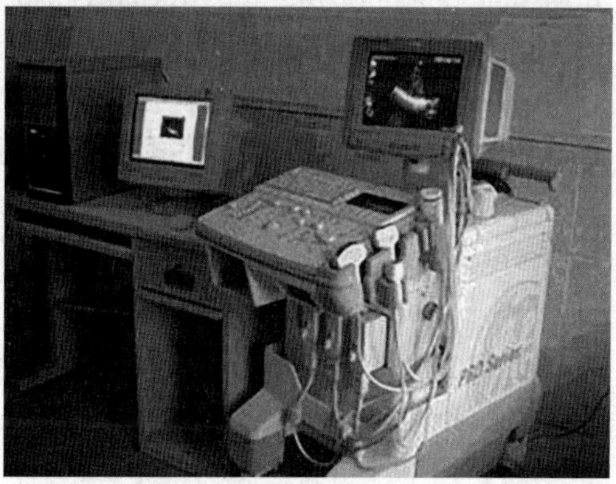

Fig. 3-11 Colorful B-mode ultrasonic apparatus
(Yang, 2009)

ultrasonic waves is characteristic of beam feature or pointing.

(2) Reflection, refraction, transmission, scattering and diffraction　When the ultrasonic propagates in the medium and meets the acoustic interface formed by two substances with different acoustic impedances, part of the acoustic energy returns to the original medium from the interface, which is called reflection. Another part is transmitted to the next medium through the interface with the direction changing, which is refraction. When ultrasonic is incident on the interface of two medium vertically or obliquely, some of the acoustic energy enters the next one through the interface, which is called transmission; When the ultrasonic encounters an obstacle less than half the wavelength of itself, it will continue to propagate around the edge of this obstacle, which is called diffraction.

(3) Absorption and attenuation　The intensity of ultrasonic decreases gradually as the distance of propagation increases when propagating because the substance absorbs some energy. When propagating in the same substance, the higher the frequency of wave, the stronger the absorption. While it absorbs most when it propagates in a gas, weaker in a liquid, and least in a solid when the wave frequency is settled. And attenuation is that the energy decreases with the increase of propagation distance.

(4) Doppler effect　Doppler effect is the phenomenon—frequency changing—that the received vibrational frequency is different from that emitted by the vibration source when existing relative motion between the source and the reflection (receiver).

2. Basic principle of ultrasonic imaging revisited

There are specific acoustic impedances and attenuation characteristics in the structures of various organ and tissues including pathological tissues, thus the structure of animal body is a complex medium for ultrasonic. The ultrasonic will go through organs and tissues from surface to deeper parts with different acoustic impedances and attenuation characteristics after being radiated into the body, leading different reflections and attenuation, which is the basis of the ultrasonic image. According to the intensity of the received echo, it can be displayed on the fluorescent screen with different light spots and then the cross-sectional ultrasonic image of the muscle that can be also called sonogram is formed.

3. Types of ultrasound diagnosis

Ultrasound diagnosis is a physical examination method based on the physical characteristics of ultrasound and the acoustic characteristics of animal tissue structure. It can accurately detect different tissues and organs of the body. At present, there are several ultrasonic diagnostic techniques for animal pregnancy diagnosis.

(1) Type A ultrasonic diagnosis　Type A ultrasonic diagnosis instrument adopts amplitude modulation display. The received echo is displayed on the oscillograph screen with different amplitude and density. The intensity of the echo is reflected by the amplitude. If there is no echo, there is a flat segment. The flat section can be divided into liquid flat section and solid flat section. The liquid is a uniform medium, which does not reflect the ultrasonic wave. If the instrument increases the transmission gain, it still does not reflect.

This kind of flat wave is called liquid flat wave. Substantive organs and muscles are also uniform media and do not reflect ultrasonic waves. However, when the instrument increases the transmission gain, the tissue structure will produce small reflected waves. This kind of flat section is called solid flat section. The image composed of wave amplitude is called echo map (Fig. 3-12).

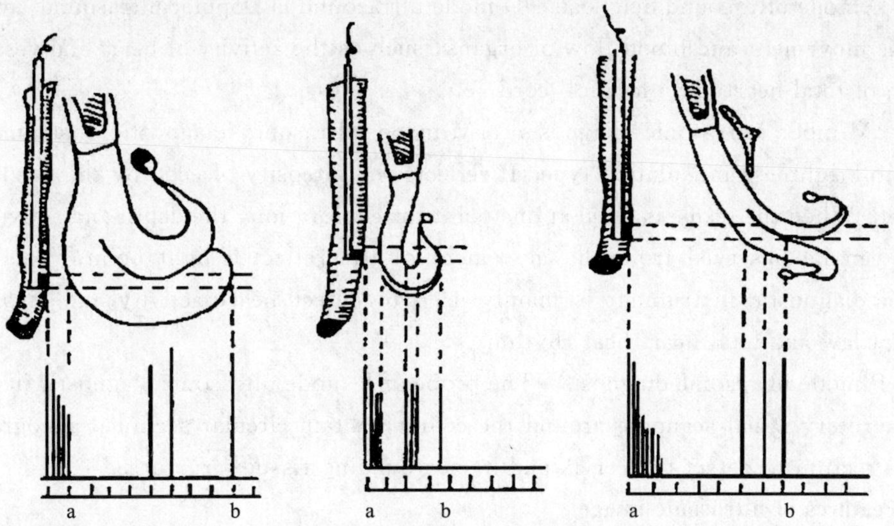

a: the wave into uterus; b: the wave out uterus; a-b: The horizontal section of fetus

Left: The echo-image at two months and seven days gestation

Middle: The echo-image of un-pregnant uterus in pelvic

Right: The echo-image of un-pregnant uterus sank into abdominal cavity

Fig. 3-12 The echo-images of uterus in pregnant and non-pregnant cattle (Yang, 2010)

(2) B-mode ultrasonic diagnosis B-mode ultrasonic diagnostic instrument belongs to brightness modulation type. It causes the vibration of piezoelectric crystal of ultrasonic probe through pulse current and emits multiple ultrasonic beams at the same time. It detects on a section, and uses the reflection of sound wave to convert it into pulse current signal through the probe. Light spots with different brightness are formed on the screen to display a sectional image of the detected part. Due to the different acoustic resistance values of tissues, there is a difference in the intensity of acoustic reflection. When the parts without reflection (liquid) and strong reflection (dense tissue) are detected, the black without echo and the white of strong echo are displayed respectively (which can be changed due to different scanning display methods). Therefore, B-mode ultrasound diagnosis is also called ultrasonic tomography. The imaging speed of this method is fast. It can immediately display the activity state of the examined organ, which is called real-time ultrasound diagnosis. B-mode real-time ultrasonic tomography instrument can be equipped with linear array scanning probe and sector scanning probe, which are called real-time linear array scanner and real-time sector scanner respectively. In addition, some B-mode ultrasonic tomography instruments can be matched with linear array and sector probes. The image scanned by linear array is

rectangular, while the image scanned by sector probe is fan-shaped. Both real-time linear array scanner and sector scanner can realize the imaging of dynamic tissues and organs, such as heart beating, fetal movement and uterine contraction. They are most suitable for pregnancy diagnosis of animals.

(3) D-mode ultrasound diagnosis D-mode ultrasound is Doppler ultrasound, which can detect the movement and blood flow of organs, such as the activity of heart and vessels, the pulsation of fetal heart and umbilical cord, etc.

(4) M-mode ultrasonic diagnosis M-mode ultrasonic diagnostic instrument also belongs to brightness modulation type. It reflects the intensity of echo by the brightness of light spot. When the probe is fixed at one point for exploration, the depth and activity of the reflector can be observed from the movement of the reflected light point. This type of ultrasonic diagnostic instrument is mainly used to detect heart activity and display fetal movement law and fetal heart beat rhythm.

(5) P-mode ultrasonic diagnosis The probe of P-mode ultrasonic diagnostic instrument can make rotary radial scanning around the center, obtain circular sectional sonogram, and enter the rectum to detect the cervix and its surrounding tissue structure.

4. Features of ultrasonic image

(1) Cross-section diagram It is sectional images that are formed with light spots according to the echo signal reflected from the animal body. This kind of image is similar to the anatomical structure of the body, so it can not only show the size, shape and internal structure of organs, but can distinguish the substantial, liquid or gas-containing tissues.

(2) Different shades of gray reflect the intensity of the echo The ultrasonic will go through organs and tissues from surface to deeper parts with various acoustic impedances and attenuation characteristics after being radiated into the body which lead to different reflections and attenuation because of the diverse densities in the various animal tissues and the different acoustic impedance of the two adjacent tissues. Thus, on the screen light or dark spots will be produced.

(3) Real-time display The real-time dynamic image can be formed by pictures that are scanned repeatedly and rapidly because of the fast propagation and imaging of ultrasonic. Therefore, the motor function of the heart, fetal heart rate and gastrointestinal peristalsis can be observed in real time.

(4) Susceptibility to interference from gas and fat affecting the quality of the image The acoustic impedance of air is 0.000 428 and that in soft tissue is 1.5, with the difference between them about 4 000 times, which causes the strong interface reflectivity. In order to prevent the sound energy from reflecting largely on the body surface and the deficiency of the sound energy to reach the detected site, there must be no gap between that probe and the body surface of the animal in the clinical detection. While with reduction of the far-field resolution, the detection of structures such as lymph nodes and extra-serosal tissues can be disturbed by gas and subcutaneous and abdominal fat, degrading the image quality.

(5) The small range of displaying Ultrasonic images only show the range that the probe has covered and the depth is limited.

5. Description of ultrasonic image

(1) No echo The field where the ultrasound passes and no reflection exists becomes a dark and non-echo area (black shadow). Reasons: in fluid sonolucent area—homogeneous liquid, there is no difference or little difference in acoustic impedance of them, which can not form the reflective interface and develop the fluid sonolucent area, e. g. blood, bile, amniotic fluid, urine et al. , pathological fluid: ascites, hydronephrosis and cysts containing fluid. In attenuation sonolucent area—tumor, the absorption of ultrasound in the tumor leads to obvious attenuation and deficient echo. In substantial sonolucent area—parenchymal organ sonolucent area, the difference of their acoustic impedance is small, and an echo-free dark area can be discovered. Normal tissues such as renal parenchyma and spleen, and renal cell carcinoma and transparent lesion tissues may appear as parenchymatous dark areas.

(2) Low echo The internal echoes are uniformly dotted in parenchymatous organs such as the liver. The acoustic impedance where occurring acute inflammation with exudation is smaller than normal tissue and acoustic transmission increases causing a low echo area (gray shadow).

(3) Slightly stronger echo Tissue is dense and reflection is enhanced appearing as a dense light spot or light mass (gray and white shadow).

(4) Stronger echo The structure is dense and there is difference between adjacent soft tissue or fluid of acoustic impedance, causing strong reflection. Mainly seen in bone, stone, calcification and performed as white shadow.

(5) Strongest echo Organs containing air such as the lungs and the inflated gastrointestinal tract are showed as extremely bright bands.

The description of each form is seen in Table 3-3.

Table 3-3 The description of each form

Echo form	Description
Light spot	The pointy echo which is thin and round
Facula	The pointy echo which is slightly larger
Echogenic mass	The echo which can be formed into a mass
Schistose echogenicity	The echo which can be formed into a slice
Striation	The echo which is thin and long
Echogenic band	The echo which is wider than striation
Echogenic ring	The echo is annular with the bright edge and dark medium, usually seen around the cancer nodule
Reticular echogenicity	The echo which is formed by a plurality of ring echoes
Cloudy echogenicity	Seen in contrast echocardiography

Table 3-3 continued

Echo form	Description
Acoustic shadowing	Due to the attenuation, reflection or refraction of acoustic energy at the acoustic interface (between soft tissue and gas, between soft tissue and bone or stone), the acoustic energy can not reach the non-echo area (dark area) which is under the extra-strong echo

From Yang (2009)

6. Exploration method

The main methods of pregnancy examination with ultrasonic diagnostic instrument include external abdominal exploration, vaginal exploration and rectal exploration. The appropriate exploration method can be selected according to the size of the animal and the different parts of the explored tissues and organs, so as to achieve the ideal exploration effect (Fig. 3-13).

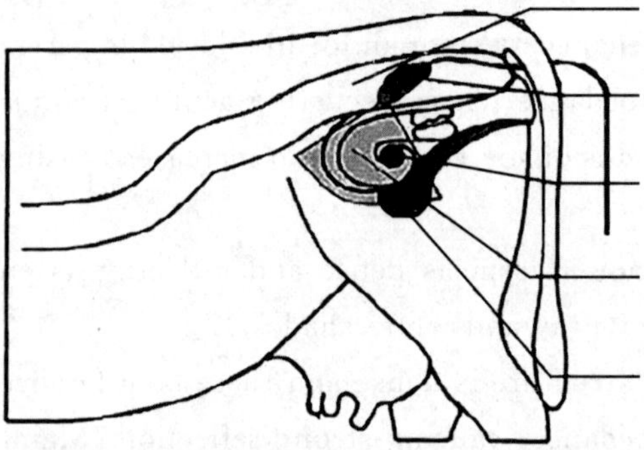

Fig. 3-13 Exploration method using B-mode ultrasound (Yang et al., 2009)

(1) External abdominal exploration It is mainly suitable for detecting the uterus and fetus of small and medium-sized animals, such as pig, sheep, dog, cat, rabbit and other wild animals. The site of exploration can be selected according to the position of uterus and fetus in the abdomen. The exploration site of pigs can be selected on both sides of the last pair of nipples. The probe is aimed at the entrance of pelvic cavity and moves forward gradually during the gestation. The exploration site of sheep can be the glabrous area on both sides of the breast or the straight front part of the breast. For the exploration of dogs and cats, the probe can be placed in the midline of the abdominal floor, moved from back to front, and explored to the left and right at the same time. In order to ensure good exploration effect, the hair of the exploration part should be scraped off, and coated with coupling agent. The probe angle shall be adjusted slowly during the exploration to make the exploration range into a sector. The probe should be a 3.5-MHz or 5-MHz sector scanning probes.

(2) Vaginal exploration It is mainly used for bovine uterus and ovary examination. During exploration, slowly send the probe into the vagina to reach the lower part of the fornix, and then adjust the angle of the probe down, left and right, so as to clearly observe the ovary and uterus.

(3) Rectal exploration It is mainly applicable to large animals, such as horses, cattle, camels, etc. (Fig. 3-14), as well as pig and sheep. During exploration, the short rectal probe can be brought into the rectum by hand, and the crystal side of the probe can be close to the uterus or ovary across the rectal wall. Fine scanning images of the ovary and corpus luteum, follicles, pregnant uterus, fetal body and fetal heartbeat can be obtained. Long handle rectal probe can be used on sheep. The pregnant uterus or fetal body can be detected by slowly inserting the rectal probe into the rectum to a certain depth and adjusting the probe angle in the left and right directions. Because the rectal exploration method can move the probe in the abdominal cavity across the rectal wall and accurately fix it on the explored tissues and organs, a clear and ideal image can be obtained by using a high-frequency probe (5. 0 MHz or 7. 5 MHz).

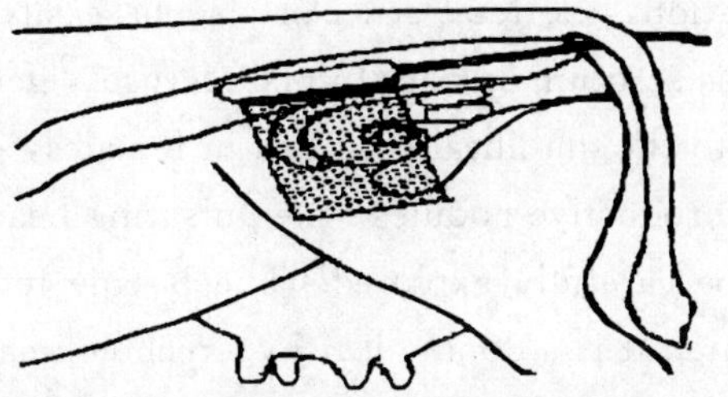

Fig. 3-14 The diagram of B-mode ultrasound in rectal of cattle (Yang, 2010)

7. Application of ultrasonic diagnosis technology

(1) Early pregnancy diagnosis Using ultrasound diagnosis technology, especially through rectal exploration, you can carefully observe the development of embryos or fetuses in early gestation, and make accurate pregnancy diagnosis according to the image characteristics of embryos or fetuses in different stages of development. Ultrasound diagnosis technology has been widely used to diagnose early pregnancy in animals in Britain, United States, Japan and Germany. The earliest time that when B-type ultrasonic diagnostic instrument can distinguish embryos in cow is 10-17 d after mating. At 13 d after mating, 93% of cows have fetal bodies in the uterus, and all embryos are in the uterine horn on the same side of the pregnant corpus luteum. It is reported that the accuracy of pregnancy diagnosis with 3 or 5 real-time ultrasound scanner is more than 80% at 20 d and 100% at 26-33 d after cattle mating. Ultrasonic early pregnancy diagnosis technology is more widely used in small and medium-sized animals. After mating, at 22 d of pig, 40 d of sheep, 20 d of dog and 8-9 d of rabbit, Doppler ultrasound or B-mode ultrasound diagnostic instrument can be used for pregnancy diagnosis with high accuracy.

(2) Observation of fetal development Using B-mode ultrasound diagnosis, you can observe the changes of fetal external structure (uterus, placenta, fetal membrane and umbilical cord), fetal shape (fetal inherent shape, limbs and external genitalia) and fetal internal structure (heart, lung, internal organs and bones), so as to monitor fetal development. Pregnancy, early fetal death, fetal age and sex can be diagnosed by monitoring fetal development.

(3) Estimate the number of fetuses Using ultrasonic technology, you can estimate the number of fetuses, which is mainly used for small and medium-sized animals with multiple fetuses such as pig, sheep, dog, cat and rabbit, so as to feed reasonably and ensure the good development of fetuses. It is reported that the number of fetuses can be estimated by real-time ultrasound diagnosis at 25 d after pig mating and 21 d after dog mating.

(4) Prediction of fetal sex Ultrasonic image detection is a fast, simple and effective method among the various methods of fetal sex identification in horses and cattle. The sexual characteristics of fetal development during pregnancy constitute the morphological basis for ultrasonic detection of fetal gender. In ultrasonic detection, the sex can be determined according to the scrotum of male fetus and breast of female fetus. Using this method, the accuracy of sex identification of fetuses at 50-120 d of gestation can reach 94%-100%; After 60 d of gestation, the fetal sex can also be easily distinguished by detecting whether the bovine fetus has scrotum or not. In male fetuses, genital nodules move from the beginning of the hind legs to the umbilical cord and in females, genital nodules move to the tail. In order to identify reproductive nodules, the pulsating heart, umbilical cord, hind leg and caudal region should be carefully explored. The bovine fetal reproductive nodule is a protuberant bilobar structure. it is a double leaf hyperechoic image with approximately oval shape in each leaf. On the 48th-49th day of pregnancy, the genital nodule is located between the two hind legs, and then the genital nodule of the male fetus gradually moves to the umbilicus and reach the tail of the umbilical cord at the 56th day. The reproductive nodule of female fetus moves under the tail at the 53th day. Under production conditions, the best time to detect fetal sex is 50-68 d of gestation. The ultrasonic image of equine reproductive nodule is similar to that of cattle, which is a double lobe strong echo structure. At about 80 d of gestation, it begins to show a three-lobe shape in some male and female fetuses. At the 63th day of gestation, the reproductive nodule moves to the tail of umbilical cord in males and under the tail in female fetuses. The best time to identify the sex of equine fetuses is at 59-68 d gestation. The accuracy of fetal sex determination can be 97% at 50-99 d gestation.

(5) Prediction of gestational age and delivery date The uterine diameter, chorionic sac diameter, fetal head diameter and fetal internal structure are closely related to fetal age. Using real-time ultrasound diagnosis technology to observe the above indexes, you can predict fetal age and delivery date. 21-24 d before delivery in dog, the length of fetal hip is about equal to the transverse width of placental belt; 8-11 d before delivery, the fetal thorax, heart cavity, stomach, lung and bladder are observed for the first time. The biparietal diameter (BPD, unit: mm) of the fetus of llama is measured by 5 MHz linear array ultrasonic scanner, and then the gestational age (GA, $GA = 18 + 3.7\ BPD$) is calculated by the formula. The date of delivery can be predicted by calculating the gestational age.

8. Research progress in ultrasonography diagnosis

In the 1990s, American scientists developed a three-dimensional ultrasound scanning technique that allows doctors to study the internal organs of patients as if they had opened a window on the body. "It makes the current ultrasound technology obsolete", said Olaf Ram, a co-inventor and director of the Center for Engineering and Research in Emerging Cardiovascular Technologies at Duke University in North Carolina.

The 3D ultrasonic technology analyzes many reflected waves in real time using parallel computing and produces images quickly. Biomedical engineers at Duke's Pratt School of Engineering have created a new three-dimensional ultrasound heart imaging probe, which can produce images of the whole heart by inserting into the esophagus. The time that it takes to form images is the time that current ultrasound technology takes to visualize a cross section of a single heart. It can not only assess the function of the heart, but also guide therapeutic devices. Furthermore, it can also be used to visualize the esophagus, rectum, colon and prostate. A related article on the study was published in the *Utrasonic Imaging*. Three-dimensional ultrasonic imaging can be classified into three modes: surface imaging, transparent imaging, and multi-planar imaging (or section imaging). The uterine and ovarian diseases are often diagnosed by surface imaging. It is reported that the diagnostic accuracy of three-dimensional ultrasonography in uterine malformation is 100%. The uterine volume is considered as an index in three-dimensional ultrasound imaging when diagnosing endometrial cancer. It can not only clearly display the external morphology of ovary and the internal structure (characteristics of nipple, diaphragm, wall) (Fig. 3-15) but its volume can be calculated to facilitate differential diagnosis of ovarian disease.

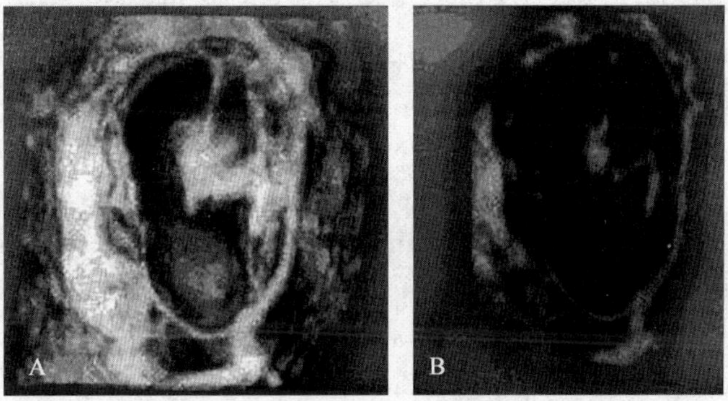

A. Three-dimensional reconstruction imaging of ovarian cysts, which clearly shows the three-dimensional structure of the cyst wall and the papillary protrusions within the cystic cavity
B. Two-dimensional sectional images

Fig. 3-15 3D imaging of liquid-containing structures

3.3.1.4 Hormone assay

In 1958, Short found that changes in the concentration of progesterone in plasma could reflect the maternal reproductive status. The concentration of progesterone in milk was detected with radioimmunoassay (RIA) for the first time by Laing et al. (1971) and they confirmed that the estrous cycle stage reflected by the progesterone concentration in milk was consistent with the results of the ovarian assessment. And then it was proved that the concentration of progesterone in blood was parallel with milk. Many studies on diagnosis of early gestation, by detecting the concentration of progesterone in blood and milk with RIA, reported that the concentration of progesterone could be used in the diagnosis of early pregnancy in cattle, pigs and sheep since 1970s. In the 1980s, the enzyme-linked immunosorbent assay (EIA) was discovered. And then researchers investigated the sensitivity, accuracy and repeatability of RIA and EIA in a large number of experiments, making it more convenient and accurate.

The level of progesterone in the body fluid is significantly higher than that in the non-pregnant one due to the presence of pregnant corpus luteum. Therefore, the concentration of progesterone in body fluid can determine whether the animal is pregnant or not. Blood or milk (saliva, hair and feces) was collected once every two days on 18-24 d after cattle mating, then the progesterone level in the sample was detected by RIA or EIA. This method is most widely used in cattle, and others can also be applied referencing that. It is reported that the accuracy rate of pregnancy diagnosis by measuring progesterone level is 85%-94% in 21-24 d after cattle mating and 16-26 d after sheep mating, and the diagnosis accuracy rate of non-pregnancy is 100%.

Furthermore, detecting estrone and human chorionic gonadotropin (HCG) with RIA and EIA can also be used for the diagnosis of pregnancy. The detection of estrone, mainly detecting the concentration of estrone sulfate, is used in early pregnancy diagnosis of pigs. The level of estrone sulfate, a metabolite of estrone secreted by blastocysts, in the plasma of pregnant pig begins to increase from 21 to 25 d after mating, reaches the peak at 25-30 d, and decreases from 31 to 35 d. However, in non-pregnant pig during estrous cycle, it is always lower than 0.4 mg/mL and in 25-30 d it is significantly different. Therefore, the concentration of estradiol sulfate in plasma of pig at 25-30 d after mating can be used for early pregnancy diagnosis. HCG, a hormone synthesized and secreted by trophoblast syncytial cells, plays an important role in maintaining pregnancy. Human chorionic gonadotropin β subunit (β-HCG) is also a sensitive index for early pregnancy diagnosis. Detecting the concentration of HCG with radioimmunoassay (RIA) is characteristic of good sensitivity, specificity and accuracy and it is mainly used in the diagnosis of human early pregnancy. If the concentration of HCG is higher than the standard, positive will be evaluated as pregnancy.

3.3.1.5 Vaginal examination

During gestation, the vaginal mucosa at the cervix is in a state similar to that in the luteal phase, the viscosity of secretions increases, and the mucosa is pale and dry. Vaginal examination is based on these changes to determine whether the animal is pregnant or not. However, this method has some disadvantages. The indexes examined are easy to be misdiagnosed because of the great differences among individuals. If the examined female animals have persistent corpus luteum or a dead fetus, it is easy to be confused with pregnancy signs and misjudged as pregnancy. When there are pathological processes in the cervix and vagina, pregnant animals are often judged not to be pregnant because of the absence of pregnant signs. Vaginal examination can not determine the date of pregnancy, especially it can not make a definite diagnosis of early pregnancy. Therefore, vaginal examination is generally used as an auxiliary method for large animals such as cattle and horse, and can not be used as the main diagnostic method.

Vaginal examination is generally carried out after an estrous cycle after mating. If there is no pregnancy at this time, the corpus luteum will degenerate, and there are no signs of pregnancy in the vagina. If there is pregnancy, changes will occur in the vagina due to the continuous secretion of progesterone by the pregnant corpus luteum. During vaginal examination, preoperative preparation and disinfection are the same as those of estrus identification. If the sterilization is not strict, it will cause vaginal infection; If the operation is rude, it will also cause abortion of pregnant animals, so vaginal examination has to be careful.

1. Cattle

During pregnancy, the vaginal mucosa becomes white, dry and dull (except the end of pregnancy). In the late stage of gestation, the vaginal wall is soft and thick. The position of the cervix moves forward (varies with the time of gestation), and is often biased to one side, with the cervix tightly closed. During gestation, the main characteristic of vaginal mucus is that it contains a large number of pale blue mucus cells and columnar epithelial cells, which are not found in non-pregnant stage. There is no change in the mucus of the cervix, but the amount of mucus was significantly higher than that of non-pregnant stage. At the first month of gestation, there is thick mucus at and near the cervix, but the amount is small;It is obvious after 3-4 months, and becomes sticky, gray or gray yellow; after 6 months, it is thin and transparent. Sometimes it can be discharged out of the vulva and adheres to the vulva and tail.

2. Horse

After 3 weeks of gestation, the vaginal mucosa turns pale, showing no luster, but there are obvious small blood vessels and surface is dry. When the vaginal opener is taken out, the mucus attached to the front edge is thick, opaque, gray white or gray yellow in color and small in quantity. It is weakly acidic when checked with litmus test paper. The above phenomena also occur before the disappearance of the corpus luteum. Therefore, the above phenomena are not unique to pregnancy. At the 4th-8th month of gestation, due to uterine prolapse, the cervix is pulled to the abdominal cavity, so the length of the vagina increases, but the number of intestinal contents will affect the position of the uterus and the length of the vagina. At the beginning of gestation, the cervix and vagina are relatively soft, and the uterine orifice is not tight even if there is contraction. After 1 month of gestation, the uterus and vagina are pale and the uterine orifice is closed in contraction. The uterine thrombus is not obvious until 4-5 months of gestation. The position of the cervix is often shifted to the left or right, not in the center of the vaginal fornix.

3. Sheep

After 3 weeks of gestation, when the vagina is opened by instrument, the vaginal mucosa turns white and pink after a few seconds, which is the pregnant sign. The mucous membrane of non-pregnant sheep is pink or pale, and the speed from white to red is slow. Vaginal contraction becomes tight, and there is resistance when inserting the opener. If the vaginal mucus is small, transparent, and begins to thin, but thickens after 20 d and can be pulled into a line, it is determined as pregnancy. If the amount is large and thin, the fluidity is strong, the color is gray and purulent, it can be determined as not pregnant.

3.3.1.6 Other diagnostic methods

1. Immunological diagnosis

During gestation, the fetus, placenta and maternal tissue can produce some chemicals (hormones or enzymes, etc.), whose contents increase significantly in a certain period of pregnancy. Some of them show antigenicity and can stimulate the body to produce immune response. For example, immunizing rabbits or turkeys to produce high titers of specific antibodies can specifically bind with antigens and introduce immune reactions, such as agglutination reaction and precipitation reaction. The pregnant animals can be diagnosed by observing these reactions (directly with the naked eye or with the help of instruments). Taking cattle and sheep as examples, the immunological diagnosis methods are briefly introduced as follows:

(1) Cattle (Agglutination inhibition test) The serum sample is diluted with 1 : 50 000, and is put on a plate or glass plate, then hyaluronidase antiserum (the amount must be determined after titration) is added. Two drops of sensitized sheep red blood cells were put in and incubated at 37 ℃ for 60 min. a large amount of hyaluronidase appears in pregnant cattle serum, which can bind with specific antiserum, and the hyaluronidase adsorbed on red blood cells is still in an unbound state, so red blood cells do not agglutinate. In non-pregnant cattle, hyaluronidase adsorbed on red blood cells binds with specific antiserum to agglutinate red blood cells because there is little hyaluronidase in serum. It is reported that this method is not reliable if blood samples are taken before 20 d after fertilization, and the accuracy rate probably is only 67%, while the accuracy rate can reach 95% after 21 d after fertilization.

(2) Sheep (Red cell agglutination test) Early pregnant sheep contain specific antigen, which can be detected from the blood of some pregnant sheep as early as the second day after fertilization, and from the fetus, uterus and corpus luteum of all sheep on the 8th day after fertilization. The materials may be a protein hormone secreted by the fetus and combined with cells. Red blood cell agglutination occurs when the prepared antiserum is mixed with the red blood cells of sheep on the 10th-15th day of pregnancy. It is reported that the positive and negative accuracy rates of pregnant diagnose for 28th-60th day pregnant sheep by this

method are 90% and 75% respectively.

2. Biosensing technology

Recently, biosensing technology is highly valued in the fields of biomedicine, life science, clinical chemistry, etc. The concept of the biosensor was first introduced by Clack and Lyons in 1962. Since the 1980s, the application and development of biosensing technology have advanced dramatically due to the great progress in life science.

Enzymes have been used as analytical reagents to detect specific substances as early as the 1940s. As enzymes selectively catalyze the reaction of the specific substance with proteins that have excellent molecular recognition, they are preferred as sensitive materials for sensors that respond specifically to organic substances. In 1962, enzymes and electrodes were firstly combined for the determination of enzyme substrates by Clark. In 1967, Updike and Hicks adopted the enzyme immobilization technique in biotechnology on the basis. Additionally, besides enzymes, other substances also can be immobilized on the membranes as sensitive elements of the sensor in organisms, such as antibodies, antigens, hormones, etc. For example, the first enzyme electrode (sensor)-glucose electrode was assembled by a glucose oxidase immobilized on an aqueous membrane. Furthermore, immobilized cells, cell organelles, and sections of animal or plant tissues all serve for a similar purpose. Therefore, such sensors are usually referred to as biomolecular sensors or biosensors which are made up of molecules of organisms like enzymes, antigens, antibodies, and hormones, or the organisms themselves like cells, cell organelles, tissues, etc.

HCG can promote gonadal development, which is a hormone assembled by protein, produced by placental chorionic trophoblast cells. It consists of two chains, α and β. The α peptide chain of many gonadotropins is the same, while the β chain is the one with specificity, which immunoassay detection antibodies direct against. HCG detection is important for the diagnosis and differential diagnosis of pregnancy, pregnancy-related diseases, and tumors. Presently, enzyme-linked immunoassay (ELISA), radioimmunoassay (RIA), latex agglutination method, and colloidal gold immunochromatography are usually used in analyses. However, the latest is simple to operate with low sensitivity and just suitable for qualitative or semi-quantitative detection. The enzyme-linked immunoassay and radioimmunoassay can perform quantitative detection with high sensitivity, however, labeling is required and both are tedious and time-consuming to operate. Moreover, the isotopes labeled by radioimmunoassay are hazardous and polluting to the human body and environment.

The principle of immunosensor to determine HCG is: using an oxygen electrode as a transducer, labeling HCG with hydrogen peroxidase glutaraldehyde and fixing anti-HCG antibody on the membrane made of cellulose bromoacetate and cellulose acetate, and then fixing the membrane to an oxygen electrode polytetrafluoroethylene membrane to form the HCG enzyme immunosensor. The enzyme-labeled HCG will compete with the free HCG to combine anti-HCG antibody on the antibody membrane after inserting the sensor into a solution containing free HCG and the enzyme-labeled one. After washing away the nonspecific binding HCG and enzyme-labeled HCG, the sensor should be inserted into the phosphate buffer saturated with oxygen (pH 7.0) at 30 ℃ and record its electron current change. When the current is stabilized, the enzyme-substrate, H_2O_2, is added and the beginning rate of current increase is recorded. The relationship curve between the beginning rate of current increase and the concentration of unlabeled HCG is produced. When the concentration of peroxidase-labeled HCG is 0.4 IU/mL, HCG can be measured from the concentration of 0.02-1.0 IU/mL with an error of about 5%. When the peroxidase-labeled HCG is 4 IU/mL, HCG can be measured from 0.2 to 10 IU/mL; and when the concentration of peroxidase-labeled HCG is 40 IU/mL, HCG can be measured in the range of 2-100 IU/mL.

3. Hematological examination

(1) Blood cell examination From the 21th day of gestation to delivery, the number of red blood cells decreases in dog. In the last week before delivery, 70% of female dogs have less than 5 million red blood cells per microliter. At the same time, the content of hematocrit and hemoglobin decrease to less than 40% and 14% respectively, and the erythrocyte sedimentation rate increase accordingly. According to these reasons, examination of the dog's blood is helpful in the diagnosis of pregnancy.

(2) Platelet examination The number of platelets in peripheral blood of mice, rabbits and pigs decreases significantly from hours to 6 d after fertilization, and then gradually return to normal level. It is reported that the accuracy of pregnancy diagnosis for sows on 6th-7th day after fertilization is 78%. In early stage of gestation, the decrease of maternal platelet may be related to platelet activating factor released by fetus. Due to the method of platelet count being simple and rapid, it is expected to become a highly valuable method for the diagnosis of early pregnancy in livestock and other species.

4. Early pregnancy factor diagnosis

Early pregnancy factor (EPF) is considered to be a substance directly related to early pregnancy. Its existence is an important sign of fertilization, pregnancy and fetal survival. It participates in the suppression of maternal cellular immunity and plays an important role in maternal pregnancy recognition and maintenance. Therefore, it is also called immunosuppressive early pregnancy factor (isEPF). EPF has been found in pregnant mice, rabbit, pig, sheep, cattle and women. The immunosuppressive activity of EPF is maintained from a few hours after fertilization to the 2/3 stage of gestation. Therefore, the detection of EPF in blood or body fluids can be used as a method for the diagnosis of super-early pregnancy. At present, the main method to detect EPF is rosette inhibition test (RIT). Generally, the inhibition titer does not exceed 4. If the lymphocytes of pregnant animals are used or incubated in the serum of pregnant animals, the inhibition titer increases significantly. When the inhibition titer is greater than 14, it is generally considered that there is EPF, which is indicating pregnancy. However, the detection of EPF by RIT is an indirect reaction. It detects the immunosuppressive activity rather than directly shows the existence of EPF, so the reliability of the detection results may be affected by many factors, and RIT takes much time. In recent years, some people labeled EPF with alkaline phosphatase and urease, quantitatively and qualitatively determined EPF with ELISA, and detected EPF in milk with 3% copper sulfate method for pregnancy diagnosis of dairy cows. It is reported that the accuracy of pregnancy diagnosis is 87%-95%. However, due to the lack of sufficient understanding of the molecular structure, characteristics and physiological function of EPF, it is difficult to obtain pure products of EPF. The detection of EPF is still in the research stage. Great improvements are needed in methodology and diagnostic criteria before it can be applied to production practice.

5. Colloidal gold immunochromatography for early pregnancy testing stripe

Colloidal gold is a new type of immunolabeling technique applied to antigen-antibodies as a tracer marker with unique advantages, which is commonly used in the labeling technique. In recent years, it has been widely used in various biological investigations and keeps an absolute position clinically.

Colloidal gold is made of chloroauric acid ($HAuCl_4$) affected by reduction agents, e. g., white phosphorus, ascorbic acid, sodium citrate, tannic acid, and so on, which can be polymerized into a certain size of gold particles. It transforms into a negatively charged hydrophobic colloidal solution that keeps a stable colloidal state due to the electrostatic effect, which is so-called colloidal gold. The specific antigen or antibody is fixed on the membrane which as the strip and the colloidal gold labeling reagent (antibody or monoclonal antibody) is adsorbed on the conjugate pad. The sample moves forward through capillary action and dissolves the colloidal gold labeling reagent on the pad with reacting with each other after being added to the pad at the end of the test strip. After moving to the area of fixed specific antigen or antibody, the tested material and colloidal gold labeling reagent will be especially combined with fixed antigens or antibodies, which are later gathered on the test strip so that they can be observed by the naked eye color results. This test stripe diagnosing early pregnancy is used widely, for example, β-HCG colloidal gold test strips for humans which can detect and predict pregnancy in the serum or urine.

6. Serum alkaline phosphatase (AP) activity assay

Alkaline phosphatase (AP) is a heat-resistant alkaline phosphatase produced by the placenta, small intestine mucosa, kidney, liver, and other tissues or organs, which is increasing with pregnancy progresses. The activity of alkaline phosphatase is higher in the serum of pregnant female, pregnant livestock and the growing infants and children. Therefore, whether someone is pregnant or not can be determined by detecting changes in the activity of AP in the serum of dairy cows.

7. X-ray diagnosis

This method can be used to diagnose pregnancy through the perspective of fetus on the abdominal wall of female animals. It is suitable for small and medium-sized livestock such as pig, sheep, dog and cat. Because this method is based on the observation of fetal bones, it can be carried out only in the middle and late stage of gestation (fetal bones have been formed). The effective diagnosis time is generally more than 65 d in sheep, more than 69 d in pig, more than 40 d in dog and more than 39 d in cat. It is reported that after 30-35 d of gestation in dogs and 17 d of gestation in cats, an appropriate amount of carbon dioxide gas is injected into the abdominal cavity, and the morphology of pregnant uterus can be observed by X-ray fluoroscopy.

8. Laparoscopy (endoscopy) examination

Laparoscopy is used to directly observe the pregnant uterus and corpus luteum for pregnancy diagnosis, especially for small and medium-sized animals such as sheep, dog and cat. It can be carried out after 17 d pregnancy in sheep and 18-19 d pregnancy in dog and cat. When using this method, the animals shall be anesthetized with Baoding (mild ones may not be anesthetized), and the abdominal cavity shall be inflated during laparoscopic observation to reduce the interference of abdominal omentums and other organs. The accuracy of this method is about 90%.

9. Pregnant mare serum gonadotropin examination (PMSG)

PMSG can promote the development and maturation of follicles on the ovaries of non-pregnant female animals and ovulate oocytes. It is generally used for estrus induction and superovulation in female animals. There are many functional corpus luteum on the ovaries of early pregnant sows, which can secrete a large amount of progesterone and inhibit the physiological response of ovaries to exogenous PMSG, so sows do not show estrus. It can be used for pregnancy diagnosis of sows. Sows after 14 d of mating can be identified as pregnant if pigs do not show estrus within 5 d and do not accept boar mating while injected with 700-800 IU PMSG intramuscularly. If a sow shows normal estrus and accepts boar mating, it is determined to be a non-pregnant. The accuracy of diagnosing non-pregnant sows is 100%. Induced estrus sows have the same normal conception ability as natural estrus pigs. In the process of early pregnancy diagnosis by this method, there are no abortion, normal litter size and fetal development, and it does not affect the estrus and conception of sows. The pregnant sows are observed with a large dose of PMSG, and there is no abortion in pregnant sows. Therefore, this method is simple, early and safe, and has the dual effects of pregnancy diagnosis and estrus induction.

3.3.2 Smart pregnancy diagnosis techniques

In recent years, with the gradual maturity of high-throughput technology, some new technologies, such as transcriptomics, proteomics, metabolomics and mid infrared spectroscopy (MIRS) have been applied to the pregnancy diagnosis. Some achievements have been made in these new technologies. Some pregnancy related biomarkers, such as genes, miRNA, amino acids and proteins, have been preliminarily identified by omics technology. MIRS technology has preliminarily identified the different spectral regions of several pregnant milk samples and non-pregnant milk samples or applied to the establishment of pregnancy diagnosis and discrimination model of dairy cattle. These new technologies establish a smart pregnancy prediction model and the pregnancy can be quickly diagnosed in the first estrous period after mating so as to improve the reproductive efficiency.

3. 3. 2. 1 Omics technology

1. Gene

On the 20th day after artificial insemination, the expression of ADP-ribosyttransferase 2a (ART2A) and Bovine-B (BOVB) genes in the blood of pregnant cattle is significantly higher than that of non-pregnant cattle, especially the sensitivity and specificity of BOVB gene reach 83% and 75% respectively. Blood samples are collected from cattle on the 14th-18th day after artificial insemination and quantitative real time polymerase chain reaction (qRT-PCR) are performed. The result of qRT-PCR shows that the expression of chemokine (C-C) motif ligand 8 (CCL8) and chemokine (C-X-C) motif ligand 10 (CXCL10) genes in pregnant cattle shows an upward trend, and significantly increases on the 18th day gestation, while the expression of non-pregnant cattle does not change. It can be seen that detecting the expression changes of these genes is helpful for early pregnancy diagnosis. Yoshino et al.'s study (2018) showed that 20-22 d after artificial insemination, the expression of interferon-stimulated gene 15 (*ISG15*) and myxovirus resistance proteins Ⅱ (Mx2) genes in blood samples of pregnant cattle was significantly higher than that of non-pregnant cattle, and the diagnostic accuracy could reach 80%.

2. MicroRNAs

MicroRNAs (miRNAs) are small non-coding RNAs that act as post-transcriptional gene regulators. The intrinsically high stability of miRNAs in bio-fluids together with, in some cases, their tissue-specificity, confers miRNA as a significant potential as non-invasive biomarkers of disease. So far, a plethora of studies have shown the association between specific plasma and serum miRNA signatures and human cancer, heart disease and autoimmune diseases, as well as, in a reproductive context, pregnancy progression and preeclampsia.

Plasma samples are collected from the jugular vein in tubes on days 0, 8, 16 and 60 after insemination. Cells are obtained from whole blood samples (100 μL) that are centrifuged at 1 900g for 10 min at 4 ℃. The resulting supernatant is discarded and the cell pellet is re-suspended to a volume of 250 μL. The fresh blood cell solution is then used for RNA extraction. Additional body tissue samples are collected at a local abattoir and are placed in ice-cold PBS until arrival at the laboratory. Tissues are then dissected, labelled, weighed and snap-frozen using dry ice. All samples are stored at −80 ℃ until further use. Small RNA sequencing libraries are prepared from samples using the Small RNA Library Preparation Kit and submitted to 36-base single-end sequencing on the HiSeq 2000 Sequencing System (Illumina). Raw sequencing data are analysed using the sRNA bench tool.

Across all samples, up to 389 unique miRNAs are identified to present at more than 10 reads. The 15 most abundant miRNAs in bovine plasma are listed in Table 3-4.

Table 3-4 Most abundant miRNAs in bovine plasma on 60 d of pregnancy identified by sequencing

miRNA	miRNA reads on 60 d of gestation
bta-miR-486	187 148
bta-miR-22-3p	86 953
bta-miR-192	45 163
bta-miR-27b	39 846
bta-miR-143	67 181
bta-miR-10b	30 810
bta-miR-142-5p	40 063
bta-miR-191	33 804
bta-miR-423-5p	23 760
bta-miR-451	27 249
bta-miR-21-5p	24 164
bta-miR-101	25 853
bta-miR-92a	19 306
bta-miR-103	13 565
bta-miR-30d	18 882

From Ionanidis et al. (2018)

3. Amino acid and protein

Thirty-nine differentially expressed proteins are found in the milk samples of non-pregnant cattle and pregnant cattle on the 35th day after artificial insemination. Lactoferrin and alpha-1G are further identified as biomarkers for early pregnancy diagnosis by Western blot. Milk samples are collected on the day of estrus and the 17th day after artificial insemination. Metabonomics technology (LC-Q/TOF-MS) is used to screen the foreign bodies with poor metabolism in the milk samples of the two groups. Finally, *L*-tyrosine, N-acetyl-*L*-tyrosine, *L*-proline, tetrahydro-1-phenylmethyl-1, 2, 3, 4-tetrahydroisoquinoline, guanosine and heptadecyl carnitine are identified as potential milk biomarkers in pregnancy recognition stage. Johnston et al. (2018) used LC-Q/TOF-MS technology to study the milk samples of cattle on the day of estrus and the 21st day after artificial insemination. The results showed that APOB, SPADH1, PLIN2 and LPO were differentially expressed in whey, and PIGR, PGD, QSOX1, MUC1, SRPRA, MD2, GAPDH, FOLR1, GPRC5B and HHIPL2 were differentially expressed in whey rich in extracellular vesicles. These amino acids and proteins may become potential biomarkers for pregnancy diagnosis.

3.3.2.2 MIRS technology

MIRS is a technology for qualitative or quantitative detection of substances based on the vibration of various chemical groups at specific wavelengths in specific areas. As a new modern analytical detection technology, MIRS technology has the advantages of non-destructive, fast, high-throughput and low cost, and has been widely used in large-scale detection of milk. At present, the MIRS technology is mainly used to predict the content of nutritional components in milk, such as fatty acids, mineral elements or proteins, and to judge whether milk is adulterated or whether there are antibiotic residues. There are few studies in the diagnosis of cattle pregnancy. A potential application of MIR for pregnancy diagnosis at advanced stages after insemination is for detection of late fetal abortion. Mid-to late-term fetal losses are often detected using careful observations because of a vaginal discharge or expulsion of placenta. However, not all abortions are detected, especially those that occur before 180 d of pregnancy, and therefore, MIR could be used as a complementary tool, provided that it can be used for slightly earlier diagnosis than 151 d. This would need to be confirmed using a larger data set. Adding data collected from other analyses such as metabolomics could further improve the prediction accuracy. Also, using a much larger data set, it is worth exploring whether deep learning techniques can help provide a better prediction.

Lain et al. (2017) showed that pregnancy had the greatest impact on the 968-1 577 cm^{-1} region of the mid infrared spectrum. Alvarado (2018) and other researchers have shown that when fat, protein, casein, lactose, parity and lactation days are used as predictors to predict whether a cow is pregnant with a general linear model, adding MIR data to the prediction model as a predictor will improve the accuracy of the prediction model. Delhez et al. (2020) collected MIR spectra and insemination records from 8 064 Holstein cows in Australia and confirmed that little ability to detect the pregnancy status as the area under the receiver operating characteristic curve was 0. 63 and 0. 65 for cross-validation and testing. The models developed using data recorded after 150 d of pregnancy showed promising prediction accuracy with the average value of area under the receiver operating characteristic curve of 0. 78 and 0. 76 obtained through cross-validation and testing. If this can be confirmed on a larger data set and extended to somewhat earlier stages after insemination, the model could be used as a complementary tool to detect fetal abortion. These results potentially implied that pregnancy was detected using different spectral wavelengths or with different weights at different stages after insemination, with an evolving pattern from the beginning toward the end of the pregnancy. These studies have laid a foundation for the application of MIRS technology in pregnancy diagnosis of dairy cattle in the future.

Assignments

1. What are the methods of early pregnancy diagnosis?
2. What are the physical (including behavior, reproductive tract) and physiological (endocrine) changes of cattle during gestation?
3. What is the commonly used pregnancy diagnosis method for large animals?
4. What is the principle of B-mode ultrasound pregnancy examination?

(Zhou Zhengrong, Huang Pan)

Chapter 4 Parturition Control

Summary

Code 4-1

The control of parturition, also known as induced parturition, refers to the use of exogenous hormone preparations during a certain period of the pregnant end to control the parturition of female livestock within an artificially determined time range and produce the healthy newborn animals. It is a measure of reproductive management to control the time and process of parturition. This chapter first introduces the principles, application and significance, then focuses on the methods of hormone treatment for parturition control.

Parturition control techniques include induced labor and induced abortion. Induced labor, also known as induced parturition, refers to use external hormone to treat female livestock to complete labor within a certain time in the end of pregnancy. Induced abortion is adopted if the appropriate time for inducing labor is extended, regardless of whether the fetus is alive or dead at the time of birth and whether the fetus has the ability to live independently after birth. The concept of induced abortion also includes the artificial interruption of pregnancy before embryo differentiation is complete. Thus, both induced labor and induced abortion involve the artificial interruption of pregnancy and the expulsion of the pregnant animal. However, the parturition control technology we are referring to here is induced labor.

4.1 Principle of parturition control

The pregnancy and parturition of animals are coordinated or interacted by a variety of hormones (Fig. 4-1). Hormone treatment can be used to terminate pregnancy and initiate parturition, so as to achieve the purpose of artificial abortion and induced parturition. Commonly used hormones include prostaglandins, oxytocin (OXT) and corticosteroids. When using adrenocorticotropic hormone (ACTH), attention should be paid to the pregnant stage of livestock. If ACTH is used prematurely (such as before 130 d of pregnant sheep) or the fetus has died, it can not cause the response of fetal adrenal cortex, resulting in the failure of induced parturition. The use of therapeutic doses of water-soluble short acting glucocorticoids can make the pregnant body deliver after 2-5 d. The combined use of a small amount of 17β-estradiol may help to improve the effect of induced parturition, but usually 17β-estradiol is not used alone or in large doses, otherwise it will cause excessive edema of uterus and birth canal, leading to dystocia. OXT is generally not recommended for cattle and sheep, especially when the cervix is not opened, the application of OXT can lead to uterine rupture. Prostaglandin (PG) $F_{2\alpha}$ can contract smooth muscle and dissolve corpus luteum, which is the most convenient, safe and effective hormone to induce parturition. When using $PGF_{2\alpha}$ analogues, the dosage should be increased or reduced according to their efficacy.

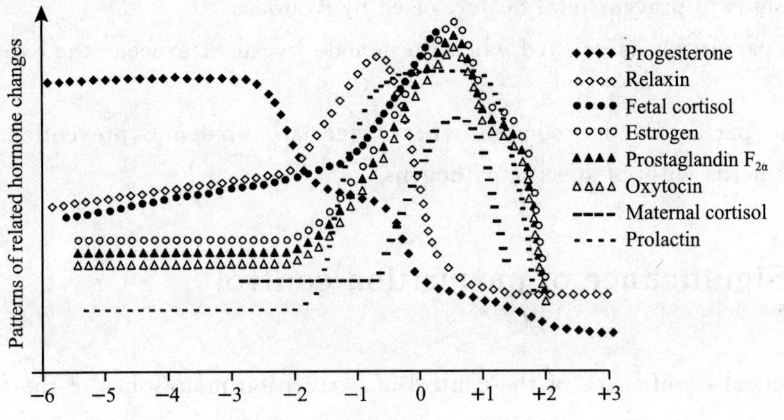

Fig. 4-1 The changing trend of related hormones in cattle before and after parturition

(Imitated from Bearden et al., 2003)

4.2 Application of parturition control

On the basis of understanding the mechanism of parturition, the control of parturition uses exogenous hormones to simulate the changes of hormones that initiate parturition, to adjust the process of parturition, and to achieve the purpose of early and centralized parturition. However, there are still some side-effects that can not be completely overcome in current control technologies of parturition, which have a certain impact on female and newborn livestock. It needs to be overcome from two aspects of improving drugs and treatment methods. The control of parturition can be used as a technology in special cases, and whether it is advisable to widely use it in production as a universal technology depends on the future development. Recently, some scholars suggested the control of parturition could be used in the following cases.

(1) The female livestock is small and the fetus grows fast, so as to avoid dystocia

during the full-term pregnancy or when the fetus can not be borne due to illness or injury during the late pregnancy.

(2) The female livestock is forced to slaughter or diagnosed with hydramnios and normal fetal growth, so as to save the pregnant fetus.

(3) The parturition is induced within 10 d in advance in order to obtain lamb fur with more beautiful patterns; or the fetal samples are collected for their later growth study in order to avoid killing the mother and taking the fetus.

(4) The parturition is required during the day by the large pregnant group for facilitating midwifery and reducing mortality.

(5) The parturition is induced since the labor pains of female animals are weak during parturition, so as to prevent fetal death caused by dystocia.

(6) The parturition is delayed when the female livestock exceeds the expected delivery period.

(7) The parturition is required when veterinary epidemic prevention requires the production of herds without specific pathogens.

4.3 Significance of parturition control

The practical significance of the control of parturition mainly lies in the following four aspects.

First, the control of parturition can enable the parturition of female animals to be carried out in batches, and the care of female and young animals to be taken intensively, so as to save manpower and time, and make full and planned use of delivery room and other equipment.

Second, the control of parturition is used to prepare the nursing work on the premise of predicting the delivery time, so as to prevent possible casualties of female and young animals.

Third, the control of parturition cooperating with the synchronization of estrus is conducive to establishing the production mode of industrialized animal husbandry, and to exchange, nest and foster newborn animals between parturient females.

Fourth, the control of parturition can make the delivery of female animals during working days and working hours to avoid holidays and night duty.

4.4 Methods of parturition control

The control of parturition includes many methods or treatment scheme, which can be used for various livestock. The mainly used hormones are prostaglandins, oxytocin, corticosteroids and their analogues. Drug types and doses are different for different animals.

Among them, glucocorticoid drugs include dexamethasone, flumethesome and betamethasone. The commonly used prostaglandin is cloprostenol.

4.4.1 Treatment scheme of prostaglandin

For pigs, 5-10 mg $PGF_{2\alpha}$ is injected in the last 3 d of pregnancy, and most of sows deliver litters after 22-32 h. In order to induce sows to deliver litters during the daytime, 2 mg $PGF_{2\alpha}$ intramuscular injection can be given on any day of the 110th-114th day of pregnancy, and about 3/4 sows deliver litters during the daytime, but the gestation period is not shortened. There are no abnormal changes in the live birth rate, the normal excretion rate of placenta and the milk secretion of sows. For sows treated early, the interval between injection and farrowing is slightly shorter.

For horses, the sensitivity of mares to $PGF_{2\alpha}$ is very different at each stage of pregnancy, because the sources of progesterone are also different in horses at different stages of pregnancy. Horses are very sensitive to $PGF_{2\alpha}$ within 30 d of pregnancy. Abortion and estrus occur soon after treatment. At this time, breeding can also be pregnant. After that, it needs four or more injections to miscarry. At the end of pregnancy, the horse becomes sensitive to $PGF_{2\alpha}$ again. The reliable method of inducing parturition with $PGF_{2\alpha}$ is to inject 2.5 mg $PGF_{2\alpha}$ every 12 h until parturition. When twin pregnancy is found, $PGF_{2\alpha}$ should be injected immediately for abortion.

In cattle, the corpus luteum is the main source of progesterone secretion within 200 d of pregnancy. If $PGF_{2\alpha}$ is injected at this stage, abortion will occur soon. Especially before 65-95 d of pregnancy, the histological connection between chorion and endometrium is not close. There is no rupture of fetal membrane, no bleeding of endometrium and no increase of endogenous $PGF_{2\alpha}$ after abortion, so the side-effects are limited. Therefore, this is an appropriate time to end unnecessary or unsatisfactory pregnancy. After 200 d of pregnancy, the placenta can produce a small amount of progesterone. Before 150 d of pregnancy, the corpus luteum is relatively insensitive to $PGF_{2\alpha}$, and cows do not necessarily miscarry after $PGF_{2\alpha}$ injection. The closer to delivery, the more sensitive to $PGF_{2\alpha}$. The parturition could be achieved within 2-3 d after 5-30 mg $PGF_{2\alpha}$ was injected at 275 d of pregnancy.

For sheep, injection of 15 mg $PGF_{2\alpha}$ at 141-144 d of pregnancy can make ewes lamb a lamb within 3-5 d. The placenta of sheep produces progesterone from the mid-gestation, which becomes insensitive to $PGF_{2\alpha}$, so it can not induce parturition with $PGF_{2\alpha}$.

For goats, the induced parturition is similar to that of sheep. However, goats rely on the corpus luteum to produce progesterone throughout the pregnancy. Therefore, pregnant goats injected with 1.2 mg $PGF_{2\alpha}$ can miscarry or deliver within 1.5-3 d.

For dogs, the corpus luteum within 10 d of pregnancy is resistant to most drugs with luteolytic effects. After 40 d of pregnancy, 25-250 μg $PGF_{2\alpha}$ is injected twice a day until abortion with continuous injections.

For cats, $PGF_{2\alpha}$ has no luteolytic effect. But during the late stage of pregnancy, when it is in a state of stress or has been injected with ACTH in advance, 1.0-1.5 mg $PGF_{2\alpha}$ per kilogram of body weight every day can cause abortion for two consecutive days, which may be the result of smooth muscle contraction induced by $PGF_{2\alpha}$.

4.4.2 Treatment scheme of oxytocin

For pigs, the parturition can be achieved after a few hours, if 20 IU OXT is injected after 15-24 h of $PGF_{2\alpha}$ injection, which can accurately control the delivery time. Alternatively, 100 mg/d progesterone is injected for three consecutive days, and then $PGF_{2\alpha}$ is injected on the fourth day, the parturition can be achieved after approximately 24 h, which can also control the delivery time in a short range.

For horses, 40 IU OXT injection can make the mare foal in about 30 min at the end of pregnancy with the normal preposition, orientation and position of fetus, when there are more colostrum in the breast, the cervix is relatively soft, and 1-2 fingers can be extended to the external opening of the cervix. The safer way of drug administration is to inject 5 IU OXT at an interval of 15 min, and then increase the dosage to 10 IU each time after three consecutive injections until the initiation of delivery. If the cervix of the horse is not loose, OXT can be injected after 12-24 h of estrogen injection. Another direct examination is conducted after 10-15 min of OXT injection, which can ensure the normal fetal position and potential.

For cattle, glucocorticoids and $PGF_{2\alpha}$ are commonly used drugs to induce parturition, and estrogen can be used as adjuvant drugs. Therefore, ACTH can be used to induce parturition because it can stimulate the secretion of endogenous glucocorticoids.

4.4.3 Treatment scheme of corticosteroid

For sows, the parturition can be induced by 60-100 IU ACTH injection on the 110th day of pregnancy with the birth interval shortened by 25% and the reduced dead piglet rate. The ideal effect of induced parturition can be obtained by injecting 75 mg dexamethasone every day for 3 consecutive days from 109 to 111 d of pregnancy, 100 mg dexamethasone every day for 2 consecutive days from 110 to 111 d of pregnancy, or 200 mg dexamethasone on the 112nd day of pregnancy.

For horses, it can make the mare deliver after 3-7 d by injecting 100 mg dexamethasone every day for 5 consecutive days from 321 to 324 d of pregnancy.

For cattle, the parturition can be induced after 30-60 h of treatment with 20 mg dexamethasone or 5-10 mg flumethasone injection intramuscularly from the 265-270 d of pregnancy.

For sheep, the parturition can be induced in most ewes after 40-60 h of treatment with 12-16 mg dexamethasone or betamethasone or 2 mg flumetasone injection on the 144th day of gestation.

For goats, the secretion control method is similar to that of sheep. $PGF_{2\alpha}$ or dexamethasone are mostly used as the drugs of induced parturition, and the treatment time should be the 144 days of pregnancy.

For dogs, dexamethasone is injected for 10 d, twice a day, 0.5 mg/kg body weight each time, which can cause fetal death and absorption in the uterus before the 45 d of pregnancy, or cause abortion after the 45 d of pregnancy.

In conclusion, it's worth noting that though the control of parturition can regulate the time and process of parturition, the effect is limited. For the reliable and safe control of parturition, the treatment time can only be a few days before the normally expected date of parturition, such as 3 d for pigs and 7 d for horses, cattle and sheep. Beyond this period, there will be some adverse effects. The earlier the time, the greater the impact. The accuracy of controlling the parturition time is that most of female treated-animals can deliver within 20-50 h after drug administration, which is difficult to control it into a very accurate time range. In addition, the clinical research on the control of parturition in dogs and cats is not sufficient to develop an accurate method.

Assignments

1. What is the significance of parturition control?
2. What are the methods of parturition control in cattle?

(Wang Zhengchao)

Chapter 5 Oocyte and Embryo Cryopreservation

Summary

Oocyte and embryo cryopreservation is a crucial step for embryonic biotechnology. It provides a long-term storage for oocytes and embryos to ensure embryonic biotechnology without time and space limitation. Therefore, the cryopreservation of oocytes and embryos has always been a research hotspot in the field of embryonic biotechnology. This chapter firstly introduces the history and principle of oocyte and embryo cryopreservation. Then, the specific methods and operation steps for embryo cryopreservation and oocyte cryopreservation are introduced in detail, respectively. Finally, we summarize the prospects and conclusions of oocyte cryopreservation.

The origin of oocytes has always been the main restrictive factor for embryo production by *in vitro* fertilization. If oocytes can be cryopreserved, high-quality oocytes could be used to produce embryos without distance limitation. In this manner, it not only reduces the production cost of embryos, but also eliminates effects of different experimental conditions on the efficiency of embryo production. In conventional embryo transfer practice, females as the donor and the recipient should be prepared at the same time. Sometimes, more embryos can be obtained from donor females due to the superovulation than the demand of recipient females, resulting in the waste of embryos. To avoid this waste, the rest of embryos can be frozen until use. Therefore, the embryo cryopreservation is crucial for the reduction of cost and the improvement of efficiency in embryo transfer. In addition, it can make embryo transfer unlimited by time and space, and improve the pregnancy rate by only using high-quality recipient. Therefore, embryo cryopreservation is an important technology for the embryo transfer industrialization.

5.1 History for embryo and oocyte cryopreservation

5.1.1 Oocyte cryopreservation

Studies on oocyte cryopreservation began in the 1970s. In 1977, Whittingham et al.

cryopreserved mouse mature oocytes at −196 ℃, and successfully obtained the first offspring from morulae transfer after fertilization. Subsequently, oocyte cryopreservation was widely used in many species and obtained successful cryopreservation in rabbit and bovine. At that time, the efficiency of bovine oocyte cryopreservation was very poor, with less than 1% blastocyst rate after *in vitro* fertilization. In 1989, Nakagata et al. cryopreserved mouse oocytes by vitrification. After *in vitro* fertilization, these oocytes could develop continuously, and achieve 45.8% birth rate after transplantation. In 1998, Vajta et al. cryopreserved bovine mature oocytes by open pulled straw (OPS) vitrification. There were 25% vitrified oocytes developing into blastocysts after fertilization, and pregnancies following transfer. Compared with mature oocytes, cryopreservation of immature oocytes is more challenging. Schroeder et al. (1990) cryopreserved mouse GV oocytes using different methods, and found that most oocytes can mature *in vitro* after thawing. However, only 9%-17.3% of these oocytes developed into 2-cell after *in vitro* fertilization, and there was no report for offspring.

Most studies have proved that immature oocytes are more sensitive to freezing. Data from Lim et al. (1992) showed that the rates of fertilization, cleavage and blastocyst in bovine GV oocytes were lower than those in mature oocytes. In 1992, Arav cryopreserved porcine GV oocytes by vitrification and found that the maturation rate *in vitro* reached 24.5% after thawing. Therefore, the cryopreservation conditions and low-temperature biological characteristics of GV oocytes need to be further studied.

5.1.2 Embryo cryopreservation

Studies on mammalian embryo cryopreservation began in 1970s. Whittingham et al. (1972) firstly invented slow freezing, namely conventional freezing. Using this method, they successfully cryopreserved mouse embryos and produced offspring after thawing and transplantation. In slow freezing, dimethylsulfoxide (DMSO) was selected as the major additional cryoprotectant. After being seeded, the cryoprotective solution containing embryos cooled to −35 ℃ at 0.33 ℃/min and to −80 ℃ at 1 ℃/min, and then directly plunged into liquid nitrogen for further storage. More than 80% mouse embryos had the ability to continuous development after slow freezing, which taken about 3 h. Soon after, embryo cryopreservation was performed in several animal species, including cow (Wilmut and Rowson, 1973), rabbits (Whittingham and Adams, 1976), rats (Whittingham, 1975) and sheep (Willadsen et al., 1976). So far, animal embryos from 16 species have been transferred to recipients after slow freezing and thawing, and offsprings have been successfully obtained.

In 1977, Willadsen et al. invented a rapid freezing method based on the slow freezing. Following pretreatment with cryoprotective solution, sheep embryos were slowly cooling down to −35 ℃ and then plunged into liquid nitrogen directly. Renard et al. (1984) froze 2-cell rabbit embryos using this method, and achieved 88% survival rate. Compared with slow freezing, rapid freezing was just shortened 1 h for cooling. Thus, rapid freezing also belongs to the conventional freezing. In general, procedures of conventional freezing are cumbersome and time-consuming. Moreover, an expensive programmable freezer is required to substantially control the optimal freezing rate.

In 1985, Rall et al. firstly invented vitrification and succeeded in the cryopreservation of mouse 8-cell embryo. Since then, vitrification has been continuously improved and gradually matured. Recently, vitrification has been widely used in the embryo preservation for several animal species.

5.2 Principle and technology of cryopreservation for ova and embryos

5.2.1 Technical terms

5.2.1.1 Heterologous nucleation temperature (T_m)

The temperature that ice crystals can form in the solution under the external mechanical and physical stimulation. It is usually referred as "freezing point".

5.2.1.2 Supercool

Supercool is a thermodynamically unstable state in which a solution whose temperature reaches below freezing point without ice crystals forming. Once subjected to external mechanical and physical stimulation, large ice crystals will achieve thermodynamic stability and cause mechanical damage to cells. Embryo and oocyte cryopreservation usually needs to carry out seeding at the freezing point of cryoprotectent solution, so as to prevent cells from fatal mechanical damage caused by supercool.

5.2.1.3 Homologous nucleation temperature (T_h)

This is the temperature at which ice crystals can spontaneously form in solution, usually about 35 ℃ lower than the freezing point.

5.2.1.4 Vitrification temperature (T_g)

The temperature at which the solution under supercool is transformed into a glassy state, usually around −140 ℃.

5.2.2 Changes during freezing and cryoinjuries

5.2.2.1 Ice-crystal formation during freezing

When temperature drops to the freezing point, ice crystals firstly form in extracellular fluid and enlarge with the temperature dropping. In this case, the concentration of extracellular solute is increasing. It will lead to a continuous outflow of intracellular water, resulting in an increase of intracellular solute concentration.

5.2.2.2 Solution effects during slow freezing

When the cooling rate is too slow, intracellular water will flow out fully. The concentration of electrolytes and other solutes in cells will increase extremely. It can denature the lipoprotein complexes in the cell membrane and some other functional proteins in the cytoplasm, which is generally named "solution effects". This effect can irreversibly damage cells.

5.2.2.3 Mechanical damage effects during rapid freezing

When the cooling rate is too rapid, intracellular water have no time to overflow outside the cell. Ice crystals directly form in the intracellular liquid, and cause fatal mechanical damage to cells.

5.2.2.4 Rupture injury

The frozen embryos and oocytes often have zonal pellucid or partial cytoplasm rupture after thawing. The embryos or oocytes are pre-equilibrated in a concentrated cryoprotective solution before being plunged into liquid nitrogen. When plunging into liquid nitrogen, solution state changes from liquid to solid. However, there is a rapid conversion process between liquid and solid during thawing. A fracture plane will appear owing to the difference of expansion and shrinkage rate. Therefore, it is important for solution to pass through the liquid phase transition temperature domain (−130-−110 ℃) slowly to avoid rupture injury. Generally, the thin tubes are exposed in air for about 10 s after removal from liquid nitrogen, and then are plunged into the water bath for thawing.

5.2.2.5 Osmotic shock during cryoprotectant removing

The thawed embryos or oocytes contain a certain amount of cryoprotectant. If directly transferred into isotonic solution, the cryoprotectant has no time to be replaced and the extracellular water penetrate easily into cells, leading to swelling and cell damage. Therefore, it is very necessary to replace the intracellular cryoprotectant in time. Selecting a cryoprotectant with strong permeability to cell membrane is crucial.

Sucrose is often added to the thawing solution to prevent the rapid penetration of the extracellular water. The molecular weight of sucrose is 342.3. Based on its impermeable characteristic, sucrose can form a high osmotic pressure in the extracellular fluid to promote the thawed cell contraction. During contraction, the intracellular water and the cryoprotectant are removed together, resulting in the intracellular osmotic pressure to reduce and the cell expansion limitation. However, sucrose will make the cells contract continuously after the intracellular cryoprotectant removal. The prolonged contraction also can damage cells.

5.2.3 Cryoprotectants

Cryoprotectants, one of hydroxyl rich chemicals, are generally used to protect from cryodamage during cryopreservation. It is worth noting that cryoprotectants exhibit variable actions depending on the types used and other factors, such as temperatures and cell types. Therefore, it is important to select a specific cell type and optimal concentration of cryoprotectants. According to their molecular weight and cell membrane permeability, cryoprotectants are broadly classified into two categories: permeating and nonpermeating.

5.2.3.1 Permeating cryoprotectants

Permeating cryoprotectants are small molecules typically less than 100 Daltons and with high amphiphilicity, including glycerol, ethylene glycol, propylene glycol, dimethyl sulfoxide, methanol and butanediol. These chemicals can penetrate the cell membrane readily and form hydrogen bonds with water molecules, thereby reducing intracellular ice formation. They also contribute to the increase of cellular hydration state during cooling, which can avoid the excessive accumulation of intracellular electrolytes. In addition, penetrating cryoprotectants can effectively dilute the remaining electrolytes as electrolyte substitutes to avoid "solution effects". However, excessive accumulation of intracellular cryoprotectants leads to increase of cellular toxicity.

5.2.3.2 Nonpermeating cryoprotectants

Nonpermeating cryoprotectants are macromolecule and with high hydrophilicity, including sucrose, trehalose, polyethylene glycol, polyvinylpyrrolidone, and polysucose. These chemicals remain extracellular and act by dehydrating the intracellular water. During thawing, the extracellular osmotic pressure decreases rapidly due to the water produced by ice melting. If the intracellular cryoprotectant can not diffuse out quickly, excessive influx of extracellular free water will result in cell swelling and even rupture. Therefore, nonpermeating cryoprotectants are always applied in combination with permeating cryoprotectants for avoidance of high osmotic stress and cryoprotectants toxicity.

5.2.4 Principles of cryopreservation

Presently, cryopreservation has three major categories: conventional freezing, ultra-rapid freezing and vitrification. The principles of these methods were described in detail as follows.

5.2.4.1 Conventional freezing

The first important step of conventional freezing is "ice seeding", which can induce extracellular fluid freezing and avoid supercool. The cooling rate is controlled by a programable freezer to ensure that cells have enough time to overflow the intracellular water. In this manner, cells can avoid damage caused by large ice crystals. Meanwhile, too slow cooling should be avoided to prevent the solution effect caused by excessive cell dehydration. When the T_h of intracellular fluid is close to T_g, embryos or oocytes can be directly plugged into liquid nitrogen for preservation. According to the temperature for plugging into liquid nitrogen, conventional freezing can be divided into slow freezing (−75-−60 ℃) and rapid freezing (−35-−30 ℃).

5.2.4.2 Ultra-rapid freezing

Before freezing, embryos oroocytes are dehydrated with nonpermeating cryoprotectants. In this manner, the dehydration changes of embryos or oocytes can be replaced during slow freezing. Next, the embryos or oocytes are directly placed at −30 ℃ for a period to reach a thermodynamic stable state. After equilibrium, the embryos or oocytes are plugged into liquid nitrogen for further preservation.

5.2.4.3 Vitrification

During vitrification, high concentrations of permeating and nonpermeating cryoprotectants are firstly used to make a basic equilibration between T_h and T_g of intraembryonic and extraembryonic fluid. Also, the toxicity of cryoprotectants to embryos is minimized by controlling the balance time and temperature between cryoprotectants and embryos. Afterwards, a rapid formation of transparent vitreous among intraembryonic and extraembryonic fluids is performed by controlling the volume of the embryo freezing container and cooled quickly. Finally, the embryos or oocytes are plugged into liquid nitrogen for further preservation.

5.3 Procedures for embryo cryopreservation

5.3.1 Conventional freezing

5.3.1.1 Preparation for cryoprotective solution

For preparing cryoprotective solution, an Dulbecco's phosphate-buffered saline medium (PBS) or TCM199 with low sodium bicarbonate is commonly used as the base medium, and then 10%-15% FCS (fetal bovine serum) or NCS (newborn bovine serum), or 0.4% BSA (bovine serum albumin) is added, as well as 1.5-2.0 mol/L glycerol, DMSO or ethylene glycol. Recently, 0.1-0.2 mol/L sucrose are added to the cryoprotective solution to prevent the formation of intracellular ice crystals during freezing.

5.3.1.2 Equilibrium in cryoprotective solution

If selected glycerol as cryoprotectant, embryos should be equilibrated in progressive concentration of cryoprotective solution. It can divide into four or six steps, and each step is about 5-10 min. Compared with glycerol, there are some cryoprotectants that can get access to cell membrane more quickly, such as ethylene glycol, propylene glycol and DMSO. If selected these cryoprotectants, three-step or one-step equilibrium method can be adopted. After equilibrium, the embryos are placed in a 0.25-mL plastic tube. The extracellular osmotic pressure is relatively high due to the cryopreservation solution. In this case, the intracellular water seeps out and the embryo contracts, so as to maintain osmotic pressure balance. With the penetration of cryoprotectant, extracellular water flows back into the cells, and the embryo volume gradually recovers.

5.3.1.3 Freezing procedure

The thin plastic tubes with embryos are placed in a programmed cooler, and cooled down to the freezing point or below (−7-−5 ℃) by a speed of 5 ℃/min. The tweezers are precooled in liquid nitrogen. Subsequently, the cotton plug part of the thin tube is clamped with tweezers to make the cryoprotective solution instantaneously produce ice crystals, namely "seeding". If seeding is not performed, the solution will continue to cool under the freezing point but no freezing occurs for the time being. At this time, the rate of cell dehydration is very slow. When ice crystals are formed in the extracellular fluid, the intracellular water can not remove completely and form ice crystals, resulting in the mechanical damage or even death of embryos.

After seeding, extracellular osmotic pressure is raised following ice crystals formation and solute concentration increase of cryoprotective solution, so as to form an osmotic pressure difference between intracellular and extracellular fluids. At low temperatures, cryoprotectants no longer penetrate through the cell membrane. In order to maintain osmotic pressure balance, intracellular water seeps out and cells contract, resulting in an increase of cytoplasmic cryoprotectant concentration.

After balancing at seeding temperature for 5-10 min, embryos are cooled to −35-−30 ℃ at 0.3-0.5 ℃/min, and are then plugged into liquid nitrogen. If slow freezing is selected, embryos should be cooled to −75-−60 ℃. During freezing, the ice crystals in extracellular fluid gradually increase and the solution continuously concentrate. In order to maintain the balance of intracellular and extracellular osmotic pressure, intracellular water continuously exudes and forms ice crystals, resulting in a simultaneous increase of solute concentration among intracellular and extracellular solution. Compared with sperm, the embryo has larger volume and smaller relative surface area, which leads to a certain time for intracellular water exudation at low temperature. Therefore, slow cooling is very necessary for embryos. If cooling is too fast, the intracellular water will have no time to seep out and form ice crystals, resulting into cell injury.

Slow cooling makes the intracellular solution highly concentrated, which makes no ice crystals formation in cytoplasm even if cooled to the same temperature as liquid nitrogen.

Besides, in addition, there are some concentrated solutions between the extracellular ice crystals, and these solutions will not form ice crystals. Under the rapid freezing condition, the viscosity of any solution increases sharply and forms an amorphous solid phenomenon, which is referred as vitrification. The vitrificated transition temperature of solution is about −130-−110 ℃. If this temperature range passes quickly, it is easy to cause embryo rupture. Therefore, samples should be fumigated in liquid nitrogen for several minutes to pass the vitrificated transition temperature firstly and cooled to −35-−25 ℃ (or −75-−60 ℃) slowly. After this operation, the effect is better than that of directly inputting liquid nitrogen.

The embryo's viability is decreased with the extension of cryopreservation time. However, the molecular movement will stop and can be preserved semi permanently during the extracellular fluid vitrification. Long term storage should be carried out below −130 ℃, usually used liquid nitrogen. The temperature of liquid nitrogen is −196 ℃ and that of gaseous liquid nitrogen can also maintain at −150 ℃. Under long-term cryopreservation, the viability of embryos will not be influenced after thawing.

5.3.1.4 Thawing

If rapid freezing is adopted (samples are plugged into liquid nitrogen at −35-−30 ℃), embryos, removed from liquid nitrogen, are stayed in the air for 10 s and directly thawed in 25-35 ℃ water bath. If slow freezing is adopted (samples are plugged into liquid nitrogen at −75-−60 ℃), embryos should be thawed in 25 ℃ wet air (with 25 ℃/min heating rate).

5.3.1.5 Removal of cryoprotectant

Cryoprotectant can be removed step by step from high concentrations to low concentrations according to the equilibrium of cryoprotective solution. It can also be removed by one step of directly placing thawed embryos in 0.3-0.5 mol/L sucrose. At present, 0.3-0.5 mol/L sucrose is usually injected from both terminals of the plastic thin tube, and embryos with the cryoprotective solution is located in the middle (separated by bubbles). After thawed, embryos directly accomplish the removal of cryoprotectant in plastic thin tube, and then are transplanted immediately.

5.3.2 Ultra-rapid freezing

5.3.2.1 Selection of cryoprotectant

The permeating cryoprotectant generally used in this method is 2.2-3.5 mol/L glycerol or ethylene glycol, and the nonpermeating cryoprotectant is 0.3-0.5 mol/L sucrose.

5.3.2.2 Equilibrium in cryoprotective solution

Before being inserted into 0.25 mL plastic thin tube, embryos need to be equilibrated in

1.5 mol/L nonpermeating cryoprotectant for 10 min, and then equilibrated in cryoprotective solution with final concentration for 1-2 min at room temperature or 5 min at 4 ℃.

5.3.2.3 Freezing

After equilibrium and packing, embryos are freezing in the low-temperature refrigerator with −30-−25 ℃ for 15-30 min, and subsequently are plugged into liquid nitrogen for preservation.

5.3.3 Vitrification cryopreservation procedure

5.3.3.1 Preparation of cryoprotective solution and equilibrium solution

Vitrified freezing solution, i.e. cryoprotective solution, is based on PBS by adding with high concentration of cryoprotectant. When the concentration of cryoprotectant reaches more than 50% (volume ratio) or 8.0 mol/L, it is necessary to choose a cryoprotectant with low chemical toxicity. So far, a variety of improved cryoprotective solutions have been developed, and the combination of cryoprotectant are also of diversity. In recent, most studies selected ethylene glycol as the main cryoprotectant due to its characteristics of low chemical toxicity and fast penetration rate.

In vitrified freezing solution, it is necessary to add the macromolecular substances such as polysucrose, dextran, polyvinylpyrrolidone, polyethylene glycol, fetal bovine serum and bovine serum albumin, as well as the nonpermeating sucrose and trehalose. These substances are not only with low chemical toxicity, but also contribute to the solution vitrification. Besides, nonpermeating supplements can also reduce the cryoprotectant concentration, so as to indirectly reduce the toxicity of vitrified solution. In addition, sucrose substances can limit the excessive penetration of permeating cryoprotectant into cells through promoting cell contraction, and reduce chemical toxicity indirectly.

Recently, the commonly used vitrified freezing solution is: 25% glycerol + 25% ethylene glycol or propylene glycol, 20% ethylene glycol + 20% DMSO + 0.3-0.5 mol/L sucrose, 40% ethylene glycol + 0.3-0.5 mol/L sucrose.

The composition of the equilibrium solution is the same as that of the vitrified freezing solution, but the concentration of each component is halved.

5.3.3.2 Equilibrium

Embryos are plugged firstly into the equilibrium solution and placed at room temperature for 2-5 min. After equilibrium, these embryos are transferred into the vitrified freezing solution and balanced at 4 ℃ for 1 min or at room temperature for 30 s.

5.3.3.3 Tubulating and freezing

After balancing, embryos are placed into a thin plastic tube or a stretched open tube with 0.25 mL volume, and then directly plugged into liquid nitrogen for preservation.

5.3.3.4 Thawing and cryoprotectant removing

The methods for embryo thawing and cryoprotectant removing are the same as the rapid freezing. However, there are more cryoprotectant in the vitrified freezing solution, which can lead a great chemical toxicity to cells. The embryo may die in a short time if the solution without dilution after thawing. Therefore, embryos in the thin tube should be washed out as soon as possible, and then diluted in 0.3-1.0 mol/L sucrose to remove intracellular cryoprotectant. Action with the osmotic pressure of sucrose solution, cell further dehydrates and contracts. Finally, these embryos are washed with isotonic physiological solution. Under the ultra-low temperature of liquid nitrogen, if the cryoprotectant can not reach the required concentration in the vitrified freezing solution, the frozen embryos will undergo de-vitrification, i.e. milky white, after immersion in water bath for a few seconds. This phenomenon occurs more frequently during thawing than during freezing. Devitrification produces ice crystals in cells, which directly affects embryo survival. Therefore, the temperature should be raised rapidly in a temperature range of −80-−50 ℃ to prevent devitrification.

In summary, conventional freezing is still widely used in production for embryo cryopreservation because of its relative stable effect and easy operation. Ultra-rapid freezing and vitrification have made many new breakthroughs in recent years, and embryo survival rate and hatching rate after cryopreservation are equal to that of conventional freezing. At present, these two technologies have not been popularized and applied in a large area because they do not have a large number of experimental results. However, ultra-rapid freezing and vitrification have many advantages, such as simplicity, economy and practicality. Therefore, these two methods will be the direction of embryo cryopreservation in future. The advantages and disadvantages of different freezing methods are listed in Table 5-1.

Table 5-1 Advantages and disadvantages of different freezing methods

Methods	Advantages	Disadvantages
Conventional freezing	Stable effect; Easy operation; Relatively time-saving; Relatively low cost	Need for an expensive temperature controlling equipment; Limitation in early oocyte and embryo cryopreservation
Ultra-rapid freezing	No need for an expensive temperature controlling equipment; Time saving; Easier operation than vitrification	Instable effect; Limitation in early oocyte and embryo cryopreservation
Vitrification	No need for an expensive temperature controlling equipment; Time saving; Wide usage in different types of oocyte and embryo cryopreservation	Instable effect; Difficult operation; Toxicity of high concentration cryoprotectant to oocyte and embryo

5.4 Procedures for oocyte cryopreservation

5.4.1 Slow freezing

The slow freezing for mammalian oocytes is similar to that for embryos. However, a great difference of ultrastructure exists among oocytes and embryos. As a result, slow freezing method for embryos may not be perfectly adequate for oocytes. Secondly, oocytes are more sensitive to the low temperature and the chemical toxicity of cryoprotectant. After thawing, the *in vitro* fertilization rate, cleavage rate and blastocyst development rate of oocytes are greatly affected. Thirdly, the freezing resistance of mature and immature oocytes is different. Studies have shown that several indexes of mature oocytes after slow freezing are significantly better than those of immature oocytes, including their fertilization ability, development rate and embryo quality. However, the obtained oocytes are often immature in actual production. Therefore, the application of slow freezing for oocytes is very limited.

5.4.2 Vitrification cryopreservation

Compared with slow freezing, vitrification has many advantages, including a higher rate of spindle morphology normalization, avoidance of intracellular ice crystal formation, saving of freezing time, exemption of temperature controller and so on. Therefore, the vitrification gradually replaces the slow freezing for oocyte cryopreservation. However, vitrification requires that the concentration of cryoprotectant reaches 6.0 mol/L or higher, which is much higher than that of slow freezing method (1.0-2.0 mol/L). The high concentrations of cryoprotectant have great chemical toxicity to oocytes. In order to reduce the toxicity of cryoprotective solution, vitrification generally adopts a low toxicity cryoprotectant or a mixture of various cryoprotectants, as well as the step-by-step equilibrium by a variety of precooled high concentration solutions before freezing. According to the different carriers, vitrification for oocyte cryopreservation can be subdivided into several categories, including thin tubes, droplets, electron microscopic grids (EMG), open pulled straws (OPS), glass micropipettes (GMP), cryoloop and so on. In general, these carriers can be summarized as vitrification surface carrier systems (Fig. 5-1) and vitrification tubing carrier systems (Fig. 5-2). The main purpose of these methods is to speed up freezing and thawing to improve the freezing efficiency. These methods have two advantages: one is to induce a decline concentration of cryoprotective solution, and another is to reduce cryoinjuries to low-temperature-sensitive cell structures. In the next, several different carrier vitrification for oocyte cryopreservation are listed in detail.

5.4.2.1 Thin tubes cryopreservation

The cryoprotective solution used in this method is 6.0 mol/L DMSO prepared with M2.

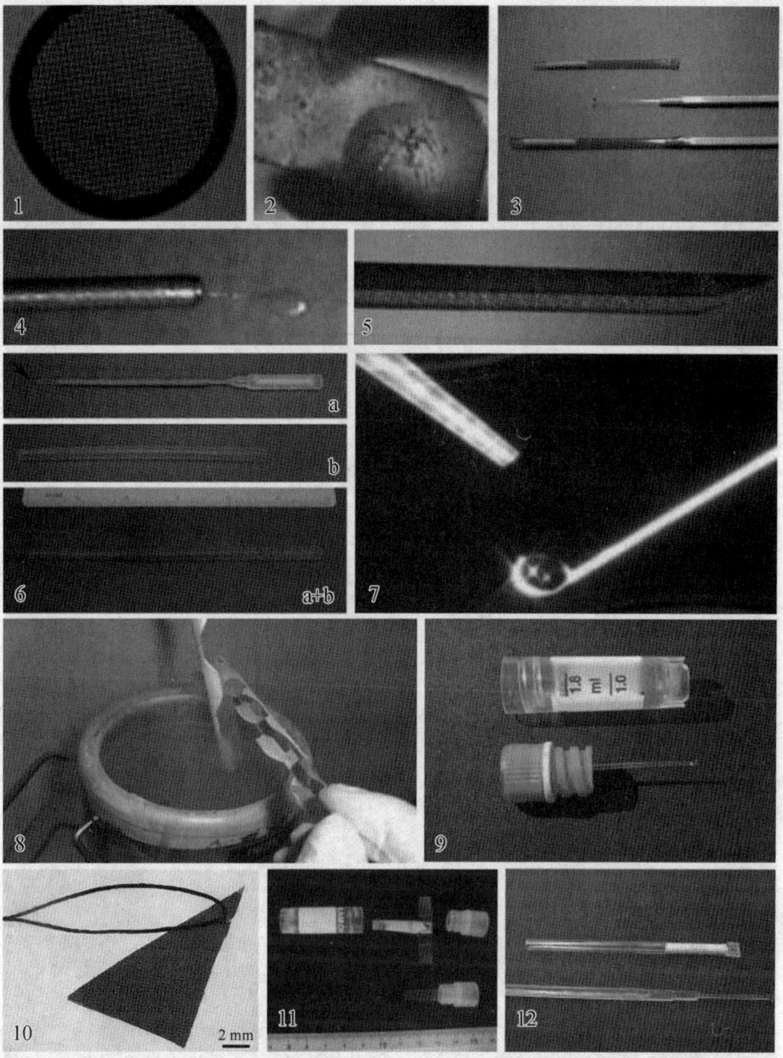

1. Electron microscope grid 2. Droplets 3. Cryotop 4. Cryoloop 5. Hemi-straw 6. Cryoleaf
7. Fiber plug 8. Direct cover vitrification 9. Vitrification spatula 10. Nylon mesh
11. Plastic blade 12. Vitri-Inga

Fig. 5-1 Vitrification surface carrier systems (Saragusty and Arav, 2011)

First, oocytes are respectively equilibrated in cryoprotective solution with 25% and 65% concentration at 20 ℃. After equilibrium, these oocytes are placed in cryoprotective solution with final concentration, and then transferred into a plastic tube with 0.25 mL volume. Subsequently, the tubes containing oocytes are plugged into liquid nitrogen for further cryopreservation after fumigating 3 min by liquid nitrogen steam. When thawing, the thin tubes are first subjected in 20 ℃ air bath for 10 s, and then heated in 20 ℃ water bath for 10 s. The morphological normality rate of thawed oocytes can reach up to 77%. After *in vitro* fertilization and transplantation, their 2-cell development rate and pregnancy rate are 91% and 79% respectively.

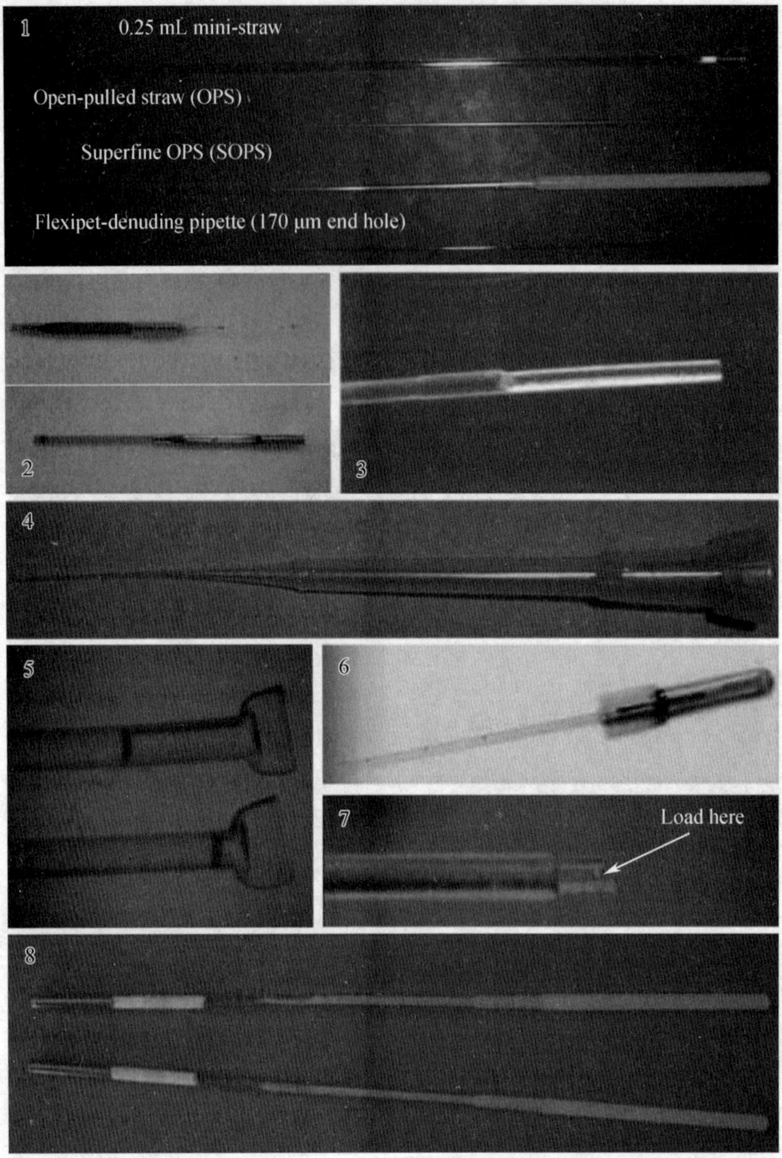

1. Plastic straw 2. CryoTip 3. High-security vitrification device 4. Pipette tip 5. Sealed pulled straw 6. Cryopette 7. Rapid-I 8. JY straw

Fig. 5-2 Vitrification tubing carrier systems (Saragusty and Arav, 2011)

If the cryoprotective solution is a combination with several cryoprotectants, i. e. , DMSO, acetamide, propylene glycol and polyethylene glycol. Before freezing, the mouse oocytes need to be equilibrated in different concentration of cryoprotective solution (25%, 50% and 100%) for 10 min respectively. Then, the final mixture is transferred into a small centrifugal tube or plastic thin tube and plugged into liquid nitrogen for cryopreservation finally. The fertilization rate and the blastocyst rate of these oocytes after thawing can respectively reach 80%-87% and 69%-78%. Porcine oocytes are cryopreserved with 40%

ethylene glycol and 18% Ficoll 70 at 3 : 2 (volume ratio). After thawing, the morphological normality rate is 97%, but 70% of oocytes show mitochondrial aggregation. In terms of livestock oocyte cryopreservation, thin tubes have not gained good result so far.

5.4.2.2 Droplets cryopreservation

The thermal conductivity of plastic thin tube is poor, which limits the maximum freezing speed of thin tube method to 2 000 ℃/min. The simplest way to improve the freezing speed is to put the mixture of oocytes and cryoprotective solution directly into liquid nitrogen without carriers, namely droplets. Firstly, oocytes are equilibrated in the cryoprotective solution for a period of time. Secondly, about 6 μL cryoprotective solution containing oocytes is directly dropped into liquid nitrogen. Finally, the frozen droplets are placed together in a small tube and then plugged into liquid nitrogen for further cryopreservation.

The thawing procedure of droplets is very simple. The frozen droplets only need to be directly put into 37 ℃ thawing solution. The oocytes are recovered and detoxified after melting. The droplets are first used to freeze mouse embryos, and then used to freeze bovine embryos, zygotes and oocytes respectively. The experimental results show that this method has a high freezing efficiency.

However, the droplet freezing method also has the following defects: ①the droplets float on the liquid nitrogen surface before sinking, resulting in a reduction of the freezing speed; ②it is difficult to label the oocyte characteristics without carrier; ③the droplets are easily missed when dropping into liquid nitrogen; ④the contact area between cryoprotective solution and liquid nitrogen is large, which is easy to cause pollution.

5.4.2.3 Electron microscopic grids cryopreservation

The freezing speed is closely related to the volume of solution, thus it can be significantly improved by a reduced volume of the cryoprotective solution. According to this principle, electron microscopic grids freezing method achieves 3 000 ℃/min freezing speed through adopting a direct contract between small volume cryoprotective solution (<1 μL) and liquid nitrogen, and using the electron microscopic grids as a carrier for oocytes. In detail, oocytes are equilibrated in the cryoprotective solution for a period of time, and with solution together transferred to the electron microscopic grids. Then, the electron microscopic grids is clamped by tweezers and directly immersed in liquid nitrogen. When thawing, the electron microscopic grids are immersed in the thawing solution at a certain temperature. The oocytes are recovered and detoxified after melting.

The morphological normality rate of bovine oocytes cryopreserved by this method is close to 60%. After *in vitro* fertilization, their cleavage rate and blastocyst development rate can reach 29%-32% and 10%-15% respectively. When this method combined with −210 ℃ liquid nitrogen as cold source, the freezing speed can reach 24 000 ℃/min. Frozen bovine oocytes are thawed and fertilized *in vitro*, and their cleavage rate is up to 48%.

5.4.2.4 Open-pulled straws (OPS) cryopreservation

The open-pulled straws is made of 0.25 mL thin pipe, with a diameter of 0.8 mm and a pipe wall thickness of 0.07 mm. The experiment is carried out on a constant temperature table with 39 ℃ and on a room temperature (25-27 ℃). During freezing, the small terminal of OPS is submerged into the cryoprotective solution droplets (1-2 μL) containing oocytes. Oocytes with solution are together loaded into OPS under the action of siphon effect. Then, the OPS are plugged into liquid nitrogen for preservation. When thawing, the small terminal of OPS is immersed in thawing solution at a certain temperature. The frozen solution is thawed within 1-2 s, and the thawing solution enters into straws immediately. Subsequently, the oocytes are precipitated and moved from OPS. The OPS method can reach a very high cooling and warming rates (over 20 000 ℃/min) and a short contact with concentrated cryoprotective solution (less than 30 s over −180 ℃), so as to avoid cryoinjuries. In addition, with the freezing and thawing procedures this method is simpler than the electron microscopic grids. Therefore, the OPS method greatly improves the efficiency of cryopreservation.

5.4.2.5 Glass micropipette (GMP) cryopreservation

GMP method is derived from the principle of OPS method. A glass micropipette with 0.3 mm diameter is used to replace OPS. Compared to plastic materials, the GMP has not only a better thermal conductivity, but also a smaller diameter and a larger mass. Therefore, GMP method has faster freezing rate than OPS method, and the GMP will not float in liquid nitrogen due to its large specific gravity.

5.4.2.6 Cryoloop

The cryoloop is a nylon loop with a diameter among 0.5 mm and 0.7 mm. Before oocyte freezing, a thin film is firstly formed on cryoloop with cryoprotective solution. Subsequently, the oocytes are carefully moved onto the film, and then quickly plugged into liquid nitrogen. This method uses a little cryoprotective solution, as well as no isolation such as thin pipe. Thus, there is a fast and uniform temperature change during freezing and thawing, which can effectively reduce the mechanical and osmotic injury of cells. If oocytes are cryopreserved for a long time, the cryoloop can be inserted into the magnetic cryocap and then put into the precooled supporting refrigeration tube of liquid nitrogen. This tube can effectively hold down the magnetic cryocap. Besides, there is a vent hole at the bottom of the tube, which is conducive to the rapid entry of liquid nitrogen. Compared with OPS method, the cryoloop method requires less cryoprotective solution and can gain faster freezing rate.

5.5 Prospects and conclusions of oocyte cryopreservation

For decades, advancement in oocyte cryopreservation has rapidly progressed with satisfactory outcomes. However, there are still many challenges existing in cryobiology that aim to improve outcomes for the cryopreserved oocytes. For example, oocytes are more sensitive to cool stress and cryoinjuries and are highly dependent on species and the freezing technology, especially the porcine oocytes. In addition, the cryopreserved oocytes are generally poor in terms of the viability, fertilization rate and developmental ability when compared to the non-cryopreserved oocytes. The reasons may include the following aspects: ①Oocyte is single cell with a large volume and small surface area relatively, and also with low membrane permeability to water and cryoprotectant. Intracellular ice crystals are easily produced during freezing, so as to cause physical damage of oocyte. ②The substances contained in oocytes are sensitive to low temperature, for instance, a large number of lipid droplets in porcine oocytes and many vesicles in sheep oocytes. ③A series of cytoskeleton related activities occur during oocyte maturation, including organelle rearrangement, polar body release, microvilli movement and so on. The cytoskeleton structure can be destroyed during freezing. Cryoprotectants can also cause serious damage to the cytoskeleton structure. ④Freezing terminates the connection between oocytes and cumulus cells, therefore delaying oocyte maturation. ⑤Freezing causes the damage of cell membrane system, which affects the physiological and biochemical activities of cell membrane.

Assignments

1. Why are the oocyte and embryo cryopreservation so important in animal production?
2. What is the principle of oocyte and embryo cryopreservation?
3. What are the advantages and disadvantages of conventional freezing, ultra-rapid freezing and vitrified freezing?
4. How to improve the oocyte and embryo cryopreservation?

(Wu Guoyun)

Chapter 6 Sex Control in Mammals

Summary

Sex control techniques can predetermine the sex of offspring by altering the normal reproductive process. In livestock production, many economic traits are restricted by sex (such as lactation traits) or affected by sex (such as meat and wool traits), differences in type and sex can translate into significant economic benefits, including the alleviation of severe food shortages. Using sex control technology to obtain the offspring of specific gender can obtain greater economic benefits. However, commercial applications of sex control techniques are currently available only for cattle. In this chapter, we introduce a brief history of sex determination, and the common methods used for sexual manipulation. The information presented in this chapter may inform future research on expanding the scope and use of sexual manipulation techniques in livestock.

Code 6-1

In the animal production, important economic traits such as meat, milk and eggs production are different between sexes, and some of them are limited traits. Under normal circumstances, the male to female sex ratio of natural animal groups is basically close to 1 : 1. Sexual manipulation or sex control (SC) can selectively predetermine the sex of animal offspring by interfering with normal reproductive processes that enables female animals to reproduce the required offspring through human intervention and according to people's wishes, and it can also resolve animal welfare-related issues such as castration, and play important roles in addressing food shortages.

Sex control is closely related to gender identification because sperm separation combined with artificial insemination results directly in specific gender, while gender identification combined with embryo transfer results in specific gender. Therefore, sex control can divide into two main methods: sperm separation before fertilization and gender identification after fertilization.

6.1 Brief history of studies on sex determination

The theory on the determination of the sex of an individual was elevated from myth to science in the early 1600s. William Harvey (1651) compared the reproductive organs of several animals and stated that "all animals, even fetal animals, including humans, are produced by the egg". Later on, Antonie Van Leeuwenhoek invented a microscope and observed sperm cells in human and dog semen, which became a turning point in the study of sexual manipulation.

In the early 20th century, Mcchung (1902) put forward the chromosome theory of sex determination when studying locust sperm cells. Nettie Maria Stevens (1905) discovered that unlike females, male *Phyllostomus flavus* has one chromosome named "Y" chromosome that is smaller than the others, and it was responsible for sex determination along with the larger chromosome named "X" chromosome. Subsequently, many studies confirmed that the normal sex of mammals is determined by a pair of sex chromosomes. The sex of offspring depends on the type of sex chromosomes carried by the female and male gametes at fertilization.

In the 1950s, two important discoveries were made in the study of sex determination. One was that Eichwald and Silmser (1955) discovered the male specific minor histocompatibility-Y antigen (H-Y antigen), after that, studies about the use of H-Y antibody in embryo sex identification combined with embryo transfer obtained 80% to 90% female sex identification rate in cattle, pigs, mice and sheep. The other was that Welshons and Jacobs (1959) put forward the theory that Y chromosome determines male, which caused people to look for sex-determining genes from chromosomes. Therefore, in the following decades, efforts were devoted to the identification of a single testicular determinant factor (TDF) in humans. Sequencing showed that an open reading frame (ORF) encodes a single exon gene in a "sex-determining" region called the "Y chromosome" (SRY in humans). The discovery of SRY sequence is a major breakthrough in sex determination theory, and the identification of embryonic SRY sequence by molecular biology then entered the stage of practical application.

6.2 Sex control techniques in mammals

Since the discovery of the mechanism of sex determination in mammals, the study of sex control techniques has received attention from a variety of studies. Methods of mammalian sex control include sex control before fertilization, gender identification of embryos, prenatal sex identification and other biotechniques.

6.2.1 Sex control before fertilization—the separation of X sperm and Y sperm

Methods of sperm separation include physical isolation, immunological separation and flow cytometry. Physical separation is based on certain physical differences between X sperm (X chromosome) and Y sperm (Y chromosome), including specific gravity, volume, surface charge, surface antigen, and DNA content, etc. Since the role of sex chromosomes in sex determination was clarified, numerous attempts (sedimentation, centrifugation, filtration, electrophoresis, H-Y antigen, etc.) have been made to separate X- and Y-chromosome-bearing spermatozoa. However, most physical separation methods impair sperm viability. Therefore, flow cytometry and immunological separation are the main sperm separation techniques currently used. Flow cytometric sorting for the preselection of sex has progressed considerably, which has allowed the production of pre-sexed offspring in a multitude of species and become a commercial success in cattle around the world.

6.2.1.1 Physical separation

1. Deposition

The principle of this method is that X sperm is slightly heavier than Y sperm, so in a certain solution, the sedimentation of X sperm is slightly faster than Y sperm. It was reported that human semen placed in 12 Percoll gradients of different densities and slowly centrifuged for 30 min resulted in more than 90% of X sperms collected eventually. The differences in the center of gravity between X and Y sperm prompted X sperm more easily through the interface between the various density gradient. However, neither sperm size, density nor activity difference between X and Y sperm can explain this phenomenon, and this method lacks repeatability and confirmation in other animals.

2. Free flow of electrophoresis

This method is based on the different surface charge in the X and Y sperm. X sperm has more net negative charge, and moves to the positive electrode faster than Y sperm. Kaneko et al. (1984) used this method to separate human semen into two distinct peaks, so that X sperm and Y sperm could be collected respectively. However, when others repeated their experiment, satisfactory results were not obtained.

3. Laminar flow separation

This method is based on the fact that the X and Y sperm has different way and behavior of swimming. It was reported that the X sperm were swimming in a kind of column flow of fluid different from Y sperm. When sperm were collected along the fluid column, most X sperm concentrated in the beginning of the liquid column, while Y sperm concentrated in the far end. This method has been proved to be successful by Y-specific probe examination, but needs to be confirmed by repeated tests and experiments in other animals.

4. Albumin liquid column chromatography

The principle is that Y sperm swim faster than X sperm in a sticky albumin column. This method was considered to be more effective than other physical separation methods mentioned above, so it has been widely used in the separation of human sperm. However, this method is only suitable for the separation of Y sperm, and only a small amount of Y sperm is collected, and its efficiency is low. This is because Y sperm are easier to penetrate the viscous discontinuous albumin gradient interface than X sperm.

6. 2. 1. 2 Immunology identification in sperm separation

Immunology sexing identification is to identify and isolate the sexually related proteins, which is mainly done by using H-Y antibodies to detect H-Y antigens on the plasma membrane of Y-bearing spermatozoa. Immunology separation includes direct input of H-Y antiserum, immunoaffinity chromatography and immunomagnetic beads.

1. Direct input of H-Y antiserum

It was reported that H-Y antiserum injected into the rabbit vagina inhibited Y sperm activity and selected X sperm, artificial insemination was carried out 15 min later, and 74% female offspring were obtained.

2. Immunoaffinity chromatography

In this method, the isolation column is constructed by immunoaffinity chromatography using agarose beads coupled with anti-mouse IgG secondary antibody. The separated X sperm and Y sperm are pre-cultured with specific H-Y antibody, and then the pre-cultured sperm are washed to remove excess antibodies. The washed sperm are treated with immunoaffinity chromatography. The Y sperm will bind to the column until the elution, and X sperm will not bind to the column.

3. Immune magnetic method

This method is to use H-Y monoclonal antibody combined with H-Y antigen in a Y sperm, which are cultured with the secondary antibody coated by magnetized polymer small oscillation, and Y sperm-monoclonal antibody complex combined with magnetic beads, when magnets are used to process samples, the magnetization of the magnetic beads (including Y sperm) adheres in the test tube wall, and the sperms in the supernatant fluid is X sperm.

Sperm from mice are successfully separated by combining immune sexing and swimming methods. In 2019, three Japanese scientists discovered that Toll-like receptor 7/8, which was encoded by the X chromosome, could affect sperm viability. Since Toll-like receptor 7/8 bound to R848, a synthetic agonist that bound TLR8 and TLR7, sperm motility was reduced upon incubation with R848. However, Y sperm motility was not affected. Thus, spermatozoa were selectively separated by the upwelling method, in which Y spermatozoa with high motility swam to the upper layer of the medium, while X spermatozoa sunk to the lower layer of the medium. As a result, the upper layer sperm produced 77 blastocysts, of which 68 were XY embryos and 9 were XX embryos, and the lower layer sperm produced 83 blastocysts, of which 58 were XY embryos and 25 were XX embryos (Umehara et al., 2019).

6.2.1.3 Sperm sexing in flow cytometry

Flow cytometry (FCM) is a new technology for quantitative analysis of cells or cell particles and classification of cells. Flow cytometry, also known as fluorescence activated cell classifier (FACS), is a high-tech product integrating fluid jet technology, laser optical technology, electronic technology and computer technology.

The principle of this method is that the DNA content of X sperm is higher than that of Y sperm (Table 6-1). The first surviving offspring were rabbits born by intrauterine surgical insemination (Johnson, 1989). Cambridge Animal Biology Co., Ltd. first succeeded on bovine IVF with isolated sperm in 1993. Since then, flow cytometry-based sperm sexing has produced millions of pre-sexed offspring of different species, including rabbits, pigs, cattle, sheep, horses, dogs, cats, etc. In general, modern sperm sorting procedures produce offspring with a sex ratio deviation of 85% to 95% (de Graaf SP et al., 2009).

Table 6-1 The differences of DNA content between X and Y sperm %

Item	Human	Cattle	Pig	Horse	Sheep	Rabbit	Dog
Difference	2.8	3.8	3.6	4.1	4.2	3.0	3.9

The detailed method is that sperm are co-cultured with fluorescent dyes (37 ℃, 45 min), and the dyes quantitatively bound to DNA which is located by the searcher and excited by the laser beam. The signal is amplified, analyzed and distinguished by the instrument and computer system to separate X and Y sperm. X sperm has a strong fluorescence signal.

Flow cytometry has been used commercially in cattle, and the quality of sperm used in flow sorter depends on the species. Thus, sorting indices for different species are obtained by combining sperm with X or Y. The difference in DNA content is calculated by multiplying the area of the flat profile of the sperm head. The sorting index indicates that bull and boar sperm are well suited for separation in the flow sorter (Garner et al., 2006).

However, the use of flow cytometry for sperm sorting in pig production is highly unlikely. This may be due to the fact that boar sperm are more susceptible to flow cytometry, as the sorting process damages the boar sperm membrane, resulting in reduced viability, fertilization capacity, storage capacity and subsequent processing. In addition, the sperm used for intrauterine artificial insemination per pregnancy are usually significantly higher in swine production than in bovine production.

Initial studies focused on identifying the steps that led to sperm damage in flow cytometry. It was earlier thought that optimizing sperm dilution rates as well as varying dye type and laser power could reduce sperm damage. However, these methods did not significantly increase sperm viability by sex sorting.

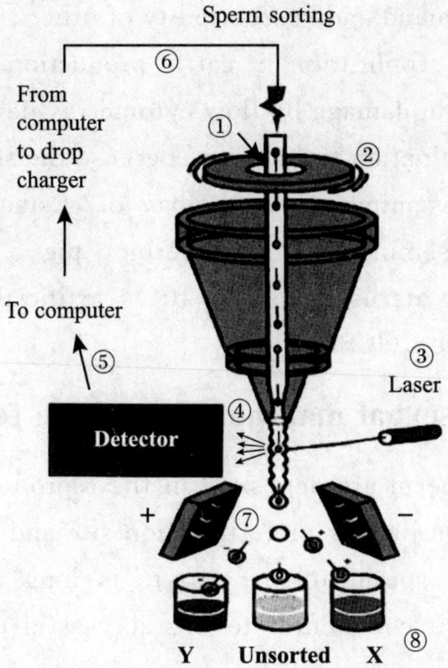

①The sperm are pumped after staining with a DNA-binding dye ②A piezoelectric crystal vibrator causes 90 000 droplets to form each second ③A UV laser illuminates the sperm as they flow by the beam ④X sperm fluoresce with 4% more intensity than Y sperm ⑤The signal detected is sent to a computer to categorize sperm as X, Y, or uncertain ⑥Negative, positive, or no charge is applied to droplets as they emerge from the stream ⑦As the charged droplets pass between continuously charged plates, they are deflected ⑧Sperm are collected in 3 containers: X, Y, uncertain or droplets without sperm.

Fig. 6-1 Sperm sexing in flow cytometry (Garner et al., 2001)

Although these improvements have increased the motility of the separated sperm, subsequent optimization is limited by current mechanical technology. In addition, improving production speed and sperm quality remains a major challenge, thus shifting attention to related technologies. In a study aimed at characterizing the effect of flow cytometry on the glycocalyx structure of bovine sperm by using lectin histochemistry, the effects of flow cytometry and energy acquisition on bovine sperm were similar, which may explain the low conception rates observed after flow cytometry (Umezu et al., 2017). Another study combining AA-2G supplementation and deep insemination to enhance fertility of sex-sorted sperm in cashmere goats revealed significant protective effects on sperm quality and function (Qin et al., 2018). Although some researchers prefer artificial insemination techniques, laparoscopic insemination (LI) in the uterine horn can achieve high fertilization rates through low sex-sorted sperm doses (five times less than intrauterine insemination) (Fantinati et al., 2005). However, improvements in the output of the flow sorter, aspects related to synchronization protocols, ovulation time and artificial insemination (AI) or laparoscopic insemination (LI) at the uterine horns deserve optimization in future studies.

In summary, flow cytometry is the most widely used and established technique for sperm separation in livestock production, with an isolation accuracy of more than 90%.

However, due to the high demand and susceptibility of other species to sperm, this strategy has only reached commercial application in cattle production. Therefore, future research should focus on reducing sperm damage by flow cytometry, developing related technologies, and optimizing artificial insemination techniques. Because the sperm separation speed is slow and the cost is high, the insemination technology of a small amount of sperm will be adopted. Such as surgical insemination (intrauterine, pig 5×10^7 in fallopian tube), laparoscopic insemination (in uterus, pig 3×10^6), artificial insemination (deep uterus, cervix), IVF and microinjection (ICSI).

6.2.2 Internal female sexual manipulation of the female reproductive tract

Millions or billions of sperm are deposited in the reproductive tract after mating, but only a few sperm arrive at the jugular or fertilization site and only one sperm fertilizes the oocyte. The journey of the sperm to the oocyte is long and dangerous. The female reproductive tract environment contributes to this successful transport by providing sperm transport vehicles, helping to remove dead sperm and other pathogens, and exerting strict selection pressure to ensure that only those sperms of the highest quality reach the site of fertilization. In addition, in many parts of the female reproductive tract, sperm interact with the epithelium and luminal fluid, which can affect sperm viability and function. Although we know little about this complex interaction between sperm and the female reproductive tract environment, many studies have attempted to reveal the influence of mammalian females on the sex of their offspring. By using bilateral laparoscopic insemination to introduce sperm with X or Y into two separate oviducts of a single female pig, researchers demonstrated that the oviducts could screen for sperm by altering the oviductal environment (Almiñana et al., 2014).

In addition, it was shown that both the timing of ovulation and the maturation status of the oocyte affected the sex ratio of the animal's offspring (Gutierrez-Adan, 1999; Dominko, 1997). This led to the study of the effect of follicular fluid composition on embryo sex, which was confirmed in subsequent studies (Grant, 2005; 2008).

Various studies have reported the effect of pH on sperm viability, motility and acquisition. However, pH does not affect *in vitro* bovine sperm bearing X or Y. Therefore, extensive studies are needed to examine the effect of external pH on X- or Y-bearing spermatozoa. It's also possible that pH may affect X- or Y-bearing sperm by altering other physicochemical factors in the bovine oviduct.

Adjusting the pH value of the insemination environment may affect the sex of offspring. Y sperm have poor tolerance to fatigue and acid. Xiangqian et al. (1996) conducted AI with semen at pH 7.2, resulting in 72.3% male calves, and 39.1% male calves at pH 6.8. Duan et al. (1991) injected different doses of arginine through vagina to change the pH of reproductive tract before insemination, resulting in 39.1% female rabbits at pH 5.22, and 61.9% at pH 9.83.

Several papers have suggested that various glycoproteins and factors from the female reproductive tract affected sperm survival, including *L*-selectin (Yu et al., 2018), PMCA4 (Al-Dossary et al., 2013), fibronectin (Osycka-Salut et al., 2020), and SBG (Teijeiro et al., 2008). Thus, the manipulation of offspring sex by regulating the expression levels of these glycoproteins and factors is possible and may be a new direction for future research.

Most importantly, although sexual manipulation in the female reproductive tract is the most succinct strategy proposed to date, it has the potential to address previously puzzling and seemingly contradictory findings. Ongoing research will allow for a better understanding of female influence on sex determination and may have a dramatic impact on existing sexual manipulation strategies.

6.2.3 Mating with semen by RNAi injection into testis

The method of RNA interference (RNAi) of specific genes is expected to realize the sex control of animals simply and quickly at a low cost. Zinc-finger Y (*ZFY*) is located in the short arm of the Y-chromosome and is a highly conserved gene that plays an important role during spermatogenesis on Y-sperm formation, while *ZFX* plays a significant role in the process of X spermatogenesis. RNAi is an effective method of silencing *ZFX* or *ZFY* mRNA expression which is a potential useful approach to control sex of offspring. It was reported that RNAi injection into testis to silence the *ZFX* produced 94 red deer, including 68 males and 26 females, and the male rates (72.34%) were significantly higher than the control groups (Wei et al., 2020). Injection of *ZFY* siRNA into bull testes resulted in 72.0% female offspring, and was greater than the 49.4% female calves in the control (Xi et al., 2019).

6.2.4 Sex identification of pre-implanted embryo

The main stages for embryo transfer are the morula or blastocyst stage. The blastomeres of the mammalian embryo are totipotent and can develop into complete animals. Each blastomere contains all genes required for animal development, including genes involved in sex determination, such as *AMG*, *ZFY* and *SRY*. In this way, firstly, a small number of blastomeres are extracted or cut from the preimplantation embryos. Then, thesex of these blastomeres are identified. Therefore, a number of techniques have been developed for the sex identification of preimplantation embryos, such as sex chromatin identification, karyotype identification, X-linked enzyme activity assay, male specific antigen identification, SRY-PCR identification and loop-mediated isothermal amplification (LAMP) etc.

6.2.4.1 Sex chromatin identification

It is the first method to identify the sex of embryos in the world. Garner et al. (1968) took cells from rabbit embryos and analyzed their sex chromatin or chromatin bodies. This method is complex and has great damage to embryos, and the success rate is very low. At the same time, chromatin bodies are difficult to find and observe in many livestock embryonic cells. However, for a small number of embryos with the recognizable sex chromatin (temporarily inactivated X chromosome), the accuracy can reach 100%.

6.2.4.2 Karyotype identification

Karyotype identification is a classical but time-consuming method. Two experienced technicians need about 5 h to identify 12-15 embryos. It is difficult to apply in production and can only be used to verify the accuracy of other methods. The detailed process is as below: a small number of embryonic cells are taken and cultured in the culture medium containing colchicine to keep cells staying in the middle mitosis stage; the cells are expanded with hypotonic solution, the cell membrane is broken, the chromosomes are released, then fixed, stained with Giemsa and observed under the microscope. Because mitosis is blocked in metaphase, chromosomes are shortened, regular in shape and have specific banding patterns. The sex of embryo is identified by karyotype analysis, and the Y chromosome not be paired with X chromosome is found.

6.2.4.3 X-linked enzyme activity assay

One of the two X chromosomes in the early female embryo is inactivated. During the short period of embryonic genome activation and X chromosome inactivation, both X chromosomes can be transferred. The concentration and activity of X-related enzymes (glucose-6-phosphate dehydrogenase, hypoxanthine nucleic acid transferase, adenine phosphate transferase) in the female embryo are twice that of the male embryo.

Prior to Barr vesicle formation or detection of H-Y antigens, the methods used were based on cytogenetic chemical analysis and the measurement of X-linked enzyme (glucose-6-phosphatase dehydrogenase, G6PD) levels. However, these methods have been replaced by molecular assays due to high embryo requirements, low accuracy, the potential for embryo damage and the unknown time of X chromosome inactivation.

6.2.4.4 H-Y antigen identification of embryo

Mammalian male embryos express a cell surface H-Y antigen, which is the basis for identifying the sex of embryos by immunological methods. H-Y antigen identification of embryo includes embryo cytotoxicity analysis, indirect immunofluorescence analysis and blastocyst formation inhibition method.

1. Embryo cytotoxicity analysis

The principle of this method is that in the presence of complement, H-Y antibody can bind with $H\text{-}Y^+$ embryo to dissolve one or more blastomeres, or make the blastomeres irregular and block the development of male embryo. It was reported that when normal embryos were transferred to synchronized recipients, 86% of the offspring were female. However, this method is rarely used due to the weak antigen of H-Y antigen, the low specificity and titer of the antibody, furthermore it is at the expense of destroying embryos.

2. Indirect immunofluorescence analysis

This method uses H-Y antibody as the primary antibody and goat anti-mouse IgG/FITC as the secondary antibody, which are co-cultured with embryos in turn. The result is that fluorescent embryos are male and non-fluorescent embryos are female. This method is non-destructive to embryos, but the fluorescence intensity is subjective. It was reported that 78% of those judged as male embryos gave birth to male rats, and 81% of those judged as female embryos gave birth to female rats. The accuracy of female identification was 83%-89% in cattle, 81% in pigs and 85% in sheep.

3. Blastocyst formation inhibition method

This principle of this method is that H-Y antibody can reversibly inhibit the formation of blastocyst. The morula is co-cultured with the H-Y antibody, and female embryo will continue to develop to form blastocyst, while male embryo continue to culture after washing off the H-Y antibody. It was reported that 79.1% (38/48) female rats were produced after ET, and 91.3% (21/23) male rats were produced after ET. Utsumi et al. (1993) cultured bovine morula in medium with rat H-Y antibody, and Karyotype analysis showed that 80%-90% of female embryos had XX karyotype and 80% of male embryos had XY karyotype. The viability of identified embryo had no difference from the normal embryo, therefore, it is a practical method for sex manipulation of embryo.

6.2.4.5 SRY-PCR identification

The in-depth study of Y chromosome structure and the development of DNA recombination techniques, especially polymerase chain reaction (PCR) and fluorescence in situ hybridization for specific gene sequences in the sex-determining region, provide accurate, efficient and rapid new molecular biological methods for early preimplantation embryo sex identification. Several molecular biology detection techniques, including the use of nucleic acid probes, fluorescence in situ hybridization (FISH), polymerase chain reaction (PCR), and loop-mediated isothermal amplification (LAMP), can be used to identify male-specific genes.

1. Nucleic acid probes

Y chromosome-specific DNA fragments (TDF, SRY) are labeled with radioactive or non-radioactive substances as probes to hybridize with the homologous sequences of DNA in embryonic cells. The existence of Y chromosomes is determined according to the speck signals displayed, so as to identify the sex of embryos. Among nucleic acid probes, isotope-labeled probes are highly accurate but require a large number of embryonic cells and time, whereas biotin-labeled probes have a short detection time but can isolate only a small number of embryos.

2. FISH

FISH is a combination of molecular hybridization and histochemistry including the preparation of fluorescein probes, denaturation and hybridization of probes and target DNA, observation and identification. It has high accuracy and rapidity due to the advent of multicolor fluorescence in situ hybridization and cells can hybridize at interdivision and division stages. However, FISH is expensive and complex to operate, so PCR and LAMP become the goal mainstream methods of embryo sex determination in pre-sex manipulation.

3. PCR

PCR is a method of *in vitro* enzymatic synthesis of specific DNA fragments which includes three stages: denaturation at high temperature (95 ℃), annealing at low temperature (37-55 ℃) and extension at moderate temperature (72 ℃). DNA from embryonic cells are exacted and specific primers are designed, and specific DNA genes are amplificated as mentioned above in a complete system. PCR products are usually revealed by agarose gel electrophoresis. Amplification of specific repeats of Y chromosome or *SRY* can greatly increase the sensitivity and improve the accuracy of identification. This method was first successful in identifying the sex of bovine and sheep embryos in 1991 (Herr et al., 1991). However, since the method amplifies only one pair of sex-specific primers, the lack of amplification would correspond to a negative result. Therefore, this method was replaced by multiplex PCR, in which a pair of sex-specific primers was designed with at least one pair of internal primers that amplified housekeeping genes. However, the method was error-prone due to the large number of cells required for conventional PCR methods and the limited sensitivity of a single amplification. Therefore, nested PCR and multiplex PCR were later combined to improve the accuracy of the assay results. The combination reduced the number of cells required to 3-8 for the accuracy of the test.

4. LAMP

This method is simple, accurate and cheap, and has many incomparable advantages so far. The product is amplificated at a constant temperature (65 ℃), no DNA denaturation is required, and it is fast, efficient (less than 1 h), with high specificity and high sensitivity. The result is usually used to detect the white precipitate (the by-product produced by the combination of pyrophosphate and magnesium ions precipitated by dNTP) by the turbidimeter.

Although significant advances in embryo sex determination have been made in PCR and LAMP due to their high sensitivity, the impact of external environmental contamination on the accuracy of results remains unavoidable. In addition, both methods are difficult to use in livestock production due to large sample sizes and the poor transfer pregnancy rates of recent embryo transfer techniques. Therefore, future research should focus on improving tolerance to the external environment and the efficiency of these techniques.

It is feasible to use modern molecular biology technology for rapid and accurate sex identification of early embryos. However, some cells need to be removed from embryos for early embryo sex identification, which will cause damage to embryos, and only a small number of embryos survive after freezing-thawing treatment, resulting in the decline of the success rate of ET, which restricts the commercialization process of providing livestock embryos with predicted sex.

6.2.5 Sex control by ultrasonography identification

The sex of the fetus is detected at a certain stage after implantation, and then selective abortion is carried out. With the commercial use of artificial insemination and embryo transfer techniques, the research to find methods for prenatal sex determination has become a focus of interest. Obstetric ultrasonography is the most common imaging technique used in veterinary reproduction, including confirmation of pregnancy, identification of congenital fetal defects, and determination of fetal sex, because it is simple, reliable and noninvasive. However, the sex of mammalian fetuses is not detected by ultrasonography until several weeks after conception (e.g., 55-85 d in cattle, 59-68 d in horses, 46-55 d in goats).

Amniocentesis can also be used to determine the content of amniotic fluid hormone (testosterone), analyze the chromosome, detect the cells obtained from amniotic fluid with Y chromosome specific DNA probe, and then judge the sex of the fetus.

6.2.6 Other biotechniques

Other biotechniques such as cloning, embryo bisection, parthenogenesis can also be used to control sex.

Cloning, also known as nuclear transplantation. Cells of known sex are used as donors and transferred into enucleated oocytes to develop into new individuals for specific gender.

Embryo bisection can get individuals with the same gender. Firstly, to identify the sex of the embryo, then divide the embryo with the specific gender, culture the splitted embryo, and transfer it to the recipient to obtain new individuals of the same gender.

Micro fertilization is a method in which *in vitro* fertilization is completed under microscope, by injecting sex-identified sperm into an oocyte.

Parthenogenesis includes parthenogernesis and patrogenesis. Higher animals mostly reproduce sexually to ensure genetic diversity. Genomes that can distinguish between sperm and oocyte, called genomic imprinting, evolved in mammals specifically. These genes make certain alleles of a fertilized egg express with one and silence with the other during normal fertilization. By imprinting gene modification (methylation and demethylation) or knockout of haploid embryonic stem cells (ESC) or oocyte and using the cells to perform complex embryonic manipulation, parthenogenetic mice with complete normal development, behavior, metabolism and fertility, as well as androgenetic male mice, were obtained (Li et al., 2018; Wei et al., 2022).

6.3 Application prospect of sex control techniques

With the continuous development of science and technology, artificial insemination, embryo transfer, *in vitro* fertilization and sex control are the four magic weapons of the new technology of livestock breeding. If the four magic weapons are combined and applied to livestock breeding and animal husbandry production, the development of animal husbandry production will play an inestimable role and have a profound influence.

Sex control is of great significance to the development of efficient and high-quality animal husbandry, for example, to allow for greater economic benefits for both sex-restricted and sex-influenced production traits, enhance the selection intensity of improved variety breeding and speed up the progress of genetic improvement, and exclude harmful or undesirable genes (e. g. , to avoid the birth of offspring with sex-linked hereditary disease). In terms of production, dairy farmers hope to breed more heifers from their high-quality and high-yield core group to renew the dairy herd; beef cattle farmers want to breed more male calves. For descendant determination, gender control can save at least half of the time, energy and cost than no gender control.

Sexual manipulation techniques are valuable biotechniques that have the potential to revolutionize animal production. The successful method is the separation of X sperm and Y sperm by flow cytometry. However, the slow separation speed and sperm motility affect the conception rate after AI. That is to say, different sexual manipulation strategies have their own strengths and weaknesses, and the choice of a particular system depends on the expertise of the individual researcher rather than the weaknesses of one of these strategies.

Overall, sex determination is a complex process that is influenced by a variety of internal and external factors. Further exploration of sex determination mechanisms, continuous improvement in the accuracy, sensitivity, and economic efficiency of sex manipulation techniques, and the establishment of a novel sex manipulation system that integrates knowledge from multiple fields could greatly facilitate livestock production and address food shortages.

Assignments

1. Design several non-invasive methods of fetal gender identification.
2. Describe the application of flow cytometry in various farm animals.
3. What are the advantages and limitations of immunology identification in sexual manipulation?
4. Briefly describe the application prospect of gender control techniques.
5. Discuss the difference and connection between gender control and gender identification.
6. What do you think the most promising gender control technology is? Why?

(Wang Ziyu)

Chapter 7 Embryo Transfer

Summary

Embryo transfer is an important assisted reproduction in which embryos are placed into the uterus of a female with the intent to establish a pregnancy. Embryo transfer has proved to be a powerful technology in genetic improvement of farm animals, primarily to propagate the genes of females of superior pedigree. Embryo transfer was first established in rabbit in 1890. Successful embryo implantation requires synchronization between the acquisition of implantation competency by the blastocyst and a receptive state in the uterine endometrium. Embryo transfer in farm animal is mostly done in cattle, small ruminants such as sheep and goat. Although the protocols or criteria of embryo transfer vary in those domestic animals, the primary procedure of embryo transfer is similar, including recipient and donor selection, donor superovulation and artificial insemination, recipient synchronization, embryo collection and quality control, and embryo transfer in recipient.

Embryo transfer refers to a step in the process of assisted reproduction in which embryos are placed into the uterus of a female with the intent to establish a pregnancy. This technique is widely applied to humans and farm animals, in whose situations the goals may vary. From commercial point of view, embryo transfer attained attention in the early 1970s. It was firstly intended to augment the spread of desirable phenotype among animal populations. Later, with advancement in the concept of multiple ovulation and embryo transfer (MOET), the potential of embryo transfer for the rapid propagation of desirable genetics was unearthed. The capability of embryo transfer program to multiply and proliferate the desired genetical goals in less time has a revolutionary role in animal production programs.

An embryo transfer program is a complex process that involves various interdependent factors and procedures. It involves selection and management of donor and recipients, superovulation, synchronization, embryo production, evaluation and transfer. Failure in achieving optimal targets in any of the aforementioned factors would result in unsuccessful embryo transfer program. Hence, sound knowledge and technical expertise are mandatory to successfully perform embryo transfer.

7.1 Research history of embryo transfer

7.1.1 Development of fresh embryo and embryo transfer

A look in the history revealed that the concept of embryo transfer in mankind exited well Before the Common Era (B. C. E). However, based on scientific traces, Walter Heape (1855—1929) was crowned to be pioneer of successful mammalian embryo transfer. He auspiciously managed to transfer two four-cell Angora rabbit embryos to an inseminated Belgian doe in 1890. Consequently, the recipient Belgian doe gave birth to four Belgian and two Angora young (Mapletoft, 2013). In this regard, a series of substratal, though unsuccessful, investigations by George John Romanes (1848—1894) were divulged to provide baseline for fruitful embryo transfer by Walter Heape. There were no reports of further success in mammalian embryo transfer until the 1920s, when several investigators again described embryo transfer in rabbits. Scientists then extended the scope of embryo transfer to other farm animal species such as sheep, goat and bovine until the first half of the 20th century. For example, Warwick group made a great success on embryo transfer in sheep and goats in the 1930s and 1940s. And Umbaugh reported the first embryo transfer in cattle in 1949. He produced four pregnancies from the transfer of cattle embryos, however, all pregnancies were terminated before the full term. In 1951, the first embryo transfer calf was born following a surgical transfer of a day-5 embryo by Willett and his colleague in Wisconsin, USA (Willett et al., 1951). Birth of calves and pigs following nonsurgical, transcervical embryo transfer were reported in the 1960s. However, it was not until the mid-1970s that transcervical embryo transfer replaced surgical embryo transfer as a routine procedure in cattle. In 1974, the first foal produced by equine embryo transfer was born in Japan. Transport of embryos over long distances was first accomplished in the early 1970's by placing pig and sheep embryos in the oviducts of rabbits, which were used as biological incubators, and in 1976, long-distance transport of equine embryos was firstly reported in England.

7.1.2 Development of frozen embryo transfer

In 1972, researchers reported the birth of the mouse offsprings that were derived from embryos that had been frozen, thawed and subsequently transferred. In 1973, the first calf was born following the transfer of a frozen-thawed embryo announced. In 1976, the first successful transfer of a frozen-thawed equine embryo was reported in Japan. Technique for successful freezing of equine embryos was reported in 1987, Colorado, USA.

7.1.3 Development of combination of micromanipulation and embryo transfer

The 1970s was remarkable era of micromanipulation and early attempts of *in vitro* fertilization (IVF). The first calf produced from an embryo that had been biopsied and the sex determined from the biopsy specimen was born in 1975. With the rapid development of reproductive biotechnology, Robert Edwards applied *in vitro* fertilization and embryo transfer (IVF/ET) to treat infertility in women with non-patent oviducts. His work resulted in the birth of the first "test tube" baby in July 1978. Till now, more than 4 million babies have been born using the IVF/ET. After a period of about 33 years, the 2010 Nobel Prize in Physiology or Medicine was awarded to British physiologist Robert Edwards for his pioneering contributions to *in vitro* fertilization, and a new specialty of assisted reproduction has been established with its own professional societies.

The first calf produced from IVF was born in 1981. This was followed by the birth of IVF pigs in 1983 and lambs in 1984. The birth of the lamb "Dolly", the first animal born following nuclear transfer or cloning, occurred in 1996. In 1991, the first foal produced by IVF was reported in France.

7.1.4 Embryo transfer organizations contributed to the technique development

Lionel Edward Aston Rowson (1914—1989) played pivotal role in commercialization of embryo transfer technique in farm animals in early 1970s. International Embryo Transfer Society (IETS) was established 1974, and Betteridge (2003) was honored as with the title of Founding Honorary President to reward his support to Rowson group on ET in farm animals. The IETS had 82 charter memberships, representing researchers, academics and veterinary practitioners from around the world and proved to be vital scientific platform of regular discussion and publications to address issues regarding embryo transfer technology. Afterwards, as research and development expanded, various organizations like American Embryo Transfer Association (AETA, 1982), Canadian Embryo Transfer Association (CETA, 1984), Brazilian Embryo Technology Society (SBTE, 1985) were established during 1980s (Rubin, 2005). The aforementioned organizations interacted with all stakeholders including researcher, practitioners and farmers and played key role in shaping up of embryo transfer technique and its commercial implementation. Member's interests related primarily to cattle, horses and small ruminants, but also included swine, companion animals, laboratory species and wild and endangered species.

7.2 Rationale of embryo transfer

Embryo implantation involves the intimate interaction between an implantation-competent blastocyst and a receptive uterus, which occurs in a limited time period known as the window of implantation. Successful implantation requires synchronization between the acquisition of implantation competency by the blastocyst and a receptive state in the uterine endometrium.

7.2.1 Gestational development of reproductive organs after estrus

Physiological changes in the organs of the donor and recipient of the same species are the same, providing similar physiological environments for the implantation of donor embryos into the recipient. In most spontaneously ovulating livestock, the series of changes that occur in the reproductive organs after estrus are the same whether or not they are mated, or whether or not they conceive after mating. This includes the formation of the corpus luteum after ovulation of the follicles, the secretion of high levels of progesterone by the corpus luteum, the proliferation of tissue in the endometrium and increased secretion function. All these changes are suitable for the survival and growth of the early embryo, providing suitable environmental conditions for development and attachment of possible embryos and preparing for pregnancy. When the female is not pregnant, the endometrium will undergo degenerative changes and secrete prostaglandins (PG), which dissolves the corpus luteum and prompts the arrival of the next estrous cycle, in turn reproducing the same changes in the reproductive organs in preparation for fertilization and pregnancy.

7.2.2 Pre-implantation status of an early embryo

Embryos are in a free state during the early stages of development, which provides the possibility for the collection of embryos. Embryos are in a free state early in development (before attachment) and have not yet established a substantial tissue connection with the mother, and the nutrients needed for development and metabolism rely mainly on the nutrients stored byitself. Therefore, after the embryo leaves the mother, it is easy to survive for a short period of time and can be cultured *in vitro* for a short time; When placed back in the same environment as the donor's physiological condition, it can continue to develop.

7.2.3 Immune tolerance of the maternal uterus to the embryo

The recipient basically does not have immune rejection to the foreign embryo transferred into the uterus, which provides the possibility for the survival of the embryo in the recipient. The reproductive tracts (uterus and fallopian tubes) of the recipient female livestock are for those with foreign antigenic substances and fetal membrane tissues generally do not have immune rejection. Therefore, when an embryo is transferred from one individual to another, it can survive and continue to develop.

7.2.4 Pregnancy signals are emitted by the embryo

The donor embryo could establish normal physiological and tissue contact with the recipient uterus. The transferred embryo develops in the recipient's womb for a certain period of time, which can send a pregnancy signal and establish contact with the recipient female animals, thus ensuring normal development in the future.

7.2.5 Stability of embryonic genetic material

The genetic characteristics of the donor embryo are not affected in any way during pregnancy. The genetic characteristics of the embryo are not affected by changes in the maternal environment, and the maternal environment only affects the physical development of the fetus to a certain extent. Therefore, the embryo transfer offspring still maintain their original genetic characteristics and inherit the excellent production performance of their donor dams.

7.3 Principles of embryo transfer

7.3.1 Physiological basis of embryo transfer

7.3.1.1 The species identity of donor and recipient

The taxonomic consistency of donors and receptors, that is, the relatives of donors and recipients are relatively close, and the general requirement is the same species. In general, different species with less kinship can only survive for a short time due to the difference in the tissue structure, developmental conditions (nutrition, environment) and speed (time of attachment and gestation period) of the embryos.

7.3.1.2 The physiological identity of the donor and recipient

The physiological consistency of the donor and the recipient, that is, the recipient and the donor are in the same stage of the estrous cycle. For IVF embryos, the time of estrus ovulation in the recipient animal and the time of IVF of oocytes are generally referenced.

7.3.1.3 Identification of anatomical sites of donor and recipient

The embryos should be in a similar spatial environment before and after transfer, and the site of embryo transfer should be the same as the site of collection (fallopian tube or uterus). For IVF embryos, the age of the embryo can be used to determine where it is in the body during development.

The above-mentioned principle of homogeneity is followed for embryo transfer because embryo development is very sensitive to the maternal uterine environment. Embryos at different stages of development require a corresponding uterine environment, and the uterine endocrine environment is in dynamic change under the action of SGS. In general, fertilization and corpus luteal formation begin almost at the same time after ovulation, and subsequent embryonic development and endometrial development are synchronized. As the embryo develops, it moves from the fallopian tube to the uterus and the environment of the fallopian tube and uterus changes accordingly. Thus, each stage of embryonic development requires a specific physiological environment and conditions for survival. If something goes wrong, the embryo will die, so there is a high degree of coherence between embryonic development and the environment of the reproductive tract.

7.3.2 Time frame for embryo transfer

Physiologically, the duration of embryo collection and transfer (the age of the embryo) should not exceed the life span of the cyclic corpus luteum, and should take place at the latest few days before the recipient's cyclic corpus luteum degenerates, and even less at the time when the embryo begins to attach. Therefore, embryos are usually collected on the 3rd to 8th day after the donor is mated, and the recipient also receives the embryo transfer at the same time. If the duration of the transfer is later than the luteal phase of the cycle, the embryo is not viable due to a degenerative change in the receptor's intrauterine environment (preparation for estrus). For example, when the embryos are removed 18-20 d after the cattle are bred, the receptor itself is also in a state of pregnancy, and the embryo itself is in the same stage of development. Although in theory this could work, there are technical difficulties in practice (embryo harvesting and transfer), as the embryo has initially established a physiological and tissue connection with the mother at this time, and the time limit for the embryo to signal pregnancy to the mother has passed (usually in cycle 15). When the embryo is removed from the mother, the interaction between the embryo and the mother will be disrupted, resulting in difficult survival after the transfer.

7.3.3 Strict quality control of the embryo transfer process

As the embryo is relatively fragile in its early stages of development, it is susceptible to the influence of external environmental factors, and any inadvertent manipulation will lead to a decrease in embryonic viability and may result in malformation of the individual after birth. Therefore, in the process of embryo collection, culture and transfer, it is necessary to create a sterile environmental condition suitable for embryo survival, and avoid the adverse effects of physical and chemical factors such as temperature and light on embryos. To prevent the spread of viruses and microorganisms, embryos are also washed with trypsin before transfer.

In conclusion, according to the above principles, the transfer of an embryo is only a change of spatial location, not a change of physiological environment, which does not affect the growth and development of the embryo, let alone endanger its life. Therefore, embryos can be transferred from one dam to another, or transported and re-transferred, and survive and develop successfully until they are finally produced by the mother and grow into new individuals.

7.4 Factors affecting the success of embryo transfer

Successful implantation requires a receptive endometrium, a functional embryo and a synchronized dialogue between them. The crosstalk between the blastocyst and the uterus can only occur during a brief period, namely the "window of implantation" (Ma et al., 2003). During this window, the blastocyst can attach to the endometrial epithelial cells and invade the endometrial stroma and vasculature. This process can only occur when the endometrium is receptive. However, it remains largely unclear how the blastocyst escapes maternal immune surveillance at the time of implantation. Since embryo implantation is a dynamic developmental process that integrates many signaling molecules into a precisely orchestrated program, it is important to understand the hierarchical landscape of the pathways governing these processes to generate new strategies to correct implantation failure and improve the outcomes (Zhang et al., 2013).

7.4.1 Maternal hormonal environment required for embryo implantation

In the majority of eutherian mammals, implantation occurs in a fixed interval of time after ovulation when the corpus luteum is fully formed. In rodents, it is in the diestrous phase of the estrous cycle, and in the mammals, which has long-term estrous cycle, such as bovine, sheep and human, it is during the luteal phase. Embryo implantation occurs at the end of the luteal phase. This stage is when progesterone and estrogen are secreted. It is generally accepted that progesterone is required for implantation nearly in all the animals studied, while the role of estrogen surges at the proestrous and luteal phase prior to embryo implantation remains controversial (Day et al., 2004). The requirement for ovarian estrogen in implantation is species-specific. In species such as guinea pig, rhesus monkey, rabbit and golden hamster, progesterone alone is adequate for implantation, however, participation of estrogen in implantation of these species may not be completely excluded (Heap et al., 1981). And estrogen synthesis enzyme aromatase is detected in the blastocyst of hamster and rabbit, while such an aromatase is absent in mice.

7.4.2 Embryonic preparation for implantation necessitates "blastocyst activation"

Acquisition of implantation competency by the blastocyst is a prerequisite for successful implantation (Paria et al., 1993). In mice, the blastocyst escapes from its zona pellucida and attaches to the uterine epithelium at day 4.5 of pregnancy. However, this sequence is interrupted in delayed implantation, occurs naturally as an obligate delay in nearly 100 mammalian species. During the delayed implantation, the blastocyst is metabolically dormant and incompetent to initiate attachment in the uterus. Under appropriate conditions, the blastocyst is rapidly activated to resume its development with attachments and invasion. The active blastocyst shows distinct morphological differences from the dormant blastocyst. Delayed implantation is often associated with the failure of implantation in the receptive uterus, highlighting the notion that the blastocyst's state of activity determines the "window of implantation". The signals participating in blastocyst activation and uterine preparation of receptivity include estrogenic derivatives, Cannabinoid signaling, Wnt signaling, and embryo-derived signals such as Erbb etc.

7.4.3 Uterine receptivity

In placental mammals, the uterus is receptive to blastocyst implantation during a spatiotemporally restricted "window" when the uterine environment is favorable for blastocyst implantation (Yoshinaga, 1988). In mice, this period is limited to 4 d of pregnancy (Wang and Dey, 2006), and in human, the receptivity window opens between 20 and 24 d of a regular menstrual cycle (6-10 d after ovulation LH surge) (Rashid et al., 2011). In contrast to rodents and primates, domestic animals (such as cattle, horses, sheep, goats and pigs) have a prolonged pre-implantation period (the pre-receptive phase). This period is characterized by migration of the embryo, spacing of the embryos (as in pigs), secretions from the endometrial glands, and generation of the embryonic signal for maternal recognition of pregnancy. The key to uterine receptivity is the dynamic and precisely controlled molecular and cellular events that drive blastocyst growth, attachment and the subsequent events of implantation. This dynamic process involves a variety of genes that includes cytokines, homeobox transcription factors, and developmental genes working together with ovarian steroids to specify uterine receptivity (Zhang et al., 2013).

7.4.4 Cell-cell interactions

Embryo implantation is a dynamic developmental event that involves a series of physical and physiological interactions among the blastocyst trophectoderm and various endometrial cell-types, including the luminal and glandular epithelial and stromal cells. Implantation proceeds through three stages: apposition, adhesion/attachment, and penetration. During the apposition stage, the trophectoderm comes into close proximity with the luminal epithelium. With uterine lumen closure, firm attachment between the trophectoderm and luminal epithelium is initiated. After the attachment reaction, the implanting embryo initiates penetration through the luminal epithelium into the stromal bed. Therefore, implantation is a bulk of cell-cell interactions between the blastocyst and the uterus under the primary influence of ovarian steroids.

During the preimplantation period, uterine luminal fluid provides a medium for transporting preimplantation embryos into the uterine horn, and absorption of uterine luminal fluid, maximal fluid absorption and uterine luminal closure can be achieved to facilitate blastocyst apposition. Adhesion molecules such as inegrins, trophinin-bystin-tastin complex, selectins and cadherins, play key roles in blastocyst apposition and attachment during implantation (Armant, 2011). Stromal-epithelial interactions are essential for uterine receptivity. It is reported that ovarian estrogen plays a critical role in uterine epithelial cell proliferation and differentiation, and progesterone acts both on epithelial and stromal to antagonize the proliferative response of the epithelium to estrogen, while it induces the stromal proliferation. Uterine glands are thought to be the principal source of uterine secretions that are required for blastocyst survival and implantation, as well as the establishment and maintenance of pregnancy (Gray et al. , 2001).

7.5 Procedure of embryo transfer in farm animals

The core principle of embryo transfer is rapid propagation of superior genetics to enhance the production potential. Embryo transfer technique involves the securing of viable embryos from a donor of elite genetics and their transfer in the uterus of a recipient to complete the gestation. To achieve this purpose, embryo transfer programs in most mammals include administration of drugs that stimulate a supraphysiological growth of follicles in the ovaries followed by hormonal induction of ovulation, except in horses. This approach increases the chances that multiple embryos are collected per procedure.

In farm animal embryo transfer, transferable embryos are embryos deemed of acceptable quality for transfer into recipient animals as fresh embryos or to be cryopreserved for future use.

7.5.1 Selection of donor

The objective of rapid propagation of superior genetics makes selection of donor cattle very important. Therefore, while underlying the criteria for the selection of donor, it must be able to justify the goal of producing embryos with required traits of high genetic merit. Once the genetic standard is met, then comes the reproductive performance including ability to produce viable multiple oocytes which are to be fertilized by equally superior semen/sperms. Hence, it is pertinent to mention that the genetic superiority of male animal, which contributes semen/sperm, must be ensured as per required traits.

The donor should have average body condition, obese animal wouldnot be able to produce competent oocyte and more prone to ovarian cysts. Furthermore, the donor animal must not be suffering from any clinical anomaly and should have regular estrous cycle.

7.5.2 Selection of recipient

The role of the recipient animal is to host high values and genetically superior embryo in uterus till the end of gestation. Hence, to accomplish a successful embryo transfer program, selection of suitable recipient is of prime importance as well. For selection of recipients, consideration should be given to sound general and reproductive health. Recipient animal should be adult, at least 60 d post-partum, having regular estrous cycle and proper body conformation to ensure calving ease. Recipient animal should be kept in a stress-free housing, environment and proper nutrition plan to fulfill its maintenance and gestational requirements. Cattle, pigs and other livestock, can be the appropriate donor after the postpartum 2 months.

7.5.3 Superovulation of donor

Superovulation refers to development of more than one oocyte on the ovaries of monoovular specie like cattle. This is usually accomplished using external gonadotropin hormone to get multiple oocytes fertilized in a single reproductive cycle. The objective of superovulation is to produce more transferable embryos from single donor for maximum harvesting genetic potential.

Timing of administration of gonadotropin plays an important role in generating consequent ovarian of follicular response. Studies have shown that for optimum super ovulatory response, hormone should be administered at the time of follicular wave emergence. Ovarian monitoring through ultrasound can be helpful to precisely tracking follicular wave emergence. On the other hand, some techniques can be applied to manipulate follicular wave emergence which includes follicular ablation or learned use estrogen, progesterone releasing devices.

Follicle simulating hormone (FSH) is administered to stimulate multiple follicles to develop. Due to less half-life of FSH, approximately 5 h, it is recommended to inject it twice daily for continued follicular stimulation. This treatment is continued for 4 or 5 d before the induction of ovulation. Similarly, equine chorionic gonadotropin (eCG) possess characteristics FSH and luteinizing hormone (LH). In addition, it has longer half-life, approximately 40 h, which allows a single injection of eCG to stimulate super ovulatory response. This ability of eCG attracted practitioner to use it instead of FSH for superovulation. Later on, it is noticed that with the continuous use of eCG in animals it results in the production of compromised oocytes due to sustained stimulation. Efforts are made to nullify the negative impact by intravenous injection of antibodies to eCG simultaneously with artificial insemination in donor. Another hormone, human menopausal hormone (HMG) is also used to induce superovulation in cattle. It is not very commonly used because of its high price.

7.5.3.1 Superovulation of bovine

In cattle, there are two generally accepted methods of superovulation. One method consists of administering a single injection (2 000-2 500 U, IM) of equine chorionic gonadotropin (eCG), typically on day 10 of the estrous cycle (in which day 0 is defined as the day cows are observed in estrus), followed 2-3 d later by two injections of prostaglandin $F_{2\alpha}$ (dinoprost or cloprostenol) 12-24 h apart.

Although some good results have been reported with a single administration, FSH is typically administered IM over 4-5 consecutive days, twice daily in decreasing doses. Both commercial FSH preparations can be diluted into 20 mL of saline (0.9% NaCl) solution and administered over 4-5 d. A typical 4-day FSH treatment protocol is as follows: day 1, 4 mL every 12 h; day 2, 3 mL every 12 h; day 3, 2 mL every 12 h; day 4, 1 mL every 12 h (total volume = 20 mL). Treatments typically begin on day 10 after estrus. On the third or fourth day of FSH treatment, a luteolytic injection of prostaglandin $F_{2\alpha}$ is administered and repeated 12 h later. Estrus can be expected 36-48 h later, and the artificial insemination can be scheduled based on the observed heat (12 and 24 h after onset of estrus) or by appointment (72, 84, and 96 h after administration of $PGF_{2\alpha}$) that is accompanied by administration of an agent to trigger ovulation [e. g., human chorionic gonadotropin or gonadotropin releasing hormone (GnRH)].

7.5.3.2 Superovulation of sheep and goats

As in cattle, FSH is commonly administered twice daily in a series of decreasing doses administered over 3 days (e. g., 5 mg every 12 h on day 1, then 3 mg every 12 h on day 2; and then 2 mg every 12 h on day 3). On the last day of FSH treatment, the progestagen source is removed and a luteolytic injection of $PGF_{2\alpha}$ administered. Estrus is typically detected by use of vasectomized bucks or rams; artificial insemination should take place 12-24 h after estrus is detected or 45-50 h after progestagen removal.

7.5.3.3 Superovulation of sows

Owing to the high ovulatory rate in sows, superovulation treatment may or may not be used in embryo transfer procedures. Embryos are typically collected 5-6 d after ovulation and transferred preferably to recipients that are in estrus about 24 h after the donor sow.

Superovulation of the donors is induced by an intramuscular administration of different doses of eCG 24 h after weaning. Only sows with clear signs of estrus 48 to 72 h post-eCG administration are treated with 750 IU, intramuscularly, of hCG at the onset of estrus.

7.5.4 Artificial insemination of donor

The recent commonly used superovulation protocols are designed to perform fix time artificial inseminations (FTAI) in donors. To increase fertilization rate in donor, it is recommended to delay the withdrawal of progesterone devices together with injection of ovulation inducers viz. porcine LH (pLH), GnRH, hCG and/or eCG. This results in higher ovulation synchronization permitting FTAI without any hampering the embryos both quantitatively and qualitatively.

7.5.5 Synchronization of recipient

Embryo transfer program is considered successful if calving rate from recipients is high. For a better acclimatization of embryo post transfer; physiological status of reproductive tract and environment within the recipient's uterus should be comparable to that of donor. Therefore, recipients are synchronized for estrus/ovulation alongside the donor.

Initially, prostaglandin (PG) $F_{2\alpha}$ is used for induction of estrus. Estrus can be detected after 60-72 h after administration of PG in recipient animal. The efficacy of embryo transfer could be compromised due to poor estrus detection. In order to overcome estrus detection, concept of fix time embryo transfer (FTET) is conceived by implementing ovulation synchronization. A protocol which uses combination of GnRH and PG and is known as Ovsynch is used to synchronize ovulation. This method is more efficient as it avoids estrus detection and embryo transfer is carried at fixed time. Efficiency of GnRH based protocol is compromised; it does not coincide with emergence of follicular. Hence, to address this issue, modification of GnRH based protocols are made by incorporating estrogen and progesterone releasing devices. It results in increased efficiency of estrus control protocol and higher ovulation synchrony.

7.5.5.1 Synchronous estrus in cattle

1. Progesterone vaginal embolization

Take 50-100 mg of 18-methylnorethinone, and dissolve in salad oil and soak in sponge. On any day of the estrous cycle, the vagina is dilated with an opener, a sponge is clamped with a long-handled spinner, and placed around the vaginal cervix with a string exposed outside the vulva, and the sponge plug is pulled out after 9-14 d. In order to increase the rate of estrus, it is best to inject PMSG or prostaglandins intramuscularly after removing the sponge plug.

2. Progesterone implantation method

20-40 mg of 18-norethinone and a small amount of anti-inflammatory powder are loaded into a plastic tube, and some holes are punched in the tube wall for drug release. The small tubes are embedded subcutaneously in the dorsum of the ear with a trocar, and the tubes are extrudated 9-14 d later, and chloroprostenol 0.2-0.4 mg or PMSG 500-800 IU are injected at the same time. Most cows ovulate in 2-5 d after removal.

3. Prostaglandin method

More than 70% of cattle are induced to oestrus and ovulation 3-5 d after treatment by intramuscular injection of 0.4 mg of chlorprostenol or by perioral infusion of 0.2 mg of chlorprostenol. Prostaglandin treatment is performed at an interval of 9-14 d after the first injection, or in recipient cattle with functional luteal bodies.

7.5.5.2 Synchronous estrus in sheep and goats

In sheep and goats, the method of synchronous estrus is basically the same as that in cattle, only the dose is different. At present, progesterone vaginal suppository method and prostaglandin method are mainly used in sheep.

1. Progesterone vaginal suppository method

The progesterone content of the progesterone device CIDR is 300 mg/piece, and the progesterone content of the sponge plug is 500 mg/piece. Use the same method as cattle. In addition, vaginal implantation and injection of Folligon (containing gestational horse serum gonadotropin 100-200 IU/animal) with CIDR can achieve a simultaneous estrus effect of more than 95%.

2. Prostaglandin method

Synchronous estrus with prostaglandins must be carried out during the breeding season when ewes have begun their estrous cycle. Intramuscular injection of 4-6 mg $PGF_{2\alpha}$ after the 4th day of the estrous cycle results in an estrus rate of about 70% after a single dose. After prostaglandin treatment, the estrus rate and conception rate in the first estrus period are low, and the conception rate in the second estrus period is relatively concentrated and normal.

7.5.5.3 Simultaneous estrus of pigs

The treatment of progesterone in pigs with simultaneous estrus is easy to cause ovarian cysts, and the rate of estrus and conception is not high. The rate of estrus in the same period treated with prostaglandin is also lower than that of cattle and sheep. Therefore, progesterone and prostaglandin are not suitable for inducing simultaneous estrus in pigs.

7.5.5.4 Equine estrus

The prostaglandin analogue ICI 81008 is effective in inducing equine synchronous estrus. Due to the longer duration of estrus in horses, the corpus luteum within 5 d after ovulation is not sensitive to prostaglandins, so the contemporaneous estrus rate after one treatment is low. The effect of reusing prostaglandins at an interval of 12-16 d is improved, but the cost is high.

7.5.6 Recovery of embryos

The process of recovering the embryos from Donor uterus is also known as flushing. Flushing is a non-surgical method of embryo recovery and it is usually performed around day 6-8 after the estrus (day=0). It is possible to perform as early as after day 4 post estrus; however, recovery rate is not optimal at this stage. Recovery of embryo beyond day 8 after estrus is hard to recognize due to its hatching from zona pellucida.

To carry out flushing, donor animal is well restrained to minimize its movements. Local anesthesia is administered and the vulvar area is thoroughly cleaned. Folly catheter is passed through cervix and its tip is placed just in uterine just beyond the bifurcation. At this point the tip is inflated, fixed and its posterior end is attached with Y-connector having inflow and outflow tubes. The flushing fluid adjusted at body temperature is passed in the horn until it feels rigid through rectum. Afterwards, the clamp from outflow tube is released to collect the flushing fluid in a container. This process is repeated in both horns 3-4 times. After the flushing process is completed, the catheter is deflated and withdrawn. The harvested flushing fluid contains embryos and it is left undisturbed for some time to allow the collected embryos to get settled in the bottom. After removing the supernatant, the remaining lower portion is collected in petri dish and observed under microscope for selection and classification for further process.

7.5.6.1 Embryo collection of cows

1. Operation method

This method is usually applied to cows before 5 d after oestrus. Firstly, make a 10-cm incision between abdominal rib and hip. Then find the oviduct through the incision and then use the hand for derivation in the umbrella to insert a catheter from tube. At last, flash the oviduct with 10-15 mL liquid.

2. Non-surgical method

Usually, a two-way catheter flusher (also known as an oocyte flusher) is used to inject the embryo flushing fluid into the uterine horn through the inner tube, and then the embryo flushing fluid (also known as oocyte flushing fluid and flushing fluid) is exported. The front end of the outer tube of the embryo flushing device is connected with an air bag. When the embryo flushing device is inserted into the uterus, the air bag is inflated to swell and block the internal mouth of the uterine neck so as to avoid the outflow of embryo flushing fluid through the cervix. Before embryo collection, 3-5 mL of 2% procaine should be injected into the first and second caudal intervertebral bodies, or 1 mL of 2% Jingasongling should be injected intramuscularly for epidural anesthesia. Generally, each side of the uterus should be rinsed with 300-500 mL of embryo flushing fluid. The embryo flushing fluid should be exported smoothly and quickly, and all injected oocyte flushing fluid should be recovered as far as possible. To this end, it is best to raise the uterine horn with the hand inside the rectum and press slightly to promote flushing fluid outflow. The more fully the rinse solution is recovered, the higher the embryo collection rate is generally.

7.5.6.2 Embryo collection of sheep and goats

Sheep and goat embryos are generally collected surgically. The operation steps are as follows: intramuscular injection of 2% xylidinothiazoline (Chinese name: Jingsongling) for general anesthesia, local use of 0.5% procaine hydrochloride for block anesthesia. The operation site is in the anterior abdominal midline of the breast, and routine disinfection is needed before operation. If the embryo is flushed 2-3 d after estrus, a catheter can be inserted from the umbrella and 5-10 mL of embryo flushing fluid can be injected from the conjoint of the uterine canal. If the embryo is flushed 6-7 d after estrus, a catheter is inserted into the base of the uterine horn, and then 50-60 mL of oocyte flushing fluid is injected from the tip of the uterine horn.

7.5.6.3 Embryo collection of rabbits

Rabbit embryos are generally collected surgically. The operation steps are as follows: general anesthesia is administered with intravenous sodium isoprenobarbital (60 mg/mL) at a dose of 40-50 mg/kg body weight. An incision between the second and third pair of nipples at the midabdominal line, 3-5 cm in length, leads to the uterine horn and fallopian tube. For example, the oviduct is flushed within 50 h after spermatoplasty. The main way to rinse the fallopian tube is to insert a silicone tube from the umbrella, and then inject 5 mL of oviduct from the connecting part of the uterine tube. If the oocyte is flushed after 72 h of sperm transfusion, the uterus should be flushed, that is, 10-20 mL of oocyte solution should be injected into the base of the uterus from the conjunctive part of the uterine canal.

7.5.7 Embryo quality control

As per guidelines laid in manual of IETS, morphological examination is performed under stereomicroscope at magnification of (50-100)× to evaluate embryo quality and subsequent classification. The viability of embryos is assessed based on characteristic of embryonic developmental stage relative to the principle that it should be on a particular day after ovulation. An illustration of blastocysts stage embryo is given in Fig. 7-1. Ideal characteristics that an embryo should possess are given in Table 7-1.

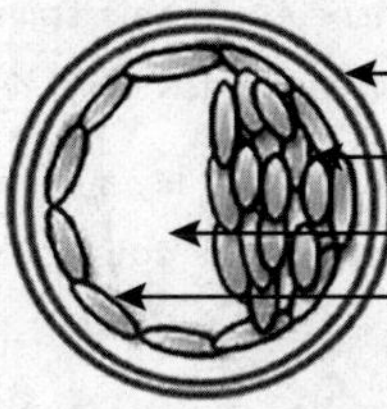

Fig. 7-1 Illustration of blastocyst stage bovine embryo

Table 7-1 Ideal characteristics of an embryo

Embryonic character	Ideal parameters
Shape	Compact and spherical
Diameter	150-190 μm
Zona pellucida	Thickness: 12-15 μm, it should be uniform; neither cracked nor collapsed and contain no debris on its surface
Blastomeres	They should be similar size with even color and texture
Cytoplasm	It should not be granular or vesiculated
Perivitelline space	It should be clear and contain no cellular debris

Embryos are classified on the basis development stage and their quality are described in Table 7-2. In order to attain maximum output in term of pregnancies the embryo should be at compact morula or blastocyst stage of development whereas qualitatively it should be excellent or good. Furthermore, for freezing point of view embryo of quality grade 1 is considered more suitable as it can bear freezing and subsequent thawing procedure. The embryos of quality grades 2 and 3 are preferred to be transferred without freezing.

Table 7-2 Classification of embryos based on evaluation performed after collection from donor on day 6-8 post estrus

Embryonic stage	Code	Description
Embryo classification on the basis developmental stage		
1-cell	1	It represents the unfertilized oocyte and it should not be considered for further processing
2-cell to 12-cell	2	It represents delayed development and perhaps dying embryo and not suitable for transfer or cryopreservation
Early morula	3	It is a loose group of 16 blastomeres, it is also not suitable for transfer
Compact morula	4	It has mass of cells which are tightly compacted and cells reside in 60%-70% of perivitelline space
Early blastocyst	5	It is characterized by development of a small cavity that contains fluid and called blastocoele. At this stage few trophoblast cells can also be seen. Cellular mass resides in 70%-80% of the perivitelline space
Blastocyst	6	It contains a well-defined layer of trophoblasts, blastocoele and inner cell mass. Thickness of zona pellucida is normal with negligible perivitelline space
Expanded blastocyst	7	It has increased diameter, 1.2-1.5 times higher than normal diameter. Zona pellucida is thickened up to 1/3rd of its original thickness. Perivitelline space disappears at this stage
Hatched blastocyst	8	It is characterized by hatching of embryos from the zona pellucida. If it is hatched completely then embryo is seen to be collapsed
Expanded hatched blastocyst	9	It is similar in appearance to an embryo which is completely hatched (code 8) and has larger in diameter. It is usually found if flushing is performed after day 8 after the estrus
Embryo classification on the basis of quality		
Excellent or good	1	It is characterized by an embryo which has is spherical and contains blastomeres of same size, color and density. Embryonic development stage should be in accordance with that which is expected on particular day after onset of estrus. At least 85% of the embryonic cells are an intact, viable and cohesive mass and no more than 15% of embryonic cells are extruded into the perivitelline space. The zona pellucida should be spherical and smooth
Fair	2	It represents moderate irregularities existing in the overall shape of the embryonic mass or in the size, color and density of individual cells comprising the embryo. Mild unevenness of cytoplasmic pigmentation is within individual blastomeres. At least 50% of the embryonic cells should be an intact, viable and cohesive embryonic mass

Table 7-2 continued

Embryonic stage	Code	Description
Poor	3	It has major abnormalities which are evident in the embryonic shape and size. There is variation in the color and density off cellular mass. Pigmentation of cytoplasm varies among cells and approximately 25% of the cells are intact and viable
Dead or degenerating	4	It contains cytoplasm which is highly dark, disintegrated cellular membranes and other prominent abnormalities. It does not reach the expected development stage in accordance with days passed after estrus

7.5.8 Embryo transfer to recipient

Selected embryo is drawn in the 0.25 mL sterile straw with the help of 1 mL syringe. First of all, a 3-cm column of medium is aspirated followed by a 0.5-cm column of air. Then a second 3-cm embryo containing column of medium is aspirated followed by another bubble of air. Embryo containing straw is passed through cervix to uterine horn in similar way used for depositing semen in artificial insemination.

It is necessary to secure already synchronized recipient in a cattle clutch to restrict unwanted movements. Afterwards, a thorough cleaning of the vulvar area is performed, and in order to prevent defecation and straining epidural anesthesia may be administered. A quick and smooth insertion of embryo transfer gun in to desired uterine horn should be ensured. Trauma to the luminal mucosa should be avoided as it may hamper fertility and could be damaging for embryo being implanted. Embryos are usually placed deep in the uterine horn which is ipsilateral to corpus luteum.

7.5.8.1 Embryo transfer to cattle

Selected embryo is drawn in the 0.25 mL sterile straw with the help of 1 mL syringe. First of all, a 3-cm column of medium is aspirated followed by a 0.5-cm column of air. Then a second 3-cm embryo containing column of medium is aspirated followed by another bubble of air. Embryo containing straw is passed through cervix to uterine horn in similar way used for depositing semen in artificial insemination.

It is necessary to secure already synchronized recipient in a cattle clutch to restrict unwanted movements. Afterwards, a thorough cleaning of the vulvar area is performed and in order to prevent defecation and straining epidural anesthesia may be administered. A quick and smooth insertion of embryo transfer gun in to desired uterine horn should be ensured. Trauma to the luminal mucosa should be avoided as it may hamper fertility and could be damaging for embryo being implanted. Embryos are usually placed deep in the uterine horn which is ipsilateral to corpus luteum.

7.5.8.2 Embryo transfer to goats and sheep

After gentle traction out of the uterus, the operator locates the ovary on the ovulating side and observes the number and development of the ovarian corpus luteum. Only recipients with a well-developed corpus luteum should be used for embryo transfer. At the time of transfer, the uterine horn of a recipient with only one corpus luteum on the ovary usually implies the transfer of one embryo, and the uterine horn with two or more corpus luteum implies the transfer of two embryos. Depending on the stage of development of the embryo and the principle that recovery coincides with the transfer site, the embryo is usually injected into the anterior part of the uterine horn on the same side of the corpus luteum or through the abdomen of the fallopian tube using a transfer tube. The uterus is then quickly repositioned and the surgical site is sutured.

7.5.8.3 Embryo transfer in pigs

1. Surgical method

As with the egg collection procedure, the procedure involves anesthesia, preservation, sterilization and finally the implantation of the embryo into the fallopian tube or the tip of the uterine horn.

2. Non-surgical method

After anesthesia with sodium pentobarbital, a 10-mm outside diameter catheter is inserted vaginally into the cervix, then a 5-mm diameter catheter is inserted into the horn of the uterus, and finally a third 2-mm catheter containing the embryo is inserted into the horn of the uterus. With laparoscopy, good results are also obtained.

7.5.9 Post-operative care of donor and recipient

7.5.9.1 Post-operative care of donor females

After donor recovery of embryos, intramuscular injection of cloprostenol is required around day 9 of the oestrus cycle to dissolve the corpus luteum and promote recovery of the donor reproductive organs. To prevent secondary infections caused by surgery, donor females that have undergone embryo transfer by surgical methods are generally given continuous intramuscular injections of penicillin and streptomycin for 3 d to reduce inflammation.

7.5.9.2 Post-operative care of recipient females

Recipient females after embryo transfer should strengthen feeding management. So pay attention to adequate vitamin and trace element supplementation, and appropriately limit energy feed intake to avoid stress reactions. At the same time, they should be carefully observed to see if they return to oestrus at the scheduled time. It is best to have an ultrasound examination after 30 d for sheep and after 60 d for cattle or to check the condition of the corpus luteum to determine if they are pregnant. Keep and manage pregnant ewes in separate groups from empty ewes. Vaccinate at the right time to prevent fetal abortion or infectious diseases.

7.6 Main problems of embryo transfer

7.6.1 Less consistent super-rowing results

Ovulation is not always as effective as it could be, the potency of different batches of ovulation medication may vary somewhat, the response to ovulation treatment varies greatly between individuals and ages, and ovulation rates are highly variable.

7.6.2 Low embryo recovery rate

Embryos located in the fallopian tube or uterine horn are often not fully recovered, and an excessive number of ovulations tends to reduce the recovery rate of embryos. The reason for this may be that the expelled eggs fail to enter the fallopian tube umbilicus and are lost. Embryo recovery rates usually range from 50% to 80%.

7.6.3 Donor reuse problems

For skilled operators, as long as the operation is performed correctly and skillfully, embryo transfer will not have much impact on the donor's own reproductive function, except for delaying the donor's breeding time by 2-3 cyclic estrus, but as the donors are all good breeding females of high economic value, the economic loss and impact caused in the event of loss of reproductive capacity can be significant.

7.6.4 Supporting technology problems

Embryo transfer is costlier and can be more valuable when combined with breeding. Abroad, it is common to use superovulation with embryo transfer program for dairy breeding, but in China there is still a clear disconnect between the two.

7.6.5 Lack of technical personnel and inadequate promotion system

At present, embryo transfer technology in China is mainly in the hands of experts from some agricultural research institutes and a few high-tech enterprises, and there is a shortage of skilled technical personnel to operate it, and no promotion network system has been established. In addition, the degree of localization of hormones, drugs and equipment required for embryo transfer has yet to be improved.

Assignments

1. Please review the development history of embryo transfer technique.
2. Please briefly describe the concept and rationale of the embryo transfer.
3. Please briefly describe the principle of the embryo transfer.
4. Please state the general procedure of embryo transfer.
5. Please describe the selection criteria of donor cow for embryo transfer.
6. Please state the embryo collection and transfer methods for cows and goats.

(Hua Guohua)

Chapter 8 Stem Cell

Summary

Stem cells are undifferentiated or poorly differentiated cells that produce progeny stem cells or directed differentiated cells through asymmetric division. Stem cells can be divided into three categories according to their origin: embryonic stem cells (ES cells) derived from the inner cell mass of blastocysts, mesenchymal stem cells (MSCs) present in mature organs, and induced pluripotent stem cells (iPS) obtained by cell reprogramming. In vitro, the three stem cells can differentiate into ectoderm, mesoderm and endoderm cells under appropriate induction and culture conditions. This chapter introduces the research history, morphological and functional characteristics, technical points of isolation and culture, main problems and application of different stem cells.

Stem cells are cells that have the ability to self-renew indefinitely or immortally and to produce at least one type of highly differentiated progeny. Most biologists and scientists believe that stem cells are from the embryo, fetus or adults with unlimited differentiation ability of self-renewal and proliferation under certain condition that could produce daughter cells with the identical phenotype and genotype of their own. In the past few decades, research on stem cell biology has been very active, and new breakthroughs have been continuously made. At present, stem cell therapy has broad prospects for development and is widely used in personalized therapy, tissue engineering, chronic disease and regenerative disease mitigation.

8.1 Research history of stem cells

Mesenchymal stem cells (MSCs) were first discovered in the bone marrow. In 1867, German pathologist Cohnheim proposed that there may be non-hematopoietic stem cells in the bone marrow when he was studying wound healing. In 1976, Friedenstein et al. confirmed that in addition to hematopoietic stem cells, bone marrow also contains spindle-shaped colony-forming fibroblast progenitor cells or fibroblast colony forming unit (CFU-

F), which are dormant *in vivo*, and can enter the cell cycle under appropriate conditions *in vitro* to form cell colonies resembling bone or cartilage fragments. Transplantation of CFU-F into host animals can form ectopic ossicles. On this basis, in 1987, Friedenstein et al. found that adherent bone marrow mononuclear cells cultured in plastic dishes can differentiate into osteoblasts, chondrocytes, adipocytes and myoblasts under certain conditions. After being expanded for 20-30 passages, these cells can still maintain the multi-directional differentiation potential. In 1911, Caplan turned bone marrow pluripotent stromal stem cells into bone marrow mesenchymal stem cells (BMSC). Since then, a series of studies have reported that these adherent-capable stromal cells isolated from bone marrow are non-hematopoietic stem cells. In 2001, Minguell summarized MSCs as: non-hematopoietic stem cell subsets that exist in the bone marrow stroma can be expanded *in vitro*, and after *in vitro* induction, they can finally differentiate into osteoblasts, chondrocytes, adipocytes, muscle cells, myotubes, nerve cells and matrix supporting hematopoietic stem cells. Many subsequent studies have shown that MSCs are also widely present in other tissues than the bone marrow. The markers expressed by MSC cells are not exactly the same in different sources, and the differentiation potential and clinical role of different MSCs need to be further studied.

ES originates from the inner cell mass (ICM) of blastocysts. Before the blastocyst stage of mammals, the blastomeres are totipotent, that is, under certain conditions, a single blastomere can develop into a complete individual. The first cell differentiation occurs when the embryo develops to the blastocyst stage, and the embryonic cells differentiate into the trophectoderm and ICM, which differentiates into the trophoblast of the placenta and any type of cell in the fetal tissue, respectively. Before the embryo implants, the ICM differentiates into the primitive endoderm, which grows close to the inner layer of the blastocoel, and finally develops into the vitelline membrane. Immediately after embryo implantation, the ICM differentiates into primitive ectoderm which differentiates into three cell layers of the embryo and further develop into the fetus.

In 1998, human ES cells were formally established, and researchers cultivated ES cell lines that can maintain pluripotency stably for a long time *in vitro*, and can differentiate into various somatic cells. Since then, the research on ES cells has achieved rapid development, and a series of progress has been made in many aspects, such as establishment of lineage, mechanism of maintaining pluripotency, and directional induction of differentiation. However, the research on ES cells is not smooth sailing, and there are many obstacles on the road of its research and development. Due to the need to destroy the early embryos to obtain human ES cells, this research has suffered huge human ethical pressure, and several countries even banned the research. In addition, since human oocytes are required to obtain early embryos, and various bottlenecks have been encountered in the maintenance of culture, these series of factors have limited the research and application of ES cells. To this end, researchers have been hoping to find a technology to replace ES cells, which can avoid ethical issues and obtain individual-specific pluripotent cells.

Induced pluripotent stem (iPS) cells are a new technology that has emerged in recent years. By introducing several specific transcription factors, somatic cells can be induced to undergo reprogramming to obtain induced pluripotency with similar characteristics to ES cells. In 2006, Yamanaka's research group established iPS cells for the first time. They used a pluripotency gene *Fbx15* as a selection marker and found that when ES cells with 24 genes specifically and highly expressed were transferred into somatic cells, the morphology and function could occur. To change the state of ES cells, they then screened these genes, and the combination of removing 1 of these 24 genes was introduced into somatic cells using retroviruses. Results showed that the combination of four key factors (Oct4, Sox2, c-Myc and Klf4) can effectively activate the expression of pluripotency marker genes in fibroblasts and reverse the cells back to a pluripotent state. The four factors were co-transfected into mouse fibroblasts by retrovirus, and the transfected fibroblasts changed in morphology, formed colonies similar to ES cells, and began to express pluripotency-specific markers. Teratomas and chimeras were then tested to verify its developmental potential. After injecting it into nude mice subcutaneously, teratomas could be formed, and section staining showed that it differentiated into three germ layers. But injecting it into blastocysts did not yield chimeric mice, indicating that the cells possessed the pluripotency of partial ES cells. However, this kind of cells can not fully reach the state of ES cells, and the promoter regions of pluripotency genes can not be completely demethylated like ES cells, and some pluripotent genes can not be expressed. Subsequently, multiple laboratories repeated this experiment, and obtained iPS cells that were more similar to ES cells in epigenetic modification by using *Nanog* and *Oct4*, which were more closely related to pluripotency, as reporter genes, and obtained healthy chimeric mice with reproductive system chimerism. In addition, this cell also passed a more stringent tetraploidy experiment to obtain healthy mice derived entirely from iPS cells, further confirming the developmental potential of this cell.

In 2007, Yamanaka's laboratory and Thomson's laboratory in the United States successively reported the establishment of human iPS cell lines. Yamanaka et al. continued to use the four factors and retroviral vectors used in mice, and replaced the mouse-derived factors with human-derived factors, and successfully induced human epidermal cells and connective tissue cells into iPS cells. Thomson et al. rescreened the induction factors and identified four factors, Oct4, Sox2, Nanog and Lin28, which were different from Yamanaka's laboratory, and used lentiviral vectors as mediators to successfully induce the foreskin cells from neonates to iPS cell. Although different inducing factors and mediating systems were used, both research groups obtained iPS cell lines that were similar to human ES cells in morphology, epigenetic modifications, and developmental potential. Since then, it has been applied to a variety of animal studies, and there have been reports of the establishment of iPS cells from species such as rhesus monkeys, rats and pigs. These three animals are important human disease research models, and the establishment of iPS cells is of great significance for medical research.

Researches on iPS have opened a new era in stem cell and regenerative medicine research. Using this technology can effectively avoid ethical issues, obtain patient-specific pluripotent stem cells, and provide a reliable and easy-to-obtain cell source for regenerative medicine treatment. In recent years, a large number of research results and progress have been made in the field of iPS technology research. The efficiency, quality and safety of reprogrammed cells obtained by this technology have been greatly improved, making the technology closer to clinical therapeutic applications.

8.2 Morphological and functional characteristics of stem cells

8.2.1 Morphological and functional characteristics of mesenchymal stem cells

Under standard culture conditions, MSCs exhibit plastic adhesion, i.e. cell-adherent growth properties. After isolation and culture, colony forming units (CFU) and population doubling time (PDT) can be used as indicators to analyze the self-renewal capacity of MSCs. MSCs have immunomodulatory properties and can interact with immune cells such as B cells, T cells, and NK cells. In addition, MSCs also have the characteristics of low immunogenicity, strong anti-inflammatory properties, inhibition of scarring, promotion of angiogenesis, and stimulation of innate progenitor cell regeneration.

Currently, there is no universal definition for the properties of MSCs, nor standard for isolation and culture procedures. Due to the lack of cell-specific markers to identify MSCs, the progress of stem cell research has been hindered in the past period of time. Therefore, the International Society for Cell Therapy (ISCT) established three basic parameters to determine the basic characteristics of MSCs, namely cell adherent growth, differentiation capacity and surface marker molecules. However, unlike human MSCs, which already have identification standards, there is no literature on the identification standards of swine, sheep and other animal MSCs by relevant international organizations.

8.2.1.1 Surface markers

Regardless of cell origin or isolation procedure, MSCs must express specific cell-specific markers while lacking the expression of hematopoietic cell surface markers. The markers expressed by MSCs are not identical from different species, as well as different tissues. Bovine MSCs from different tissues are all positive for the expression of adhesion-related mesenchymal markers such as CD29, CD166, CD105, surface enzymes (CD73), receptors (CD44), and glycoproteins (CD90). In addition, bovine MSCs also express pluripotency genes such as *OCT4*, *SOX2* and *NANOG*. Goat and sheep BMSCs express CD44 and CD166, and weakly express CD34, CD45, CD105 and CD90.

Chen et al. analyzed the biological characteristics of sheep umbilical cord mesenchymal stem cells (UCMSCs) such as multi-directional differentiation and clone formation ability through morphological observation, reverse transcription polymerase chain reaction (RT-PCR), growth curve and immunophenotype analysis. In this study, 29 passages of UCMSCs were successfully isolated and passaged, and RT-PCR results showed that UCMSCs were positive for CD29, CD44, and CD73, but negative for CD34.

8.2.1.2 Multidirectional differentiation potential

UCMSCs have the ability to differentiate into other adult cells. Therefore, UCMSCs obtained by umbilical cord isolation can be identified by assay of differentiation ability. At present, it is mainly detected after induction to differentiate into adipocytes (intracellular fat droplets), chondroblasts (smooth spheres of cells expressing type Ⅱ collagen), and osteoblasts (enhanced expression of ALP, and von Kossa or Alizarin Red staining).

MSCs are responsible for the transformation and replenishment of tissue. When the tissue needs to be repaired, these cells are stimulated to proliferate and differentiate. Under specific *in vitro* conditions, MSCs from animals such as sheep and cattle can also be induced to differentiate into a wide range of functional cells such as osteoblasts, skin cells, renal epithelial cells, and chondrocytes. In the past, it was generally believed that MSCs could only differentiate into mesoderm (adipose-like cells) cell lineages; However, in recent years, studies have shown that MSCs from species such as bovines can also differentiate into endoderm (islet B-like cells) and ectoderm (neurodermal cells) cell-like cells.

UCMSCs can differentiate into three germ layer cells *in vitro*. It has been reported that 5-azacytidine can induce the differentiation of UCMSCs into cardiomyocyte-like cells, and the differentiated cell morphology has typical cardiomyocyte characteristics. At the same time, the differentiated cells express cardiomyocyte-specific markers, namely N-cadherin, troponin, desmin, α-actin and MyHC (myosin heavy chain). In addition, sphingosine 1-phosphate (S1P) can also induce the differentiation of UCMSCs into cardiomyocyte-like cells. The differentiated cells not only change their morphology, but also express similar action potentials like α-actin and MyHC.

UCMSCs are not only induced to differentiate into the above-mentioned mesodermal cells, but also differentiate into hepatocyte-like cells after induction by growth factors, such as hepatocyte growth factor (HGF), basic fibroblast growth factor (bFGF) and FGF-4. When the morphology change significantly, it specifically express hepatocyte markers such as albumin, alpha-fetoprotein, keratin-19, connexin-32, and dipeptidyl peptidase Ⅳ. In addition, differentiated cells express hepatocyte-specific secretory proteins. It also has the functions of absorbing urea, storing glycogen and ingesting low-density lipoprotein.

Under the induction of nicotinamide or β-mercaptoethanol, UCMSCs could differentiate into insulin-secreting islet-like cell clusters and express the islet β-cell development-related gene *PDX-1*. In addition, transplantation of induced islet-like cells into diabetic mice can alleviate hyperglycemia in diabetic mice.

UCMSCs can not only be induced to differentiate into mesoderm and endoderm cells, but also can be differentiated into neuron-specific marker proteins such as nestin, β-tubllin Ⅲ and neurofilament M under the induction of bFGF and lysine. After the addition of VEGF and bFGF, UCMSCs could differentiate into endothelial cells and form blood vessel-like structures in Matrigel medium.

The above studies show that, similar to embryonic stem cells, UCMSCs are a class of stem cells with self-renewal and multi-directional differentiation potential, and are an ideal cell model for studying cell differentiation and trans-differentiation, especially in pigs, bovine, and relevant research fields of animals such as sheep, which are particularly important.

8.2.2 Morphological and functional characteristics of embryonic stem cells

8.2.2.1 Morphology of embryonic stem cells

Compared with ordinary cells, embryonic stem (ES) cells are smaller in size, larger in nucleus, higher in nuclear-cytoplasmic ratio, and have obvious nucleoli. In normal culture systems, ES cells maintain a very stable karyotype. In the cycle of ES cells, the S phase is longer, accounting for most of the entire cycle, and there is no G_1 phase checkpoint. ES cells with a karyotype of XX do not have X chromosome inactivation.

8.2.2.2 Differentiation potential of embryonic stem cells

ES cells can maintain an undifferentiated state and achieve self-renewal on feeder cells (feed layer cells) or in media containing differentiation inhibitory factors (such as leukaemia inhibitory factor, LIF). Thus, it can be immortalized *in vitro* like common cell lines, or cryopreserved. The aspect lies in the differentiation of embryonic cells in the medium of differentiation inhibitors or under the action of specific inducers. After being injected into the blastocyst and the blastomeres of the early embryos, ES cells can be differentiated into reproductive stem cells. Therefore, ES cells can differentiate into any type of cells in the body *in vivo*. Unlike early embryonic blastomeres, ES cells can not directly develop into offspring. The current research results show that ES cells can not differentiate into trophoblasts *in vivo*, e.g. ectoderm (trophectoderm) and primitive endoderm (primitive endoderm). Therefore, it can not directly develop into a fetus; However, through nuclear transfer, or polymerization with early blastomeres, or injection into the blastocoel, ES cells can develop into a complete individual or into specific tissue of the body, ES cells *in vitro* can also achieve directional differentiation under the action of inducing factors. Mouse ES cells have been induced to differentiate into neuronal cells, glial cells, neural stem cells, islet cells, liver cells *in vitro* cells, osteoblasts, adipocytes, skeletal muscle cells, vascular endothelial cells, cardiomyocytes, hematopoietic progenitor cells, yolk sac cells, smooth muscle cells, chondrocytes, melanocytes and primitive endoderm cells (Wobus, 2001). ES cells injected into genetically homogeneous or immunodeficient adult mice subcutaneously or under the capsid membrane of the kidney can form teratomas, and typically teratomas, which are gastrulation-like structures composed of epithelial cells, smooth muscle, skeletal muscle or cardiac muscle, nervous tissue, cartilage or bone which are derived from the inner layer, the middle layer and the outer layer, respectively, from the perspective of developmental biology.

8.2.2.3 Molecular characteristics of embryonic stem cells

There are stage-specific embryonic antigens (SSEA) on the surface of stem cells. Different animals have different types of antigens. For example, human ES cells contain high levels of SSEA-3 and SSEA4, whereas mouse ES cells contain only SSEA-1. Human ES cells also contain tumor rejection antigen-1-60 (tumor rejection antigen-1-60, TRA-1-60) and TRA-1-81.

All ES cells contain higher levels of alkaline phosphatase, Oct4 protein and telomerase. Oct4 protein is the expression of *Pou5f1* product, which is a transcriptional regulator. At least 7-8 target genes of Oct4 protein have been found in mice nucleoblasts, ICM cells and PGCs. When the expression of Oct4 protein in mouse ES is inhibited, ES cells differentiate into trophoblasts, and when Oct4 levels are artificially increased, ES cells differentiate into primitive endoderm and mesoderm. Therefore, Oct4 is a key regulator in maintaining ES pluripotency. Telomerase plays an important role in maintaining the telomere length of cell chromosomes, and telomerase activity and telomere length are markers of cell lifespan. Both indices are higher in embryonic cells, PGCs and ES cells than in somatic cells. The lifespan of somatic cells *in vitro* is generally 50-80 passages, while human ES cells still have high telomerase activity after 300 passages in culture. Furthermore, the nuclei of ES cells are very stable. The above characteristics of ES cells are particularly important for the study of mammalian developmental mechanisms and cell therapy.

8.2.2.4 Molecular mechanism of self-renewal of embryonic stem cells

At present, the understanding of the molecular mechanisms by which ES cells maintain pluripotency and achieve self-renewal is very limited. Existing research results suggest that ES cells maintain normal function through three signal transduction pathways.

1. LIF

There is a complex receptor on the surface of ES cells, whichis a heterodimer composed of a low-affinity LIF receptor (LIF-R) and a common subunit gp130. After binding to the receptor, LIF activates the JAK enzyme (Janus kinase), which activates genes that maintain cell self-renewal by phosphorylating STAT3 (Fig. 8-1).

2. Transcriptional regulator Oct4

Oct4 is always maintained at a certain level in ES cells, and is the switch for ES cells to maintain pluripotency and enter into differentiation. It is a DNA-specific binding protein that activates or inhibits the transcriptional activity of target genes maintaining self-renewal of ES cells. The expression level of Oct4 in ES cells must be maintained at 50%-150% of the

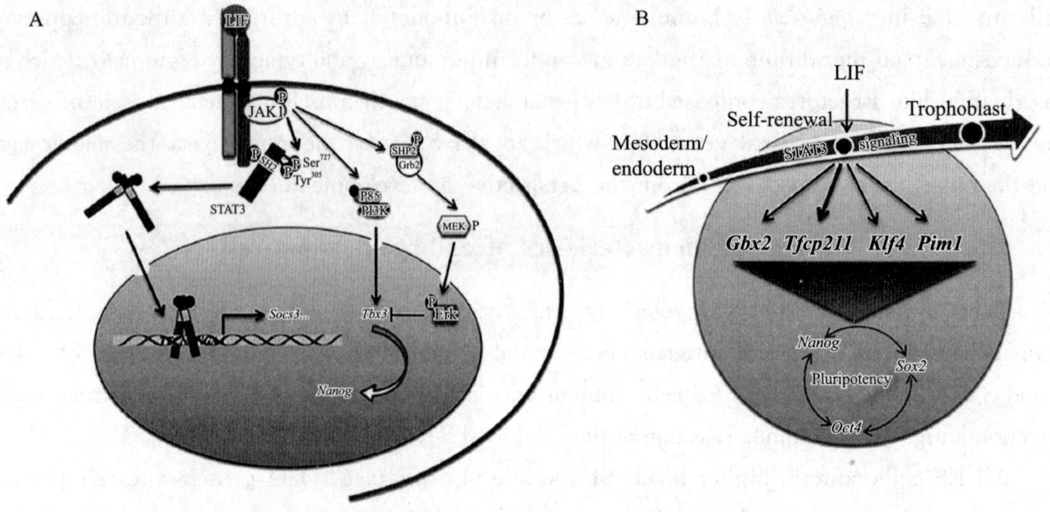

Fig. 8-1 LIF/JAK/STAT3 signaling pathway in mouse ESC self-renewal(Huang, 2015)

expression level *in vivo* in order to achieve self-renewal and maintain pluripotency. Recent findings suggest that STAT3 and Oct4 may act on the same target genes to coordinately maintain ES cell function.

3. Tyrosine phosphatase SHP2 and extracellular regulatory kinase ERK

SHP2 can remove the phosphate group on tyrosine, and acts on the intracellular gp130 receptor area. ERK is activated upon stimulation of gp130 and other membrane receptors. SHP2 and ERK negatively regulate STAT3 function, and they can transmit signals that inhibit STAT3 function. In addition, the intracellular binding protein Gab participates in the regulatory functions of SHP2 and ERK by changing its conformation.

8.2.3 Morphological and functional characteristics of iPS cells

8.2.3.1 Morphology of iPS cells

The iPS cells of sheep are smaller in size, with large nucleus, high nuclear-to-cytoplasmic ratio, obvious nucleoli, and hill-like growth; iPS cells of cattle are flat and compact multicellular clones. Growing in normal culture systems, iPS cells maintain a very stable karyotype. In the cycle of iPS cells, the S phase is longer, accounting for most of the entire cycle, and there is no G_1 phase checkpoint. The iPS cells with karyotype XX do not show the phenomenon of X chromosome inactivation.

8.2.3.2 Differentiation potential of iPS cells

iPS cells are reprogrammed somatic cells that maintain long-term proliferation and the ability to differentiate into any cells within the body (Fig. 8-2). Research on the transcriptional and epigenetic state of iPS cells is highly informative, and it might ultimately be possible to characterize newly derived iPS cell lines based on their genomic profiling alone. Before selecting the most informative markers, it is important to use *in vivo* assays to analyze the interplay between transcriptome, epigenome, and developmental potential. Jaenisch provided a detailed comparison of the different strategies for assessing developmental potential and their stringency. *In vitro* differentiation is the least stringent assay, whereas tetraploid-embryo complementation is the most stringent assay for testing developmental potential. These strategies could be used to determine the pluripotency of mouse iPS cells, but only *in vitro* differentiation and teratoma formation could be applied to test human iPS cells. Mouse iPS cells appear to have developmental potential similar to that of ES cells, as confirmed by teratoma formation capability and by a high contribution to chimera formation with germline transmission. To show the developmental potential of iPS cells as equivalent to that of ES cells, in three separate groups, mouse iPS cells ($2n$) are injected into tetraploid blastocysts ($4n$), and live mouse are created. The procedure, called tetraploid complementation, is the most stringent test for pluripotency. If the stem cells that are injected into the tetraploid blastocyst differentiate into embryonic tissues that produce a mouse, then the stem cells are considered truly pluripotent.

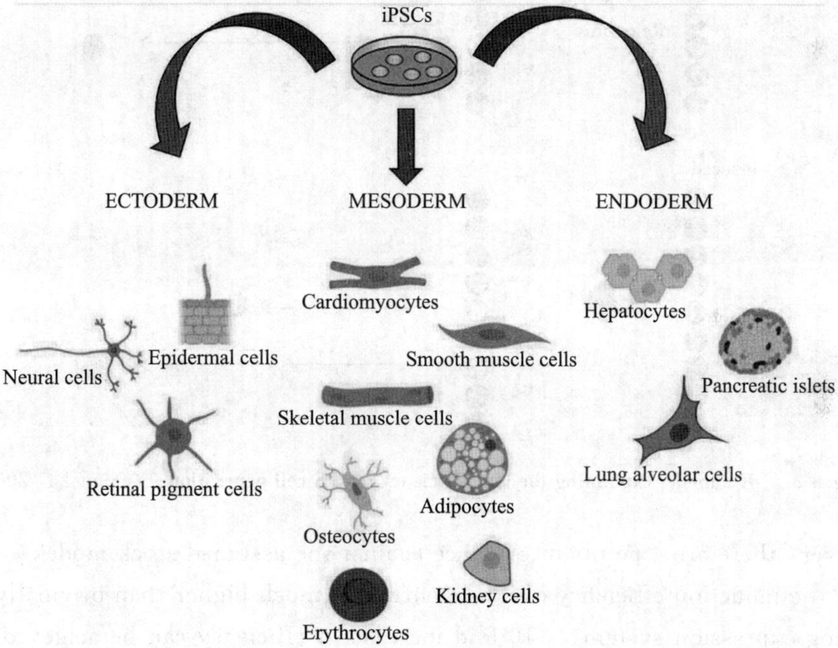

Fig. 8-2 iPSC differentiation into different lineages (Poorna, 2021)

8.2.3.3 Theoretical mechanism of iPS cell induction

Despite the continuous development of iPS technology, the efficiency of reprogramming is low, and many cells are not fully reprogrammed. Researchers have put forward some hypotheses for the cause of the iPS cell proliferation bottleneck. There are currently two main models: the elite model and the stochastic model.

1. Original model

The elite model assumes that only a few cells are capable of reprogramming. Such models can be further divided into two types: "assumed elite" models and "induced elite" models.

In the hypothetical model, a small fraction of cells is reprogrammed without retroviral transduction of the quadruple factor (Fig. 8-3). Tissue stem cells and undifferentiated cells in other regenerative tissues are good candidates for "stocks" that are capable of reprogramming. Reprogramming by somatic cell nuclear transfer, with nuclei from less differentiated primitive cells, such as neural stem cells or embryonic stem cells, yields much higher reprogramming efficiencies than with cells from terminally differentiated cells, such as lymphocytes. Likewise, the more primitive stem cells are preferentially reprogrammed with the induction of pluripotency with defined factors.

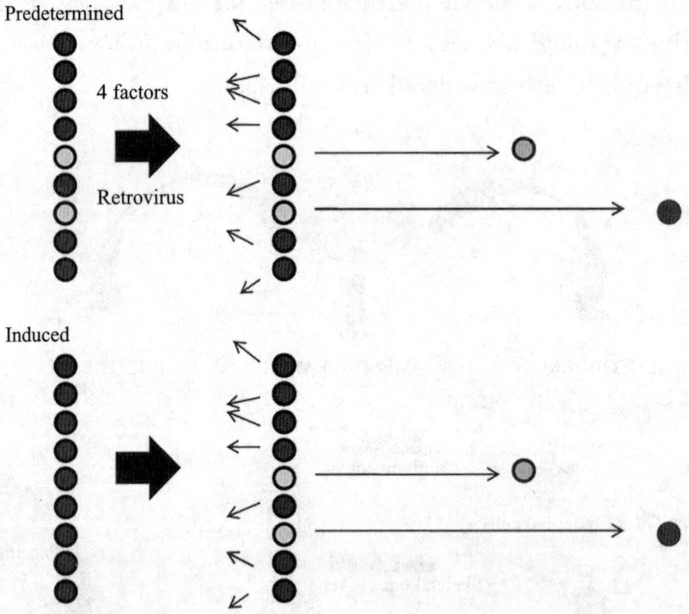

Fig. 8-3 Ete model explaining the low efficiency of iPS cell generation (Yamanaka, 2009)

However, there are 4 points of evidence against the assumed stock model.

First, the induction efficiency of iPS is currently much higher than originally reported. Using Nanog expression system, a 10-fold increase in efficiency can be achieved by simply prolonging the time of drug selection; 0.5% MEFs can be turned into iPS cells.

Second, 10% of MEFs can be reprogrammed using small molecules such as valproic acid in combination with factor 4. But in the original fibroblasts, 10% of tissue stem cells or undifferentiated cells are not present. While reprogramming factors or chemical treatments may preferentially increase proliferation of tissue stem cells or blast cells, it appears that more somatic cells are also reprogrammed to form iPS.

Third, genetic lineage trace analysis using iPS derived from the Cre/LoxP system show that most iPS derived from liver tissue are derived from albumin-expressing cells. Likewise, many iPS from the pancreas are derived from insulin-expressing cells. These data do not demonstrate that terminally differentiated cells can form iPS, as both albumin and insulin are expressed in blasts, mature hepatocytes, and pancreatic beta cells. Nonetheless, these findings unequivocally demonstrate that four factors can reprogram already differentiated lineage-committed cells that express either albumin or insulin.

Finally, Hanna et al. obtained iPS from B lymphocytes. They confirmed that iPS cells were derived from terminally differentiated cells by displaying genetic recombination of globin sites. Despite requiring ectopic expression of myeloid transcription factors, or specific knockout of the B lymphocyte transcription factor Pax5, and induction of factor 4, their data demonstrate that germline committed cells can be reprogrammed by defined factors.

In the induced elite model, certain genes in the host cell genome other than these four factors need to be activated or inactivated by viral insertion (Fig. 8-3). Therefore, only cells with specific viral integration sites are capable of reprogramming. However, there is also a lot of evidence that does not support this view.

Epithelial or tissue-derived iPS cells, such as hepatocytes, gastric mucosal cells, and skin cells, integrate less retrovirally than fibroblast-derived iPS cells. Taking advantage of this phenomenon, Yamanaka et al. used inverse PCR to detect retroviral integration sites in two iPS clones from hepatocytes and gastric cells, respectively, but did not find any defined common integration sites. Six iPS clones derived from mouse MEFs showed no common retroviral insertion.

More direct evidence came from several studies that did not use retroviruses to induce iPS. Adult mouse liver-derived iPS can be obtained by inducible expression of the four factors using adenovirus. Another study used 2 plasmids to induce MEFs into iPS cells, one of which was Oct4, Sox2 and Klf4 ligated with 2A (self-cleavable peptide) and the other was c-Myc ligated cDNA. Both groups showed the absence of adenovirus integration and plasmid insertion in the resulting iPS cells. The protein induction of iPS cells by four factors suggested that insertional mutagenesis was not necessary for iPS formation.

2. Stochastic model

In a stochastic model, most or all differentiated cells have the potential to form iPS after induction with four factors. Cell differentiation is often described as a ball rolling down an epigenetic state, first described by Conrad in 1957, as cells move from a totipotent state, through a pluripotent state, and finally to a state of lineage commitment (Fig. 8-4). During normal development, a cell's pluripotent state exists briefly. They are unable to stay on slopes, are pulled down by gravity, and rapidly diverge into various lineages. In contrast to normal development, embryonic stem cells are able to self-renew and maintain pluripotency for long periods of time. Thus, embryonic stem cells appear to be bumped or hit a snag, stuck in their characteristic epigenetic state. In this hypothesis, four reprogramming factors cooperate to push the cell to a pluripotent state (Fig. 8-4).

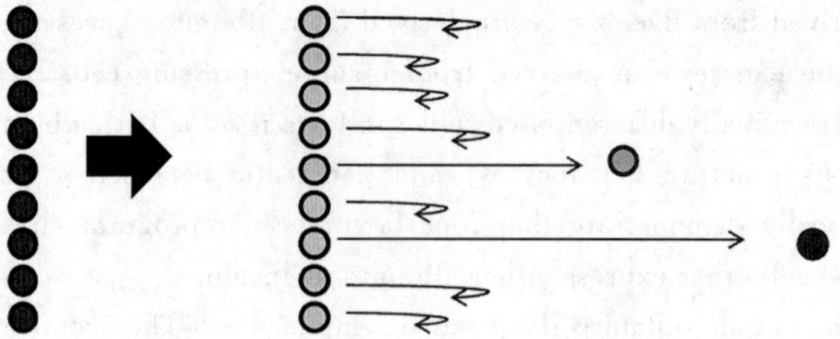

Fig. 8-4 Stochastic model explaining the low efficiency of iPS cell generation (Yamanaka, 2009)

Complete reprogramming requires at least two aspects. First, the four factors must be expressed in sufficient quantities to push the cell state in the right direction. Because the current technology can not precisely control the expression of 4 genes, this requirement can only be achieved randomly. Second, cells must reach a stage where they are trapped in an epigenetic state characteristic of ES cells, so that they can remain in a pluripotent state when foreign gene expression is lost. Since these four reprogramming factors alone can not form such an epigenetic barrier, there is still randomness in the process of iPS formation. Reprogramming by nuclear transfer suggests that DNA methylation and histone modifications likely play important roles in iPS formation.

Direct reprogramming depends on the expression, balance, continuity and silencing of the four factor transgenes. There is a high copy number of provirus in each iPS clone, indicating that strong foreign gene expression is initially required. *Klf4* is expressed in fibroblasts, but the expression of endogenous genes is not sufficient to form iPS. The endogenous *c-Myc* is also expressed in fibroblasts, but its ectopic expression results in significantly increased efficiency in the formation of mouse and human iPS.

The formation of iPS cells may also depend on a stochastic balance specific to four factors. For example, excess Oct4 and Sox2 are detrimental to maintaining pluripotency. Indeed, in those neural stem cells that also express endogenous *Sox2*, the iPS formation efficiency of the four factors is higher than that of the other three factors ectopically expresses without Sox2. The balance between *c-Myc* and *Klf4* is critical for apoptosis and senescence caused by these cancer-related genes. An imbalance of these four factors can lead to inappropriate reprogramming, senescence and even apoptosis.

Sustained expression and silencing of foreign genes is also important. During the initial 10-14 d, the foreign genes must be continuously expressed. In this regard, retroviral vectors are clearly superior to plasmids and adenoviruses. However, in order to complete reprogramming, after the initial phase, the exogenous gene must be silenced and replaced by the endogenous gene. If foreign genes can not be silenced, so-called partially reprogrammed cells have the state of embryonic stem cells and express some of the hallmark genes of embryonic stem cells, but their ability to differentiate is limited.

The introduction method of foreign genes has a great influence on their expression, balance, persistence and silencing. With retroviruses and lentiviruses, the position of integration of the provirus can be greatly affected. This may explain why only a small fraction of induced cells are able to complete the reprogramming process, as each induced cell has a unique pattern of integration. If this is the case, the use of fibroblasts containing the reprogramming components, integrated foreign genes, will significantly increase the efficiency of iPS formation.

This experiment was first done by Werning et al. They injected doxycycline-induced iPS cells into blastocysts, then isolated MEFs from chimeric embryos, and then treated these MEFs with doxycycline to form a "secondary" iPS. The same treatment was subsequently done on human cells. Human primary iPS cells were obtained by doxycycline-regulated lentivirus, and then induced to differentiate to obtain fibroblasts, and secondary iPS cells were again induced to form with doxycycline. In these experiments, 4% of the fibroblasts became iPS cells. Recently, an approach using the piggyBac transposon achieved a reprogramming efficiency of about 20%. These findings suggest that the efficiency of becoming iPS cells can be improved as long as the cells with the foreign gene inserted contain the appropriate pattern.

Even when the four factors are correctly expressed, lifting the cell to the top of the valley, the cell may roll down again without the expression of the foreign gene. Cells need to be fixed by specific epigenetic states. The endogenous sites of these four factors need to be fully activated. An important aspect is the correctness of DNA methylation. Genes associated with pluripotency are heavily methylated in their promoter regions in fibroblasts and somatic cells, but not in embryonic stem cells and iPS cells. Therefore, the DNA in these regions must be completely demethylated during direct reprogramming. Because these four factors have no inherent DNA demethylation activity, this process may require secondary effects of cell division. This may be the reason for the slow and inefficient iPS cell formation. In fact, demethylating agents, such as 5-azacytidine, increase the efficiency of iPS cell formation in this model.

The formation of iPS cells also requires the correct histone modification. In embryonic stem cells and iPS cells, genes associated with pluripotency are hyperacetylated on histones H3 and H4 in the promoter regions. In contrast, differentiated cells have a low degree of acetylation of H4. Therefore, H4 in these regions is acetylated during iPS formation. Because these transcription factors themselves have no histone-modifying activity, other factors are required. One function of c-Myc is to recruit acetylases for genes of interest. Indeed, inhibitors of histone deacetylases, such as valproic acid, can increase the efficiency of iPS cellformation, supporting the notion that histone acetylation is important during direct reprogramming.

Histone methylation is also important in preventing iPS cell formation. In ES cells and iPS cells, in the promoter region of pluripotency-related genes, H3 is methylated at lysine 4 and demethylated at lysine 9. Fibroblasts contain the opposite pattern of histone modifications. However, both ES cells and iPS cells have bivalent chromosomal structures called developmental genes that are methylated at positions 27 and 4 of histone H3 lysines. The methylation status of these histones is established for the formation of iPS cells.

8.3 Technical points of stem cell isolation and culture

8.3.1 Main points of culture techniques for mesenchymal stem cells

The bone marrow tissue adhered to the plastic matrix culture plate after enzymolysis, and the cells "crawled out" and showed the characteristics of MSC. The study by Gazit et al. showed that MSCs derived from adipose tissue were easily isolated after collagenase treatment of bone marrow tissue. Collagenase broke down collagen fibers and "served" as the connections between cells, which is one of the commonly used methods for isolating MSCs. However, the optimal concentration, incubation time and temperature of the enzyme need to be explored during the separation process.

Different experimental protocols can be used to obtain MSCs. Passaging tissue blocks adhered to a dish for 48-72 h is a common and effective practice, but this method often leads to cell heterogeneity. In order to obtain MSCs with good consistency, more stringent isolation protocols have been proposed, including the use of different media, flow cytometry sorting, and passage immediately after adhering to culture dishes for 3 h.

In order for cells to survive, proliferate, and maintain differentiation capacity *in vitro*, culture systems must mimic the conditions of the original tissue. These cells must be kept in a constant temperature incubator with 5% CO_2 in order to maintain the proper pH, oxygen content, and optimal physiological temperature in the medium.

Fetal bovine serum in cell culture medium can support the attachment and proliferation of MSCs on culture plates. Different concentrations of FBS have different effects on MSCs. At high concentrations, the immunogenicity of MSCs decreases and the proliferation rate increases; at low concentrations, the immunogenicity of MSCs increases and the proliferation rate decreases; at very low concentrations, the immunogenicity and proliferation rate all decline. Adding an appropriate concentration of serum replacement (KSR) can promote the proliferation rate of bovine mesenchymal stem cells and shorten the proliferation cycle. Using higher concentrations of KSR instead of FBS causes cell proliferation arrest.

Adding an appropriate amount of growth factors can promote the proliferation of MSCs: for example, TGF-β_1 at an appropriate concentration and time (2 ng/mL, 24 h) can reduce apoptosis, and improve the cell migration ability and normal karyotype rate of BMSCs. Under the action of TGF-β_1, the expression levels of pluripotency-related genes (*Sox2* and *Nanog*) and mesenchymal markers (N-cadherin and Fn1) are significantly up-regulated. In addition, TGF-β_1 can also activate the epithelial-mesenchymal transition (EMT) process, increase the expression levels of EMT-related genes (*Snail* and *Slug*), and maintain the cellular properties of bone marrow MSCs.

Studies have shown that hypoxic environment may maintain the characteristics of MSCs: hypoxia (5% oxygen) can promote colony formation and strengthen fiber synthesis of BMSCs, and improve cell migration ability and normal karyotype rate. In addition, the expression levels of genes related to the HIF-1α/Twist1 signaling pathway axis are up-regulated under hypoxia, thereby maintaining the mesenchymal properties of bovine bone marrow mesenchymal stem cells(bBMSCs).

8.3.2 Main points of culture techniques for embryonic stem cells

8.3.2.1 Selection and handling of blastocysts

Since blastocysts fertilized *in vivo* have more ICM cells than embryos fertilized *in vitro*, embryos fertilized *in vivo* should be selected as far as possible to isolate ES cells. For embryos from the same source, the embryos with clear color and obvious ICM should be selected. Embryos with earlier developmental stages need to be cultured to blastocysts *in vitro*, and frozen and thawed embryos need to be cultured *in vitro* to recover to blastocysts before use.

8.3.2.2 Select a culture system that inhibits cell differentiation

1. Feeder layer culture system

Feeder layer culture system is the first conventional culture medium, such as DMEM to add amino acids, serum and antioxidants, etc., to culture somatic cells. After the cells are covered with the bottom wall, γ-ray irradiation (usually the dose is 30-50 Gy) or mitomycin C treatment (about 10 pg/mL, 3-4 h) are used to stop cell division and form a special cell layer, the feeder layer, which can be inhibited by co-culture of blastocysts and feeder cells. ICM cells differentiate and promote their division and proliferation, and ES cells can be obtained by further separation. The somatic cells used to prepare the feeder layer require slow division but strong vitality. When preparing the feeder layer, commercial cell lines, such as STO, TM4, etc., can be used, and fibroblasts from mouse fetuses can also be used. Studies have shown that feeder cells can secrete differentiation inhibitory factors (DIA), such as leukemia inhibitory factor (LIF) and other substances.

2. Conditioned medium culture system

In this system, blastocysts are directly cultured in a special medium, and ES cells can be obtained without preparing a feeder layer. This medium is derived from the liquid recovered after certain cells have been cultured for a period of time. Cells commonly used to produce conditioned medium are guinea pig liver cells (BRL), human bladder cancer cells, and mouse EC cells. These cells can not only secrete LIF during growth, but also other growth factors, such as IGF-Ⅰ, can promote the growth of ES cells. Therefore, ICM cells can maintain an undifferentiated state in the conditioned medium and continue to proliferate to obtain ES cells.

3. Addition of differentiation inhibitory factor system to the culture medium

This method is to directly add the differentiation inhibitory factor LIF or interleukin-6 (IL-6) family to the culture medium of blastocysts at a certain concentration to separate ES cells. This culture system is not only simple in method, but also can avoid the interference of feeder cells and the toxic effect of mitomycin C on embryonic cells. Since the chemical composition of the culture medium is relatively clear, it is very useful for studying the differentiation mechanism of cells. However, this system has only been successful in the isolation and culture of mouse ES, and can not be used in other animals. In practical applications, in order to improve the separation efficiency of ES, a certain concentration of LIF (1 000 IU/mL) is also added to the feeder layer system to improve the separation efficiency of ES.

8.3.2.3 Isolation of ES cells using blastocysts as materials

There are two methods of obtaining ICM: immunosurgery and mechanical dissection. The basic method of the former is to let the blastocyst continue to be cultured *in vitro*. After hatching the zona pellucida, it is treated with a specific antibody. After the antibody is combined with the trophoblast cells, it is then treated with guinea pig complement to dissolve the trophoblast cells and retain the ICM cells. The latter is to allow the embryos to grow *in vitro* until the ICM protrudes from the trophoblast, and then use microneedle cutting or repeated capillary pipetting to remove the trophoblast cells, ICM is isolated. Because of its simplicity and complete removal of trophoblasts, immunosurgery is the most commonly used method for the isolation of ICM. The isolated ICMs are cultured in a differentiation inhibitory culture system for 9-16 d to form multicellular clumps, which are digested with 0.05% trypsin or pipetted with a micropipette to form small cell clumps, which are then transferred to new differentiated cells. Continue to culture in the suppression culture system for 3 weeks, and then use microneedles to select terminally differentiated cell clusters with uniform morphology and continue to culture and expand for 5 generations, and then inoculate single cells into small-cell culture plates to obtain cells grown from single clonal cell mass. After the karyotype analysis, the cell mass of the normal karyotype is continued to be cultured and expanded, and the morphology and function of the ES cells are identified to determine whether they are ES cells.

8.3.3 Technical points of iPS cell culture

The main steps of iPS technology including: ① isolation and culture of host cells; ②introducing several or more pluripotency-related genes into host cells by virus-mediated or other means; ③the virus-infected cells are planted on feeder cells and cultured in a special culture system for ES cells, and corresponding small molecules are added to the culture as needed to promote reprogramming; ④identification of iPS cells (cell morphology, epigenetics, *in vitro* differentiation potential, etc.) after the emergence of ES-like clones(Fig. 8-5).

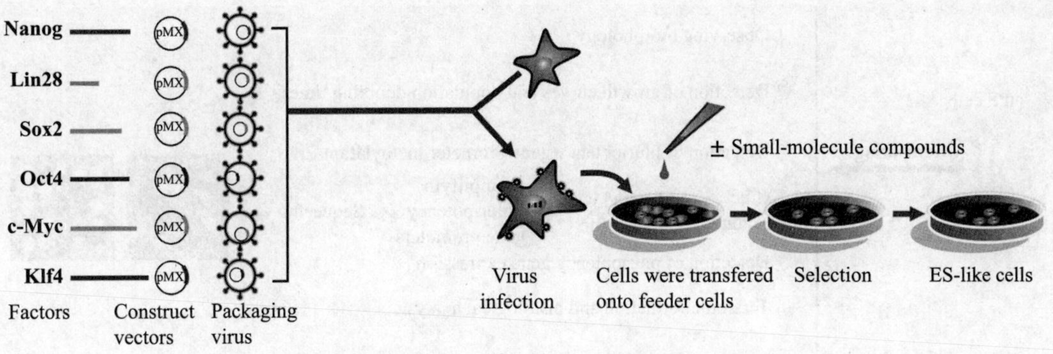

Fig. 8-5 Basic process of iPS induction

8.3.3.1 Methods to identify and characterize reprogrammed cells

The screening of iPS cell clones is mainly based on the morphological judgment of the unique characteristics of ES cells. During reprogramming of mouse and human cells, the morphological changes in the resulting clones are progressive. It is worth mentioning that some clones that are morphologically similar to ES cells but not multifunctional are also formed during the induction of human fibroblast reprogramming. These clones are a misshapen form of iPS cells that are distinct from iPS cells in that they are loose and granular in appearance and begin to form during the initiation phase of reprogramming. Certain helper labeling methods are also used to identify iPS cells, such as the isolation of Thy-1-SSEA-positive cells during fibroblast reprogramming in mice, which helps improve the efficiency of iPS cell formation and is specific to human ES cells. Staining for the antigen Tra-1-81 can help identify which iPS cell clones are derived from fibroblasts. Mouse iPS cells/ES cells prevent the differentiation of individual cells, and nascent clones immediately begin enzymatic breakdown, simplifying the process of forming cell lines. Cells must be passaged before they can be used for dissociation by digestive enzymes. Because human iPS cells/ES cells are highly differentiated, especially in the first few passages, it is especially necessary to deplete differentiated cells to prevent premature differentiation.

After the formation of ES cell-like clones, the detection of pluripotency factors can be performed at the RNA level and protein level, respectively. The genes detected by RNA level mainly include (endogenous) *Oct4*, (endogenous) *Nanog*, *Fbx15*, *Eras*, *Rex1* and *Esg1* (Fig. 8-6).

8.3.3.2 Selection of reprogramming factors

1. Oct4

Oct4 is encoded by *Pou5f1* and belongs to the POU (Pit-Oct-Unc) transcription factor family. It participates in the regulation of downstream genes by binding to an eight-nucleotide conserved sequence containing ATGCAAAT. Oct4 is mainly expressed in

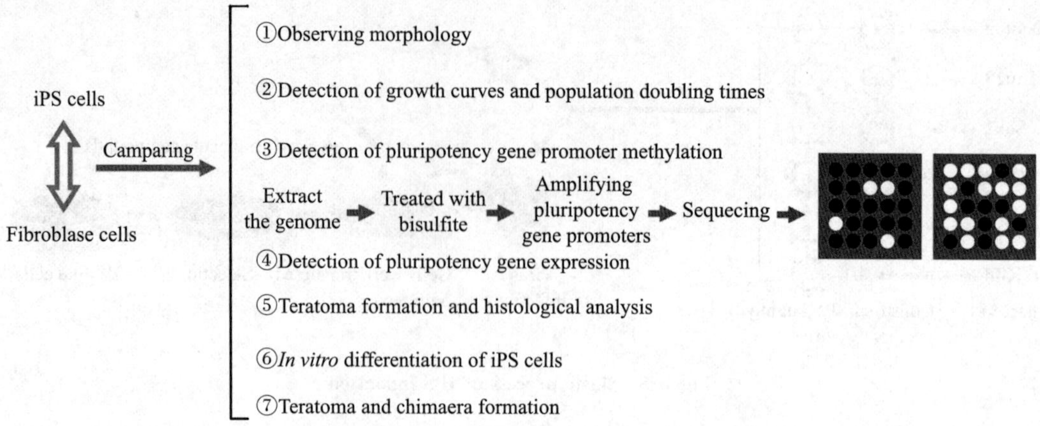

Fig. 8-6 Identification procedure of induced pluripotent stem cells

fertilized egg, morula, ICM of blastocyst, and epiblast and primordial germ cells (PGCs) of post-implantation embryo. In mice, knockout of the *Oct4* resul in embryonic death in the preimplantation stage (embryogenesis 3.5-4.5 days). The blastocyst is developmentally arrested due to its inability to form a pluripotent ICM and to implant normally. This suggests that Oct4 is necessary for the formation of pluripotent cells in early embryonic development. In ES cells cultured *in vitro*, only when the expression level of Oct4 is within a specific range can the cells remain undifferentiated. When the Oct4 expression level is more than twice the normal level, ES cells will differentiate into primitive endoderm and mesoderm; when the expression level is lower than 50% of the normal level, ES cells will develop into trophectoderm. However, Oct3/4 can not maintain the undifferentiated state of ES cells when exogenous inhibitory factors are withdrawn, indicating that Oct3/4 can not independently maintain the subpluripotent differentiation characteristics of ES cells. This suggests that Oct4 precisely regulated in ES cells, and its expression level determines the fate of ES cells. A large number of literatures on the induction of somatic cell reprogramming have shown that Oct4 is the core inducer of iPS, and there is no report that Oct4 is not involved in the generation of iPS.

2. Nanog

Nanog is an endogenous homeoprotein. During *in vivo* development, Nanog is mainly expressed in inner cells of compact morula, blastocyst ICM, epiblast and ectoderm of embryonic primitive streak, and its expression is down-regulated during formation of mesoderm and defined endoderm. *In vitro*, Nanog is only expressed in ES cells, embryonic germ cells and embryonic carcinoma cells. The *Nanog* has not been discovered for a long time, but its important role in ES cell regulation has been confirmed. This gene can maintain the pluripotency of ICM and ES cells independently of LIF/gp130. When exogenous LIF is removed, ES cells that continuously express Nanog can remain in an undifferentiated state, while ES cells that lack Nanog will differentiate into body and visceral endoderm.

Studies have shown that Nanog's function is dependent on the continuous expression of

Oct3/4. Oct3/4-depleted ES cells still express Nanog, but they can not maintain ES cell differentiation, so Nanog is functionally related to Oct3/4. Oct3/4-depleted cells express Nanog, but Nanog overexpression does not reverse the differentiation that occurs in the absence of Oct3/4. Oct3/4 is necessary to determine the following two differentiation fates of embryonic cells: one is at the morula stage and the second is in the pre-implantation stage, and Nanog is extremely critical in the decision of the second differentiation fate. Oct3/4 is not required for *Nanog* expression, but Nanog can not function without Oct3/4. Therefore, only the combined effects from Nanog and Oct4 signaling can maintain the self-renewal and multi-directional differentiation potential of embryonic stem cells. Yu et al. believe that Nanog can promote the recovery of ES cell clones derived from mesenchymal cells, but it is not necessary for the initial emergence of clones. In experiments, overexpression of Nanog in somatic cells increased the efficiency of cell reprogramming by 200-fold. The expression of Nanog can improve the clone formation rate of human ES cells, that is, increase the survival probability of reprogrammed cells.

3. Sox2

Sox2 is a member of the Sox family expressed in embryonic cancer cells. *Sox2* is also a marker gene for pluripotency in early embryonic development and is expressed in the ICM, ectoderm and germ cells. During embryogenesis, *Sox* exhibit diverse and dynamic expression patterns, and the expression of maternal Sox2 protein persists throughout developmental stages prior to implantation. *In vitro*, the *Sox2* is expressed in undifferentiated ES cells and EC cells, and the expression is down-regulated as cells differentiate. Unlike Oct3, Sox2 is also expressed in the pluripotent cell extraembryonic ectoderm. Embryos lacking Sox2 die at implantation due to poor ectoderm development. Homozygous mutants appear morphologically normal, but blastocysts do not differentiate or expand *in vitro*, producing only cell-like trophectoderm and endoderm. Deletion of Sox2 results in trophectoderm differentiation of ES cells. Therefore, Sox2, like Oct4, is required for the maintenance of pluripotency.

Sox2 often forms a transcriptional regulatory complex together with Oct4 to regulate the expression of genes. For example, Oct4 and Sox2 are also involved in the regulation of Fgf4 expression. Gene knockout experiments have showed that Sox2 is required for epiblast and extraembryonic ectoderm formation, suggesting that Sox2 and Oct4 are involved in specifying embryo implantation as the fate of pluripotent stem cells. Experiments have also shown that Sox2 is necessary for regulating multiple transcription factors that affect Oct4 expression, thereby maintaining necessary levels of Oct4 expression and stabilizing the pluripotent state of embryonic stem cells.

4. Klf4

Klf4 belongs to Krüppel-like factors (Klfs). The transcription factor Klf4 affects the overall physiological function of cells by regulating the expression of its downstream target genes. Klf4 is highly expressed in anaphase of mitosis in skin and gastrointestinal epithelial cells. Initial studies found that some conditions related to cell growth inhibition, such as serum starvation, contact inhibition, and DNA damage, can up-regulate Klf4 expression, and exogenous *Klf4* expression can inhibit DNA synthesis and cell growth. The mechanism comes from detecting Klf4 expression and the study of DNA damage which induces cell growth inhibitory responses. The results of Miller et al. further indicated that Klf4 plays an important role in cell proliferation, differentiation and cell cycle regulation by regulating the expression of some cell cycle regulation and differentiation-related genes.

The *Klf4* is not necessary for ES cell self-renewal and maintenance of an undifferentiated state, but if all Krüppel-like family genes are knocked out, ES cells will differentiate and disrupt ES cell self-renewal. The sites of *Klfs* and *Nanog* are bound together. *Nanog* and other genes are effectors of Klfs protein biological function, and Klfs genes are important mediators for ES cell identification. Shields et al. found that Klf4 mRNA expression is high in NIH3T3 growth-arrested cells, but undetectable in exponentially growing cells. *Klf4* is both a tumor suppressor and an oncogene, and overexpression of *Klf4* results in inhibition of DNA synthesis and normal cell cycle progression during cell culture. Embryos lacking the *Klf4* develop normally, but neonatal mice die within 15 h after birth and show impaired differentiation in the skin and large intestine, suggesting that KLF4 plays an important switch role in cell expansion and differentiation.

5. c-Myc

c-Myc is not only a translocatable gene, but also a regulated gene regulated by various substances, and also a gene that can enable cells to proliferate indefinitely, obtain immortalization functions, and promote cell division. *Myc* is involved in apoptosis and is related to the occurrence and development of various tumors. The *c-Myc* is mainly activated by amplification and chromosomal translocation and rearrangement, and has an important relationship with the occurrence, development and evolution of tumors in some tissues. Overexpression of *c-Myc* is associated with early tumor recurrence. Its ability to affect gene expression has in the past been thought of as a means by which it promotes tumor development, but the discovery that *c-Myc* also affects DNA replication directly suggests that there should be another explanation for some of its effects. *c-Myc* controls DNA replication by targeting DNA synthesis sites and binding to the pre-replication complex: when out of control, the same mechanism may also cause DNA damage and incorrect cell proliferation. Homozygous mouse embryos knocked out of the *c-Myc* will die between 9. 5 and 10. 5 d of gestation. Malformations in dead embryos include malformations of the heart, pericardium, and neural tube, as well as delayed or stalled embryonic development. There are more than 25 000 *Myc* binding sites in the human genome. By binding to *c-Myc* binding sites within the genome, *c-Myc* can modify chromatin structure and regulate the expression of noncoding RNAs.

Yamanaka's research group, removed *c-Myc* and used the remaining three factors (OSK) to induce iPS cells in mice. Although the probability of tumor formation was reduced, the induction efficiency was also greatly reduced, and it also showed that *c-Myc* is not essential in inducing somatic reprogramming.

6. MicroRNA

miRNAs are a large family of regulatory RNAs that inhibit the expression of target genes and are important for normal development and ESC biology. ESCs express a unique set of miRNAs required for rapid cell proliferation and cell cycle progression. Most of these miRNAs are derived from two genomic transcriptional loci: the mouse miR-302/367 cluster (containing five miRNAs—miR-302a/b/c/d and miR-367) and the human miR-290-295 cluster (miR-290, miR-291a, miR-291b, miR-292, miR-293, miR-294 and miR-295) or miR-371-373 cluster (miR-371, miR-372 and miR-373). Because ESC-specific miRNAs share a very similar "seed" sequence, they are likely to have regulatory effects on excessive overlap of target genes.

Studies by Morrisey et al. showed that a single primary miRNA-expressing miR-302/367 cluster alone is sufficient to reprogram murine and human fibroblasts. The gene expression and functional properties of the formed iPSCs are characteristic of fully reprogrammed pluripotent cells. Compared to OSKM, iPSC clone formation is more than 100-fold more efficient, in addition, reprogramming induced with miR-302/367 occurs somewhat faster than with OSKM colony in terms of iPSC colony appearance and activation of multifunctional markers. Interestingly, valproic acid is administered with the miRNA-expressing virus to reprogram mouse cells, but not human cells. The histone deacetylase Hdac2 may partially explain this species difference, and the function of Hdac2 is an important obstacle to cellular reprogramming. In-depth studies have shown that miR-302 inhibits lysine (K)-specific demethylase 1A (LSD1) and methyl CpG binding protein 2 (MECP2), etc. Epigenetic regulators cause extensive demethylation of the somatic genome.

Valproic acid can destroy Hdac2, and the expression level of Hdac2 in mouse fibroblasts is significantly higher than that in human fibroblasts. Therefore, reprogramming murine Hdac2-/-fibroblasts does not require valproic acid, and miR-302/367 alone can efficiently reprogram these cells.

Judson et al. co-transfected ES cells with specific and highly expressed miRNAs: miR-291-3p, miR-294 and miR-295, together with the three factors after removing c-Myc, and found that they could improve the induction of iPS. These miRNAs that maintain the pluripotency and proliferative properties of stem cells replace the role of c-Myc, thereby reducing the risk of carcinogenesis in iPS cells.

In 2011, He et al. revealed a miRNA family (miR34) that plays an important role in the process of somatic cell reprogramming, and confirmed that inhibiting this miRNA family can greatly improve the efficiency of iPS cell generation. miR-34 is a highly conserved miRNA family discovered in recent years, which is widely found in arthropods, nematodes and mammals. Three members of the miR-34 family (referred to as miR-34s) exist in almost all vertebrates: miR-34a, miR-34b and miR-34c. Dr. He's group has confirmed in previous studies that miR-34s plays a non-negligible role in the p53 regulatory pathway known as the "guardian of the genome". The transcription of miR-34s is regulated by p53 protein and participates in the p53 pathway. Since some studies in recent years have found that the p53 signaling pathway has a hindering effect on cell reprogramming, and inhibiting the function of p53 will greatly improve the productivity of mouse and human pluripotent stem cells. miR-34s, especially its member miR-34a, is involved in the regulation of somatic reprogramming in a p53-dependent manner. The somatic reprogramming efficiency of miR-34a-deficient mice is significantly improved and the cell dynamics are enhanced. Unlike p53 deficiency, which results in the loss of pluripotency of the generated iPSCs while increasing the reprogramming efficiency, miR-34a gene knockout promotes iPSC generation, and also the generated iPSC cells maintain their self-renewal or differentiation ability. Further analysis indicates that miR-34a may exert its reprogramming inhibitory effect by negatively regulating pluripotency genes including *Nanog*, *Sox2* and *N-Myc*. This miR-34a-induced post-transcriptional gene repression effect also affects the differentiation kinetics of iPS cells.

In addition, two other family members of miR-34s, miR-34b and miR-34c, also show similar inhibitory effects as miR-34a on somatic reprogramming. This suggests that the three miR-34s may play a synergistic effect in this process.

8.3.3.3 Inducible cells

So far, in addition to the successful induction of reprogramming in mouse embryonic fibroblasts, several types of cells are induced into pluripotent stem cells, including mouse tail tip fibroblasts and human skin fibroblasts, mouse hepatocytes, gastric epithelial cells, pancreatic β cells, terminal differentiated B lymphocytes, neural progenitor cells, keratinocytes and peripheral blood cells, further demonstrating that terminally differentiated somatic cells can be induced into pluripotent stem cells.

Li et al. transferred four factors into amniocytes that highly express some pluripotency factors such as Nestin, Vimentin, and Gata4, which can efficiently and rapidly establish iPS cells. The morphological changes occurred on the second day after the infection of the amniocytes, and the clones of human embryonic stem cells appeared on the fourth day. Suna et al. induced adipose stem cells into iPS cells, and the induction efficiency was 20 times higher than that of fibroblasts, and it could be independent of feeder cells.

When selecting target cells, the following points must be considered: ①cells that are easily infected; ②cell reproduction capacity; ③cell generation and source. Aged cells or cells that have been passaged several times may contain genetic defects that reduce the clinical efficacy of iPS cells; while primary cells obtained from organs are prone to DNA damage, such as skin cells that tend to aggregate UV-induced mutations.

8.3.3.4 Selection of inducible delivery system

So far, the generation of iPS is mainly induced by nucleic acid delivery systems, protein inducible systems and small molecule compounds. Among them, nucleic acid delivery systems can be divided into virus vectors, plasmids, and direct transfection of synthetic microRNAs.

Retroviral and constitutive lentiviral vectors were initially used for mouse and human iPS production, and later inducible lentiviral vectors were used. However, these viral systems have been questioned because of their permanent integration into the genome. To make iPS more therapeutically viable, a non-integrative approach must be pursued. At present, there are two main methods: transient transfection and adenovirus transfer, both of which have been successfully usedin the reprogramming of mouse cells, but both have the problem of low efficiency.

Although lentiviruses can infect non-isolated cells and maintain high levels of exogenous expression, their silencing in the pluripotent cell state is less than desirable, and such sustained translational expression is less desirable in reprogramming attempts. Although the induction of iPS by lentiviral vectors with continuous expression has been reported, it is not clear how cells differentiate under the condition of continuous expression of the transgene.

The drug-induced lentiviral system offers a more attractive option because it provides immediate control of reprogramming factor expression levels. Although these viruses also integrate into the host cell genome, it can not be denied that they are particularly suitable for mechanistic studies.

The use of integrative viruses is a major hurdle for iPS to move towards clinically relevant applications, as there are indications that genomic insertions may affect gene function, while reactivation of proto-oncogenes can lead to tumorigenesis. Analysis of integration sites did not reveal common targets and pathways, suggesting that genome integration is not required during reprogramming. This notion was further demonstrated by the successful acquisition of mouse iPS by transient transfection method.

Okita et al. used a plasmid to transfer Oct3/4, Sox2, and Klf4 into fetal mouse fibroblasts, while using another plasmid to transfer c-Myc to induce iPS cells with non-integrated foreign genes. It was lost after the second passage, and iPS cells without foreign genes were obtained. However, this induction method is less efficient. Later, iPS cells were also obtained using other non-integrating vectors such as adenovirus, liposome, etc. , but they all had the problem of low efficiency. Yu et al. induced human iPS cells using a non-integration episomal vector, which only attaches to the genome but does not integrate into the genome. After obtaining iPS cells, stop using drug screening, and the attachment vector will fall off from the genome, and finally iPS cells without any vector and foreign genes can be obtained.

After obtaining iPS cells using four factors, Soldner et al. used the Cre/LoxP recombination system to excise the exogenous gene to obtain iPS cells without exogenous genes, but the residual vector fragments on the genome still brought the risk of mutation. Kaji et al. used transposon to mediate four factors to induce iPS cells, and then completely excised the exogenous gene under the action of transposase. The transposon system can completely excise the exogenous inserted gene, unlike Cre/LoxP, the system leaves vector residual fragments, but the transposon system still has to undergo the process of insertion into the genome, and the impact on cells remains to be further analyzed. Vectors that do not integrate into the genome have greater safety advantages, but the process of iPS induction requires the continuous expression of exogenous factors over a certain period of time.

Ban et al. used Sendai virus vectors to remove foreign genes and vectors from iPS cells by culturing at 37-38 ℃. The Sendai virus vector does not enter the nucleus, but only replicates its own genome in the cytoplasm and produces a large amount of target protein. The genome is RNA, which is fundamentally different from chromosomal DNA. Therefore, Sendai virus vectors in principle do not present the danger of altering chromosomes in the nucleus(Table 8-1).

Table 8-1 The transport way of reprogramming factors

Methods			Advantages	Disadvantages
Integrated virus vector method	Retroviral vector based on moloney		Silence in pluripotent state	Genome integration; risk of insertion mutation; restricted to dividing cells
	Lentivirus vector based on HIV	Constitutive type	Can infect dividing or non-dividing cells	Genome integration; risk of insertion mutation; no silence in pluripotent state
		Inducible type	Immediate control of foreign gene expression	Genome integration; risk of insertion mutation; possibility of leakage expression
Non integrated virus vector method	Adenovirus carrier		Low genome integration frequency	Repeated infection of certain cell types is required; delay the dynamic process of reprogramming; some produce tetraploid cells
	Sendai virus carrier		In principle, genome integration will not occur	RNA viruses can only process RNA
Others	Transposon		Virus free; After gene expression, it will be removed from the genome	Gene insertion occurs; low efficiency
	Plasmid transient transfection		Virus free components; Low frequency of genome integration; Simple operation	Multiple rounds of transfection are required; lower than the level of integrated expression; the dynamic process of reprogramming is delayed; consolidation provides selection advantages and necessary clone screening
	Small molecule compound addition		Instantaneous control capability; Easy to operate	Controversy over toxic efficiency; uncertain or nonspecific effects
	RNA transient transfection		Virus free components; Low frequency of genome integration; Simple operation; Faster expression than DNA transfection	Multiple rounds of transfection are required
	Protein mediated		Direct delivery of transcription factors avoids the complexity of nucleic acid delivery system	Short half-life; multi round application is required; some proteins are difficult to purify

8.4 Main problems of stem cell technology

8.4.1 Main problems of mesenchymal stem cell technology

8.4.1.1 Fewer sources

At present, bovine MSCs can only be isolated from body tissues such as uterus, umbilical cord, bone marrow, adipose tissue, placenta, and amniotic fluid, and the sources are not extensive, and there is a certain age limit for the animals from which they are drawn. High-quality bovine MSCs can not be extracted.

8.4.1.2 Limited self-renewal capacity

The limited self-renewal of bovine MSCs makes the number of subcultures limited and can not be continuously sub-cultured for a long time. If there are too many passages, such as after more than ten generations, there may be genetic instability, mesenchymal stem cell function decline, automatic differentiation and other phenomena.

8.4.1.3 Differentiation potential is controversial

Studies have shown that adding dexamethasone or triiodothyronine to the culture medium can promote the differentiation of MSCs into chondrocytes. However, the conditions for directional induction of differentiation are still unclear. There is only a clear way to differentiate MSCs into cartilage, fat and other cells. There are few studies on the differentiation of MSCs into other specific tissue cells. The results are not rich enough to generalize a systematic induction pattern.

8.4.1.4 Clinical application is less researched

There is still a lack of robust clinical data to demonstrate the function of bovine MSCs. For example, some experimental and clinical data have provided support for the use of MSCs in the treatment of diabetes, but the clinical data for the treatment of diabetes need further research to accumulate results; although experiments have shown that bovine MSCs can promote the regeneration process of skeletal muscle, there is no clinical data to suggest that MSCs can be used for bovine bone repair. Signaling molecules and nutritional factors secreted by MSCs can modulate innate immune responses to treat bovine chronic arthritis, but there is still a lack of relevant data support.

8.4.2 Main problems of ES cell technology

8.4.2.1 Low efficiency of cell line isolation

At present, ES cell lines with the ability to form chimeras have been truly established only in mice and rats. Although ES cell lines have been obtained from livestock, these are not ES cells in the true sense. The main reason for the low separation efficiency of ES cells is the lack of an ideal separation and culture system. Due to the lack of understanding of the molecular mechanism of ES cells maintaining pluripotency, a reasonable isolation and culture system can not be established according to the characteristics of cell differentiation and the nutritional requirements for maintaining undifferentiated cells.

8.4.2.2 Low theoretical and technical level of directional induction and differentiation

The broadest application prospect of ES cells in medicine is to utilize their directed differentiation ability. The current theoretical and technical level is still unable to specifically induce differentiation of ES cells, that is, to differentiate into specific cell types, such as neurons or cardiomyocytes. The research and observation of ES cell differentiation is mainly accomplished through the cellular structure of chimera, teratoma and EB, which are complexes composed of various differentiated or undifferentiated cells, and the research on the mechanism of ES cell differentiation is very limited.

8.4.3 Main problems existing in iPS cell technology

Remarkable progress has been made in iPS cell technology. Breakthrough results have brought us both joy and new challenges. Cell reprogramming is expected to usher in a new wave of research. Although iPS cells have attractive application prospects, however, the future research of iPS cells also faces many problems that need to be solved.

8.4.3.1 Efficiency issues

At present, the rate of inducing iPS cells is still very low, which is related to various factors such as the way of gene introduction, integration site, and epigenetics. This has become the biggest bottleneck restricting iPS from the laboratory to the clinic. Therefore, how to improve the production efficiency of iPS cells is still a common concern.

8.4.3.2 Safety issues

At present, the induction of iPS cells usually uses retrovirus as a vector to transfer several oncogenes into differentiated cells to induce them to become iPS cells. This method may interfere with the insertion of foreign genes into the cell genome, which can cause cancer.

8.4.3.3 Mechanisms

The pluripotency of cells is regulated by the transcription factor network and epigenetics in a complex manner. Then, how to complete this complex task by only relying on a few transcription factors to induce pluripotency? What about cells? In-depth study of the molecular mechanism of somatic cell reprogramming, the regulatory mechanism and the *in vitro* directed differentiation mechanism is also one of the important tasks of future research.

8.5 Application prospects of stem cell technology

8.5.1 Application prospects of mesenchymal stem cells technology

8.5.1.1 Treatment of mastitis

Clinical mastitis has a serious impact on breast tissue and will significantly reduce milk production. The anti-inflammatory properties of MSCs can reduce the damage caused by mastitis to animals, and stem cells modified with therapeutic agents can also be used to treat mastitis. It has been reported that the introduction of the cloned bovine lactoferrin (LFcinB) gene into the pggyBac transposon vector through exogenously expressed hybrid antimicrobial peptides is a feasible treatment method.

8.5.1.2 Treatment of diabetes

Diabetes in cattle is similar to juvenile diabetes in humans, which increases the glucose levels in the blood and urine of animals, affects the synthesis of fatty acids in the liver, and causes acid-base imbalance, poisoning and dehydration, and in severe cases, coma and death. It has been found that bone MSCs can be transformed into islet-like clusters, express insulin and glucagon, and alleviate the symptoms of diabetes.

8.5.1.3 Promotion of animal reproductive performance

Bilateral ovarian malnutrition may cause endocrine disorders, developmental abnormalities and other animal reproductive disorders, and in severe cases may lead to inability to ovulate. Amniotic fluid MSCs have anti-inflammatory effects and low immunogenicity, which can repair tissue cells, which provides a new therapeutic solution for clinical tissue engineering. Studies have shown that injection of cultured amniotic fluid MSCs *in vitro* can promote follicle maturation, alleviate the symptoms of malnutrition, and restore the reproductive capacity of cattle.

8.5.1.4 Supply of ideal donor cells for nuclear transfer

Nuclear transfer (NT) is the transfer of the nucleus of a donor cell into an enucleated oocyte so that the donor cell can be activated, divided and developed into a reconstituted embryo without going through sexual processes such as sperm penetration. After the reconstituted embryos are transferred, cloned animals can be obtained. After proper modification of nuclear donor cells and partial gene recombination, the transplanted cells can treat various diseases caused by cell function defects, and can also help species reproduce and retain excellent genetic resources. Compared with fibroblasts, MSCs are less differentiated and more easily reprogrammed. Therefore, studies have successfully used amniotic fluid MSCs and adipose MSCs to clone calves.

8. 5. 2 Application prospects of embryonic stem cell technology

8. 5. 2. 1 An extremely valuable tool to study on the laws of mammalian ontogeny and development

Since ES cells can differentiate into any type of cells in the embryonic endoderm, mesoderm and ectoderm, therefore, by tracking the traces of ES cells during fetal development, it is possible to find out the temporal and spatial order of the formation of organs and tissues; to study the occurrence and development of individual higher mammals. At present, the commonly used method is to genetically mark ES cells with genes expressing special products, such as β-galactosidase gene (*LacZ*) or green fluorescent egg (GFP). After *in vitro* screening, positive cells are injected into human blastocysts. The process of ES multiplication can be traced through histochemical staining or direct observation of marker gene products in chimeric fetal tissue, and the regularity of tissue and organ formation during embryonic development can be ascertained. The formation time and developmental process of a certain organ can be studied by using the expression of the marker gene of the promoter of organ differentiation-inducing factor. Mouse ES cells have become one of the commonly used methods to study the early developmental patterns of cellular drivers.

8. 5. 2. 2 An ideal means to study on cell differentiation

ES cells can maintain undifferentiated state only on feeder cells or in culture medium supplemented with leukemia inhibitor (LIF) or interleukin-6 family (IL-6). When a differentiation inducer is added to the culture medium, ES cells will undergo directional differentiation. For example, retinoic acid can induce ES cells to differentiate into neurons and neuroblastoma cells (Villenueveetal, 1982), dimethyl sulfoxide (DMSO) can induce ES cells to differentiate into smooth muscle cells (Dinsmore et al. , 1996), and butyryl cyclic adenylate can induce ES cells to differentiate into body wall endoderm (Strichland, 1980). By studying the differentiation characteristics of ES, we can reveal the pathway of intracellular signal transduction, discover new signal receptors, and reveal the mechanism of cell differentiation and apoptosis.

8.5.2.3 The first choice to study on gene function

Because ES cells have a stable karyotype and are easy to trap exogenous genes, they are ideal cells used for gene knockout (knock-out) or directed insertion (knock-in) of foreign genes. Genetically modified ES cells are used to produce chimeras or cloned progeny are obtained by nuclear transfer technology. By observing the phenotypic changes of these progeny, the biological function and expression regulation of unknown genes can be identified.

8.5.2.4 The main method to produce transgenic animals

ES cells, like ordinary cells, can integrate foreign DNA into the genome through random or directional integration after transfection of foreign genes. After screening, positive cells are obtained, and then the positive cells are injected into the blastocoel cavity or aggregated with other blastomeres to obtain chimeric offspring. If the transfected ES cells differentiate into germinal stem cells, the offspring are transgenic positive animals. The transformed ES can also be used as the donor nucleus, and the transgenic animals can be directly obtained by the nuclear transfer technique.

8.5.2.5 Materials to perform cell therapy

ES cells can be used as cell therapy for some intractable diseases caused by cell necrosis, degeneration or abnormal function, such as Parkinson's syndrome, juvenile diabetes and Alzheimer's disease, etc. New cells may be repaired and return to normal function. The results of animal models show that special tissue cells differentiated from ES cells have successfully cured diabetes, Purkinje cell degeneration, liver failure, heart failure, Duchenne muscular dystrophy, osteogenesis imperfecta, etc (Odorico et al., 2001). With the development of ES cell isolation and culture and cell therapy technology, everyone will have ES cell lines that are identical to their own genetic material. Once the function of a certain type of cells in the body is abnormal, by inducing the directional differentiation of ES cells, the diseased cells can be repaired, and human lifespan will be greatly extended. With the development of tissue engineering technology, it is possible to cultivate artificial tissues or organs by inducing the differentiation of ES cells *in vitro*. The current clinical problems such as insufficient donor organs and immune rejection after organ transplantation will be easily solved.

8.5.2.6 Materials to study the mechanism of cell carcinogenesis and the screening of new drugs

Cell carcinogenesis is mainly due to the disorder of the normal cell cycle and the disordered proliferation of cells. ES cells can maintain a stable karyotype in a normal culture environment, and are ideal materials for studying the mechanism of carcinogen-induced cell carcinogenesis. In the process of developing new drugs, it is necessary to study the mechanism of action and efficacy of drugs. ES cells differentiate into a variety of somatic cells *in vitro* and can be directly used as experimental materials to screen new drugs. For example, in the development of drug engineering for the treatment of myocardial diseases, ES cells can be induced to differentiate into cardiomyocytes as experimental materials, and then drugs with less side effects and stable therapeutic effects can be screened. The development of this drug can shorten the development cycle of new drugs, and the therapeutic effect is highly targeted.

8.5.3 Application prospects of iPS cell technology

8.5.3.1 Cell transplantation and tissue repairment

Autologous induced pluripotent stem cells can be matched with HLA-related stem cells without HLA database matching. The obtained immune-tolerant stem cell lines are very suitable as seed cells for autologous transplantation. The problem of rejection of severe, long-term follow-up drug treatment drives the development of regenerative medicine and individualized treatment of disease.

At present, researchers have induced somatic cells with clinical application prospects, such as pancreatic beta cells, skin fibroblasts in amyotrophic lateral sclerosis, and hepatocytes. Hayashi et al. successfully induced iPS cells into primordial oocytes *in vitro* for the first time, transplanted them into mice, and fertilized with sperm to produce healthy offspring mice. iPS cells have the potential to differentiate into germ cells under the induction of granulosa cell culture supernatant. On this basis, Chen Mo et al. used cyclophosphamide and busulfan intraperitoneal injection to construct a mouse model of premature ovarian failure, and injected iPS through the tail vein. Cells, using their homing effect from both reproductive and endocrine aspects, can repair premature ovarian failure caused by chemotherapy.

8.5.3.2 Establishment of disease models

iPS cells can provide an *in vitro* disease model, which is conducive to reproducing the development process of the disease, studying the pathogenesis of the disease, replacing the distantly related experimental animals for drug development, and clinically replacing the patient for drug dosage and administration, mode adjustment and other aspects of the test. At present, patient-specific iPS cell lines have been established for a variety of genetic diseases, including amyotrophic lateral sclerosis and spinal muscular atrophy, Parkinson's disease, Huntington's disease, and Down's syndrome, etc. It has played an important role in the study of the mechanism and treatment of sexually transmitted diseases. There are certain differences between mouse models commonly used in research and actual human diseases. Compared with primate models, there are differences in the induction stage, culture environment, differentiation, behavioral cognitive evaluation and drug testing of iPS cells. There are significant limitations in long-term studies of neurological disorders.

The patient's autologous cells are an ideal source for establishing disease models. They retain the unique genotype and phenotype of the disease and are the best manifestation of the patient's autologous disease. However, the interaction between cells and the environment can not be fully displayed *in vitro*. Zhou et al. used *Ngn3*, *Pdx1* and *Mafa* to directly reprogram fully differentiated pancreatic exocrine cells into inducible β cells in mice and found that although the obtained β cells could express β cell function essential genes, remodel peripheral blood vessels and secrete insulin, however, is distributed as single cells or small clusters of cells, can not form islet structures, and lacks tissue integrity, limiting the effectiveness of cell use. In addition, pathological models of disease are complex and difficult to fully represent by a single type of cell.

8.5.3.3 Screening and pharmacological evaluation of candidate drugs

During the "screening" phase of drug development, some leading drugs are identified from a predetermined compound library using high-throughput screening (HTS), followed by the selection of several promising drug candidates and the optimization of HTS applications. Sensitivity and reproducibility are prerequisites for the screening and identification of target compounds at this stage. The iPS cell technology not only satisfies the prerequisites well, but also circumvents the interspecies differences that often hinder accurate evaluation of compounds.

At present, the most commonly used disease model is the mouse model. Mice belong to rodents, while humans are primates. There are interspecies differences. iPS cells derived from the patient's own cells are an ideal source for establishing disease models, which retain the unique genotype and phenotype of the disease and are the best manifestation of the patient's own disease. For example, Kobari et al. used iPS cells to analyze the genetic mutation of sickle cell anemia and the function of red blood cells, and used gene sequence analysis to show that all patients' somatic cell-derived iPS cells differentiated into red blood cells with mutations, which brought the research on the mechanism of the disease. For rare diseases, due to the lack of samples, the research on the disease is not in-depth, but patients with rare diseases are eager to obtain new drugs for treatment as soon as possible, and obtain the efficacy of corresponding drugs by using patient-derived iPS cells, and safety practices may soon advance the study of the pharmacological and toxicological effects of compounds in the clinical setting.

Assignments

1. What are stem cells? How many categories can stem cells be divided into according to the source ?
2. What are the morphological and functional characteristics of MSCs?
3. What are the morphological and functional characteristics of ES cells?

4. What are the morphological and functional characteristics of iPS cells?
5. What types of cells can MSCs differentiate into *in vitro*?
6. What types of cells can ES cells differentiate into *in vitro* and *in vivo*?
7. What types of cells can iPS cells differentiate into *in vitro* and *in vivo*?
8. What are the main points of the separation and culture technology of MSC cells?
9. What are the main points of the separation and culture technology of ES cells?
10. What are the main points of the isolation and culture technology of iPS cells?
11. In the process of inducing somatic cells into pluripotent stem cells, which factors can we choose?
12. What are the main problems of MSC technology?
13. What are the main problems of ES cell technology?
14. What are the main problems of iPS cell technology?
15. In which fields can MSC technology be applied?
16. In which fields can ES cell technology be applied?
17. What fields can iPS cell technology be applied to?

(Wang Chunsheng)

Code 8-1 Excerpt from the Report to the 20th National Congress of the Communist Party of China

Chapter 9 Animal Cloning

Summary

Animal cloning is the process by which an entire organism is reproduced from a single cell in a genetically identical manner. However, the term reproductive cloning usually refers to an artificial form of a sexual reproduction that enables the generation of genetically identical organisms and does not involve fertilization. Therefore, cloning is not limited only to somatic cell nuclear transfer (SCNT). It includes parthenogenesis embryos splitting, and nuclear transfer. This chapter mainly introduces the development, production and application of parthenogenesis, embryos splitting and SCNT in domestic animals, especially in pigs.

To date, generating viable cloned mammals from adult cells is technically feasible. Various mammals have been successful produced and the list will only continue to grow with time. The success of many of the proposed applications of the cloning technique obviously depends upon the health and survival of founder animals generated by SCNT. Although some altered phenotypes in cloned animals may be attributable to the cloning process, additional influences may arise from interlaboratory differences in technical proficiency. Thus, it is crucial to have the systematic longitudinal studies of SCNT and more detailed care must be exercised in dissecting the effects of these techniques.

9.1 Parthenogenesis

Parthenogenesis is a form of reproduction in which an egg can develop into an embryo without being fertilized by a sperm, also can been defined as the production of an embryo from a female gamete without any genetic contribution from a male gamete, with or without the eventual development into an adult. Discovery of parthenogenesis may trace back to the 18th century by Charles Bonnet. In 1899, the first case of artificial parthenogenesis was found by Jacques Loeb in frog eggs (Fangerau et al., 2007). In 1936, rabbit eggs were parthenogenetic activated reported by Gregory Pincus (Werthessen et al., 1997).

Parthenogenesis is an appropriate method to study some mechanisms that are also involved in the SCNT process, and an improved parthenogenetic activation of porcine oocytes would facilitate further development of SCNT.

Many methods of oocyte activation have been used in different species including physical, chemical, biological impacts or their combination. The most commonly applied techniques in pigs are using the electrical pulses, connected to later chemical treatment by the protein synthesis blocking chemical cycloheximide (CX) and by the spindle disassembling chemical cytochalasin B (CB).

In addition, the zona pellucida (ZP) that surrounds mammalian oocytes and early preimplantation stage embryos is generally assumed to be indispensable for normal development of early mammalian embryos. However, the presence of ZP is one of the major obstacles during the enucleation step in SCNT procedure, and different protocols have been devised for removing the ZP from oocytes prior to enucleation. The ZP-free somatic cell nuclear transfer method is technologically less demanding compared to the conventional micromanipulation, however, this technique requires a specially adjusted activation and *in vitro* culture system to allow ZP-free embryo development to the compacted morula or blastocyst stage. ZP-free oocytes must be cultured in an environment that does not only prevents blastomere separation, but also individually prevents conglomeration, embryonic aggregation or blastomere separation. To overcome these problems, apart from the traditional droplet culture method, special systems have also been applied including the Well-of-the-Well (WOW) system (Vajta et al., 2005), the glass oviduct (GO) (Thouas et al., 2003) culture system and the polyester mesh (PM) individual culture system. The WOW culture system has proven its value for reconstructed embryos without ZP described in previous studies in cattle, pig and mouse. Therefore, PA both of ZP-free oocytes and ZP-intact oocytes in pig are described in the following.

9.1.1 PA of ZP-free oocytes in pigs

1. Collection and maturation of pig oocytes

In vitro maturation (IVM) of oocyte is performed as described earlier. Briefly, cumulus-oocyte complexes (COCs) are aspirated from 2-6 mm follicles of slaughterhouse-derived sow ovaries with an 18-gauge needle using vacuum suction. COCs are selected according to their morphological characteristics, i.e. at least three layers of compact cumulus and even cytoplasm. After washing twice in HEPES-buffered tissue culture Medium 199 plus 0.3% heparin, 5% amphotericin and 10% cattle serum (CS), the COCs are cultured in groups of 50-60 for 42-44 h in 4-well dishes (Nunc, Danmark) with bicarbonate-buffered TCM-199 supplemented with 10% (volume ratio) CS, 10% (volume ratio) pig follicular fluid, 10 IU/mL PMSG and 5 IU/mL hCG at 38.5 ℃ with 5% CO_2 in air with maximum humidity.

2. WOWs preparation

Standard 4-well dishes are filled with 400 μL of porcine zygote medium 3 (PZM-3) and covered by 400 μL mineral oil. WOWs are made by pressing an unheated darning needle into the bottom of the dish to create microwells (approx. depth 130 μm, width 150 μm; Fig. 9-1A, B). In each well of the dish, 25-30 microwells are made.

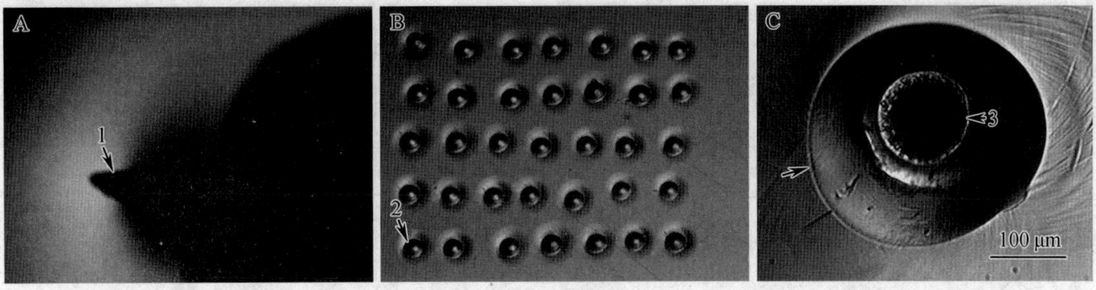

A. An unheated darning needle is pressed into the bottom of the dish to create microwells

B. WOWs made in the dish C. One ZP-free oocyte cultured in a WOW

1. A needle 2. A WOW 3. A ZP-free PA oocyte

Fig. 9-1 ZP-free PA oocyte cultured in WOWs

3. ZP digestion

The optimized activation protocol applied for ZP-free oocytes was used as described previously but with slight modifications (Booth et al., 2001). Briefly, oocytes are stripped of cumulus cells in 1 mg/mL hyaluronidase dissolved in HEPES-buffered TCM-199 and washed three times in T2 [T for TCM-199, the number denoting percentage (volume ratio) of CS supplement, here 2%]. Matured oocytes with a visible polar body, indicating successful nuclear maturation, are selected and placed in a 20-μL T2 droplet covered with mineral oil at 38.5 ℃. After a short equilibration, the ZP is removed by incubating the oocytes in 3.3 mg/mL pronase in T33 for 1 min. ZP-free oocytes are then moved into T10 medium until to activation.

4. Electrical stimulation

The electrical stimulation is delivered with an electrofusion machine to a chamber with two parallel platinum wire electrodes (BTX microslide 0.5 mm fusion chamber, model 450; BTX, San Diego, CA, USA) overlaid with activation medium (0.3 mol/L mannitol, 0.05 mmol/L $CaCl_2$, 0.1 mmol/L $MgSO_4$, and 0.01% PVA, pH adjusted to 7.8). After washing in T2 droplets at least three times and a short equilibration in the activation medium, 10-15 ZP-free oocytes are transferred into the chamber in each working round. With a 4 V/cm alternating current (AC) and 700 kHz, oocytes are aligned at the north wire of the chamber and a single direct current (DC) pulse with 0.86 kV/cm is applied for 80 μs.

5. Chemical post activation

For post-activation, the oocytes are transferred into one well of a 4-well dish in 400 μL PZM-3 supplemented with 5 μg/mL CB and 10 μg/mL CX, and covered with 400 μL mineral oil for 4 h at 38.5 ℃ in 5% CO_2, 5% O_2 and 90% N_2 with maximum humidity.

6. *In vitro* culture of ZP-free PA embryos

Four hours after activation, the ZP-free oocytes are thoroughly washed, then incubated in the WOW culture system (Fig. 9-1C) in 400 μL PZM-3 medium covered with 400 μL mineral oil for 5-6 d in the Galaxy R CO_2 incubator at 38.5 ℃ in 5% CO_2, 5% O_2 and 90% N_2 with maximum humidity, until developed to blastocysts (Fig. 9-2).

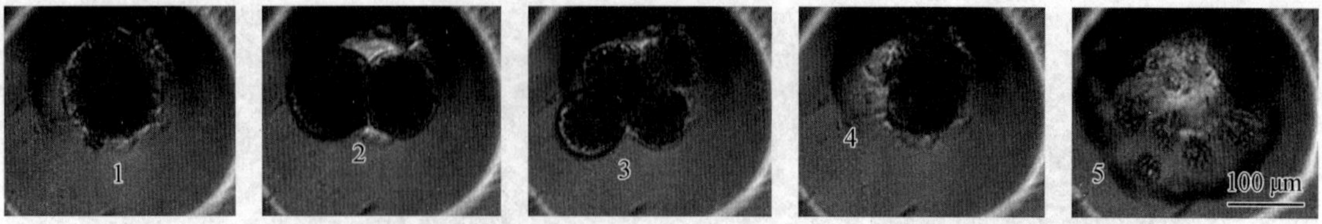

1. 1-cell 2. 2-cell 3. 4-cell 4. Morula 5. Blastocyst

Fig. 9-2 A ZP-free embryo cultured in a WOW from 1-cell to blastocyst

9.1.2 PA of ZP-intact oocytes in pigs

All of the procedures to produce ZP-intact oocytes is same as described above, but ZP-intact matured oocytes are activated for 80 μs by two different electrical pulses, 1.26 or 0.86 kV/cm, respectively and culture is performed in groups (Fig. 9-3).

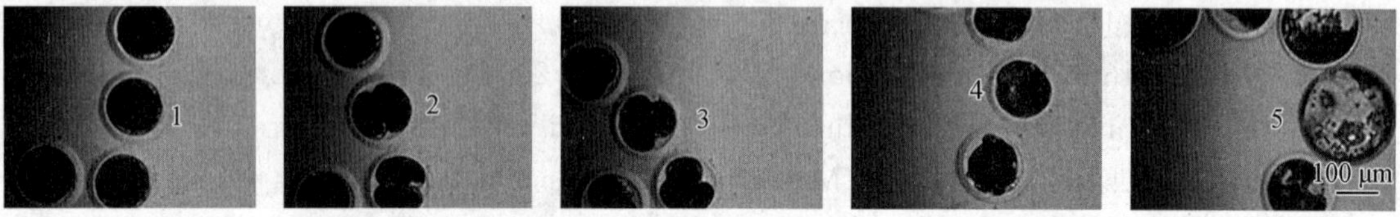

1. 1-cell 2. 2-cell 3. 4-cell 4. Morula 5. Blastocyst

Fig. 9-3 ZP-Intact embryo cultured in a WOW from 1-cell to blastocyst

9.2 Embryos splitting

The first identical twins were produced by splitting of pre-compaction embryos in sheep (Rowson et al., 1964). At that time, *in vitro* culture system of early embryonic developmental stages had not been established, therefore, the pair of the separated blastomeres from one sheep embryo was inserted into an empty ZP, respectively. Several days, the pair of blastocysts were transferred into recipients with born of the identical twins. Actually, identical bovine twins were also produced by the same procedure. With the development of micromanipulation, half bisecting of compacted morulae or blastocysts has been successfully produced by using a fine glass needle, of which the bisected halves were transferred directly to recipients for offspring production, and monozygotic twin bovine, sheep, goats, and rabbits have been successfully produced. However, micromanipulation of pre-compaction embryos is laborious and techniques experiences needed, therefore, simplified procedures were developed. When handmade cloning (HMC) procedure has been established (Vajta et al., 2003; Peura et al., 2003), splitting of embryos could be performed by a metal blade under an inverse microscope at the morula or blastocyst stage.

9.2.1 Basic procedure to split embryos

To split preimplantation embryos into two halves, ZP of the embryo should be removed firstly. Commonly, ZP of embryos to be separated could be digested as described in section 9.1.1. Without ZP, blastomeres of early developmental stages embryos can be separated directly by pipetting gently (Fig. 9-4).

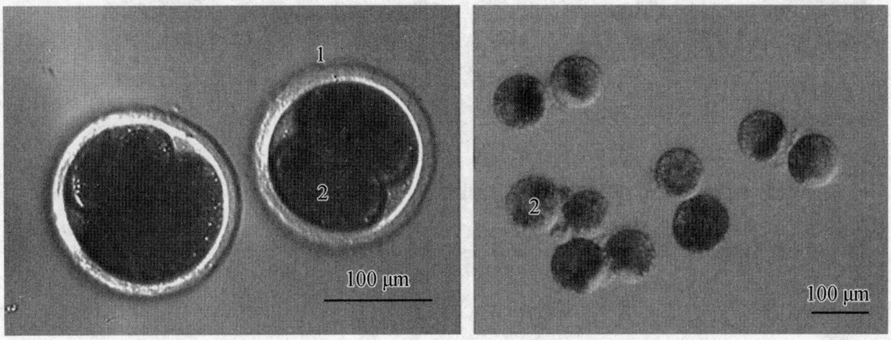

1. The embryo with ZP at 2-cell stage 2. One blastomere from the 2-cell embryo

Fig. 9-4 Separation of a 2-cell embryo

Because the necessary of ZP for the development of pre-compaction blastomeres, the ZP-free half embryo should be *in vitro* cultured with a right condition to maintain their development. WOWs as described in section 9.1.1. have developed, in which the ZP-free embryos can be *in vitro* cultured from 1-cell directly to blastocysts with the comparative embryonic developmental efficiency.

In addition, when embryos developed to blastocysts, the separation of embryos can be performed by microblade handling by micromanipulator (Fig. 9-5).

Therefore, the separation of embryos can be carried out at any stage of early embryonic developmental stages. In addition, triplet, fourths or more could be successfully produced by embryos separating at early embryonic developmental stages. For example, an 8-cell embryo could be separated into four pairs (Tsunoda et al., 1987) and then each pair could be *in vitro* cultured in WOWs for further development into blastocysts, to be followed up by the transfer of blastocysts into recipients for production of the fourths. Quarters of calves have been reported produced successfully.

9.2.2 Significance and application of embryos splitting

Genetic identical twins can be achieved by embryos splitting, therefore, the embryo bisection is an accepted and even commercially used technique in breeding of large domestic species to propagate rapidly animals with high genetic value. Splitting of embryos has been especially used under commercial field conditions to increase the production of calves by

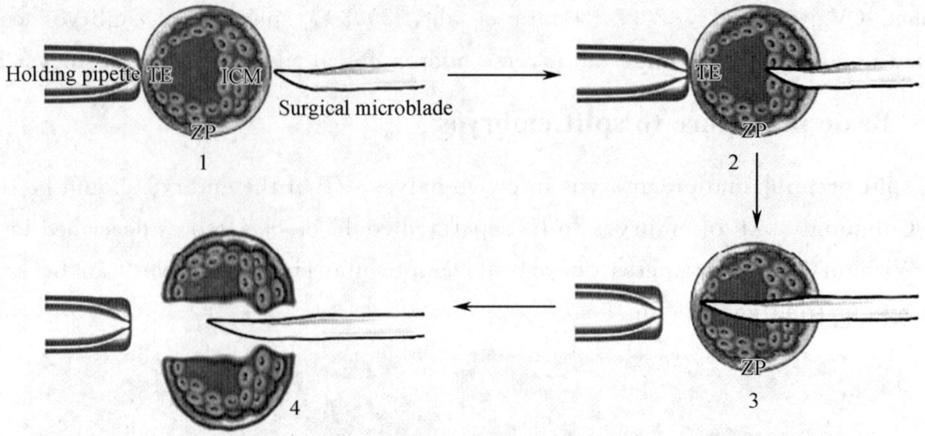

1. Holding micromanipulator 2. Blastocyst to be separated
3. Place of ZP to cut 4. Bisection of the blastocyst by micro-blade

Fig. 9-5 Sketch map of embryos splitting by micromanipulator at blastocyst stage
(Rahbaran et al., 2021)

embryo transfer. A total of 132 embryos recovered from 17 superovulated donor cows after estrus, 74 of which were selected and assigned to directly transferred (44 whole embryos) and bisected (30 embryos for bisection), respectively. Whole and demi-embryos without ZP were non-surgically transferred, that only one whole or demi-embryo was transferred to each recipient. Results showed that no significant difference between the pregnancy rates of whole embryos with 63.6% (28/44) and bisected embryos with 74.6%. Using the embryo bisection technique Forty-three calves including the 14 sets of identical twins were obtained from 30 original embryos (143.3%) separated into four pairs (Seike et al., 1989).

However, even the theoretically achievable productivity is very limited (two offspring from one embryo) and is further compromised by the low survival rates of the bisected halves. The developmental competence of one part of the fourth is very low with decreased cell number per blastocyst. In the limited available reports, the loss of post-parturition piglets obtained from separated embryos is higher than that of piglets produced after natural mating, but body weight of born kids is within the normal range of natural mating. In addition, abnormalities such as cryptorchids, atresia, and hernia formation are high in piglets that developed from bisected embryos.

9.3 Nuclear transfer

9.3.1 Brief introduction of nuclear transfer

The main procedure of nuclear transfer (NT) is to introduce the nucleus from a cell into an enucleated egg cell, whose own nucleus is removed. The donor nucleus used for NT may come from an undifferentiated embryonic cell or from a differentiated adult cell which is called somatic cell; in the latter case, the technique is called SCNT.

The concept of NT was first conceived in 1928 by German embryologist Hans Spemann (Beetschen et al., 2004), who initially experimented with transferring salamander embryonic cell nuclei into egg cells. At that time, blastomeres from undifferentiated early-stage embryos were mainly used as the nuclear donors and it was believed that only cells derived from early-stage embryos could be used as donors for NT in mammals. However, in the 1950s American scientists Robert Briggs and Thomas King successfully cloned tadpoles using nuclear transfer of frog blastula cells, which had lost some of their totipotent properties (Berardino et al., 2004). Later that decade British developmental biologist John B. Gurdon generated cloned tadpoles from differentiated frog intestinal cell nuclei. Then the possibility to generate clones through the transfer of nuclei from differentiated cells into enucleated egg cells was believed.

9.3.2 Somatic cell nuclear transfer

In 1996, a sheep named Dolly was generated by somatic cell nuclear transfer (SCNT) with adult mammary gland cell, which provided the possibility that cloning animals could be produced by differentiated adult somatic cell (Wilmut et al., 2007). However, the only cloned sheep was in doubt initially, but one year later, the birth of bovine and mouse cast away that doubt (Kato et al., 1998; Wakayama et al., 1998), and confirmed the belief that mammalian cloning by using adult somatic cells was indeed possible. Just like Pandora's opening of that evil spirit box, different species were cloned from adult somatic cells in the following years, including pig, goat, cat, rabbit, mule, horse, rat, dog, deer, monkey and so on. This research on SCNT not only stimulated renewed interest in the ability of egg cells to reprogram differentiated nuclei, but also inspired the development of SCNT, which has become an important technique in stem cell research and in understanding the cellular mechanisms controlling embryonic development and nuclear reprogramming.

All presently applied techniques of SCNT include the following steps: ① nucleus removal from a matured oocyte; ② transfer of a donor somatic cell derived from various tissues of an adult, young animal, or fetus; and ③activation of the reconstructed embryo for the following development. Depending on species to clone and equipment qualified in laboratories, all of the above three steps can be performed in various ways. But two main different enucleation techniques have been applied: traditional cloning (TC) by using a micromanipulator and handmade cloning (HMC) by using a bisection blade (Vajta et al., 2001).

9.3.2.1 TC by micromanipulation

TC by micromanipulation method can be performed by microinjection to remove nucleus in the matured oocyte and inject the donor somatic cell into the enucleated cytoplasm, following by electric and/or chemical activation. In brief, a flow diagram of the procedure is normally used in TC by the microinjection.

1. Preparation of donor somatic cells

Donor somatic cells that reach confluence which synchronizes in G_0/G_1 phase are prepared at the day of cloning. In brief, the pig fetal fibroblasts established from day 24 pig fetus are cultured in Dulbecco's modified Eagle's medium (DMEM) supplemented with 10% fetal calf serum (FCS) at 37 ℃ in a humidified atmosphere containing 5% CO_2. After 1-week *in vitro* culturing, the confluent pig fetal fibroblasts are washed twice with Ca^{2+} and Mg^{2+}-free PBS, and those cells are further incubated at 37 ℃ with 0.5 g/L trypsin. Three minutes later, 500 μL DMEM supplemented with 10% FCS is added with pipetting. Subsequently, the medium containing the cells is transferred to an Eppendorf tube for centrifugation for 3 min. Then the supernatant is discarded and 500 μL T2 is added for the suspension of fibroblasts, which is stored at room temperature (22-24 ℃) until fusion.

2. Enucleation

The matured oocytes with extruded first polar body (PB1) are collected from superovulation or from *in vitro* maturation. Subsequently, the enucleation is carried out by micromanipulation in TPCB medium [HEPES buffered medium supplemented with 2% (volume ratio) FBS and 5 μg/mL CB]: oocytes are handed by one arm of micromanipulator, while PB1 with a small amount of cytoplasm is aspirated out by a glass pipette fitted to another micromanipulator (Fig. 9-6). The rest of the oocyte is assumed as the cytoplast which will be used for cytoplasm recipient.

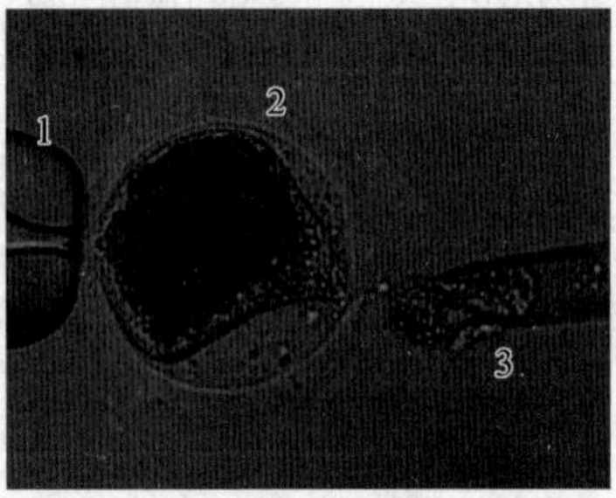

1. Holding hand 2. Oocyte to be enucleated 3. Nucleus with small part of cytoplasm aspirated out

Fig. 9-6 Enucleation in TC by aspiration

3. Microinjection for nucleus transfer

One donor cell is aspirated into the glass pipette used for enucleation, and is directly injected into the cytoplast (Fig. 9-7).

4. Activation

The reconstructed oocytes are activated by electric pulse in the F^+ activation medium, with a single DC pulse of 0.86 kV/cm for 80 μs on the electrode wire (Fig. 9-8), following the chemical post activation in PZM-3 supplemented with 5 μg/mL CB and 10 μg/mL CX for 4 h.

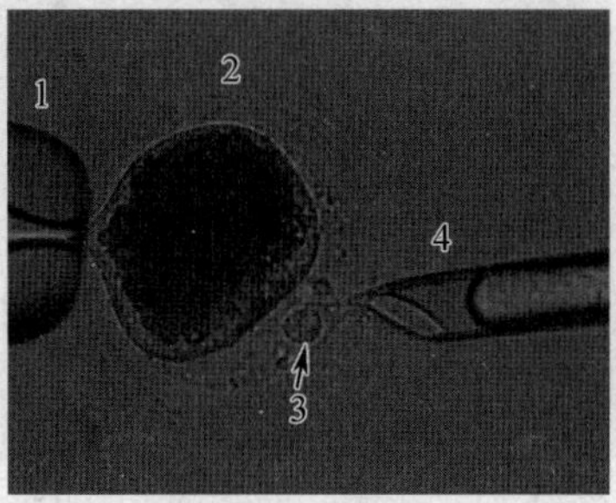

1. Holding hand 2. Enucleated oocyte 3. Donor cell being injected 4. Glass needle for microinjection

Fig. 9-7 Microinjection of donor cell

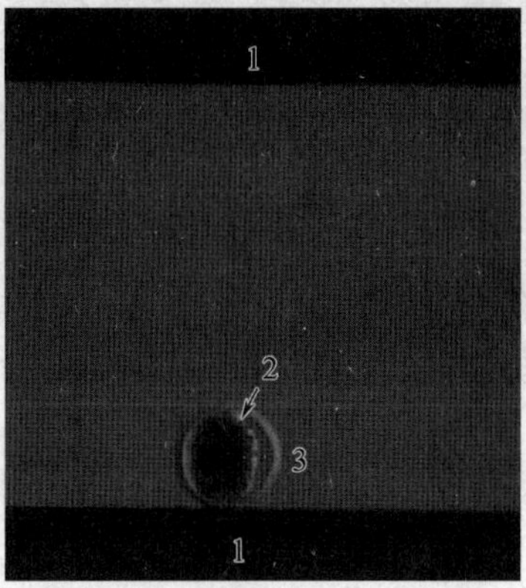

1. Two electric lines 2. The injected donor cell 3. The reconstructed embryo

Fig. 9-8 Electric activation

5. *In vitro* culture

Activated embryos in group of 30-40 are transferred into one well of 4-well dish in 400 μL PZM-3 medium covered with 400 μL mineral oil in the CO_2 incubator at 38.5 ℃ in 5% CO_2, 5% O_2 and 90% N_2 with maximum humidity, until developed to morula and/or blastocysts.

9.3.2.2 HMC

With the development of the biotechnology, another SCNT technique named HMC technology has become a novel method that eliminates the complicated operation and expensive micromanipulator.

1. Partial digestion of ZP

After the removal of cumulus cells performed as described in section 9.1.1, the ZP of the selected pig oocyte with a PB1 is partially digested in 3.3 mg/mL pronase (Fig. 9-9).

2. Oriented bisection for enucleation

Subsequently, oocytes are lined in a droplet of TPCB medium, and then oriented bisected is carried out according to the PB1 by a self-made blade (Fig. 9-10).

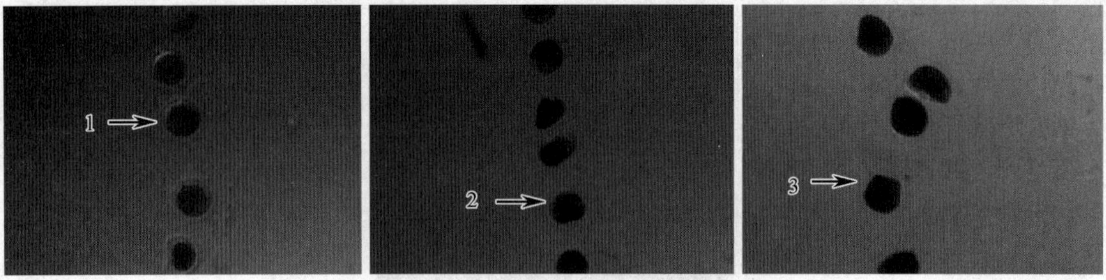

1. Light digestion 2. Right digestion 3. Over digestion

Fig. 9-9 Partial digestion of ZP

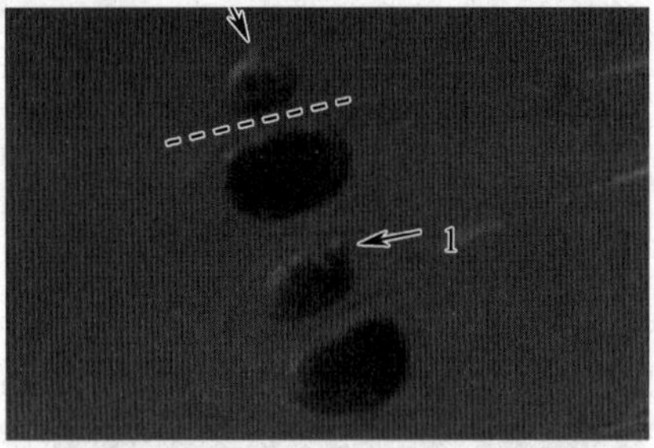

1. The position of PB1

Fig. 9-10 Oriented bisection of pig oocytes

3. Fusion and primarily activation

The semi-oocytes far away the PB1 are collected for the putative cytoplasm. After 2-5 s equilibrium in 1 mg/mL phytohemagglutinin (PHA), the cytoplasm is one-to-one adhesion to the somatic cells that has been pre-balanced in T2. Subsequently, the pair of cytoplasm-somatic cell are lined on the electricwire covered F_0 activate solution (0.3 mol/L mannitol supplemented with 1 mg/mL PHA) (Fig. 9-11), and then a single DC pulse of 2.0 kV/cm for 9 μs is performed for the first fusion of donor cell and cytoplast.

1. Two electric lines 2. Cytoplast from HMC 3. Donor somatic cell

Fig. 9-11 Fusion of donor cell and cytoplast in HMC

After 1 h, completely fused pair is selected for second fusion with another cytoplast by using an F^+ activate medium with a single DC pulse of 0.86 kV/cm for 80 μs on the electrode wire (Fig. 9-12).

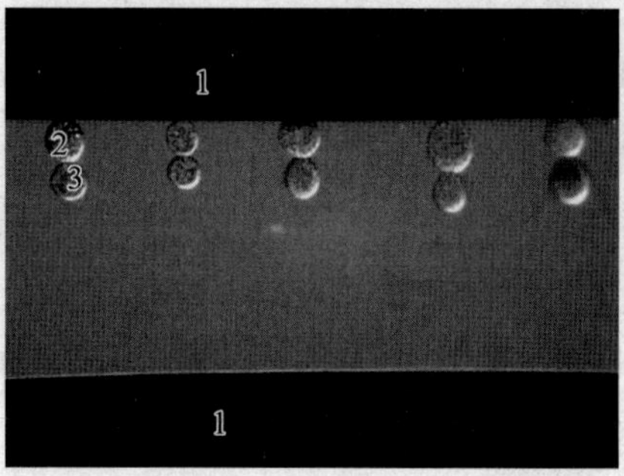

1. Two electric lines 2. Fused pairs 3. Another cytoplast

Fig. 9-12 Second fusion of the fused pairs with another cytoplast to reconstruct a cloned embryo

4. Chemical post activation and *in vitro* culture

In 15 min, the reconstructed embryos are collected for chemical activation as described in section 9. 1. 1. Briefly, reconstructed embryos are transferred into one well of a 4-well dish in 400 μL PZM-3 supplemented with 5 μg/mL CB and 10 μg/mL CX. Four hours later, cloned embryos are thoroughly washed, then incubated in the WOW culture system in 400 μL PZM-3 medium covered with 400 μL mineral oil in the CO_2 incubator at 38. 5 ℃ in 5% CO_2, 5% O_2 and 90% N_2 with maximum humidity, until developed to morula and/or blastocysts.

Unlike in other animals like sheep and cattle, multiple embryos are needed to establish and maintain pregnancy in recipient pigs. In addition, because of the generally low quality of cloned embryos, a relatively large number of reconstructed embryos are recommended for embryos transfer in pig. Normally, cloned embryos by TC selected at 2- or 4-cell stages and cloned embryos by HMC selected at morula and/or blastocysts are transferred into the oviduct of surrogates whose estrus has been synchronized.

9.4 Main problems of animal cloning

After the first success, the number of offspring is increasing rapidly, as well as the efficiency of the whole procedure. However, SCNT remains challenging, and its inefficiency in producing live offspring limits its application for animal production. Even different species have been successfully cloned, low efficiency of SCNT and high percentages of abnormalities are commonly observed in offspring born after SCNT. In sheep and cattle, high postnatal mortality, placental abnormalities, as well as the high percentages of stillbirths, miscarriages, and malformation of fetuses have been reported. Although less abnormality is reported or observed in cloned piglets, the birth weight, placenta weights and growth rates are usually reported in cloned pigs.

In addition, that many of the anomalies remain hidden at the early developmental stages, and are manifested only later. So, seemingly normal, morphologically intact embryos should be selected and are transferred into recipient animals to get healthy offspring. However, possibilities are very limited to produce exclusively healthy embryos or to select the right ones at the early stage of development, before the embryo transfer.

9.5 Significance and application of animal cloning

Regarding domestic animals, the simplest application is their high genetic value of breeding animals, such as increased production of superior breeding stock and conservation of genetic resources. Somatic cells are of great significance for the expanding propagation, breeding and rescue of endangered species of livestock. By using SCNT, the breed can be recovered and propagated rapidly in a short time, and its once-underestimated genetic values can be used for future animal breeding applications. The most practical application of SCNT is in the reproductive cloning of farm animals that have exceptional qualities, such as the ability to produce large quantities of milk. Another important application for genetic value is the conservation of endangered species or breeds. Fibroblasts derived from tissues of animals in danger of extinction can be cryopreserved and be used for centuries as donors.

Another most disputed application of SCNT is the stem cell research. As described above, donor cells could be established from the skin obtained with painless procedure, subsequently being used for SCNT to produce cloned embryos. When reconstructed embryos develop into blastocysts, some cells from inner cell mass can be isolated and *in vitro* cultured in infinitely in the undifferentiated state. Finally, embryonic stem cells have been established, if required, differentiation of these cells can be induced under certain conditions with some factors or hormones. The possible and expected application of these stem cells might be in treatment of various diseases, if legal barriers do not hamper it.

With the development of transgenic and gene editing, SCNT opens the way to get reconstructed embryos with random gene transfer, or even targeted insertion or deletion of genes. The benefits can be very diverse, including the production of new breeding of animals, production of pharmaceuticals for human medicine, or human disease models. Near-human-sized diseases models are needed to validate the methods and to facilitate further technological development. Genetic modified model animals are produced by SCNT more and more for human disease models such as transgenic mouse and pigs.

Xenotransplant is transplantation of organs, valves, or other medical products (like blood or stem cells) between species that may or may not be similar to each other. The most potentially beneficial uses of xenotransplants would be to transplant organs from animals to humans. In the start of 2022, an American man with life threatening heart disease who received a genetically modified pig heart during a groundbreaking operation was recovering well. But unfortunately, the patient died in almost two months later, due to the gene edited pig heart given to the patient infected with a pig virus. Another astonishing step has been taken that the pig's kidney was attached to a brain-dead patient who was sustained on a ventilator, with the family's consent. The transplanted kidney was obtained from a pig genetically engineered to grow an organ unlikely to be rejected by the human body (Dolgin, 2021). The "a watershed moment" surgery was hailed as a milestone. Hence, xenotransplantation and a steady supply of organs from transgenic cloned animals, which could eventually include hearts, kidney, lungs, and livers, would offer a lifeline to the transplant waiting lists, and genetically engineered pigs can potentially be a sustainable, renewable source of organs.

Assignments

1. What are the main factors affecting parthenogenetic activation?
2. What's the commercial application of embryos splitting?
3. What are the steps of animal cloning?
4. What are the pros and cons of animal cloning?
5. What are the applications of animal cloning?

(Li Juan)

Chapter 10 Transgenic Animal

Summary

Transgenic animal technology, a fast growing biotechnology, can provide a better platform to develop transgenic animals for breeding new varieties and promote the development of medical sciences, livestock production, and other fields. Here, we first introduce the conception, development and application of transgenic animal technology. And then, we introduce the main technical links of transgenic animal technology such as targeting gene selection, gene modification, expression vector construction and so on. Finally, we discuss the main question and the developmental trend in transgenic animal technology.

Transgenic animal technology, a fast growing biotechnology, have been changing the ways of science and human life in every aspect by integrating exogenous genes into the animal genome which contain new traits or beneficial expression products (Moro et al., 2020). Briefly, the foreign genes need stable and integrated in the genome and all tissue cells are integrated with foreign genes which can be inherited and expressed by offspring. Another method is to inactivate the gene in the body and make it lose its function using embryonic stem cells—the animals thus bred also belong to transgenic animals (Lanigan et al., 2020). Therefore, in a broad concept, transgenic animals can be defined as animals in which new or altered genes have been experimentally inserted into their genome by genetic engineering techniques. However, in the preparation of transgenic animals, foreign genes may only be integrated into the genome of some tissue cells of animals, and partial tissue cells integrated with foreign genes of animals are called chimera mosaic animal (Rossant and Tam, 2004). Generally, the first-generation transgenic animals prepared by embryonic stem cell method or retroviral vector method are chimeric animals, and 20% of the first-generation transgenic animals obtained by microinjection method are chimeric animals.

10.1 Point and application of transgenic animal technology

Although transgenic animal technology is widely applied in the world, there are many milestone events accompanied nearly 50 years (Fig. 10-1). In 1980, Gordon et al. first described the introduction of a foreign gene into mice using pronuclear injection into oocytes which was an approach first in transgenic animal technology. And then, Ralph Brinster and Richard Palmiter integrated growth hormone (GH) genes into fertilized mice and obtained "super mice" that grew at a rate two- to four-fold greater than control littermates, reaching a mature weight twice of normal mice in 1982, and subsequently influenced the emerging field in a most compelling manner for both basic and applied sciences. The first transgenic sheep and pigs were obtained via microinjection of DNA into pronuclei in 1985 (Hammer et al., 1985). Meanwhile, to improve the transgenic efficiency and precise control of gene expression, transgenic methods especially about vector and injection were also upgrading such as microinjection, sperm vector, intracytoplasmic sperm injection (ICSI)-mediated transgenesis, somatic cell nuclear transfer (SCNT), lentiviral vectors and primordial germ cells (PGCs). In 1996, cloning of sheep Dolly with somatic donor cells nuclear transfer using adult somatic donor cells affected dozens of mammals such as rabbits, sheep, pigs, cows, chickens, fish, cats etc (Cibelli et al., 1998). Furthermore, artificial chromosome was used to create trans-chromosomal cattle in 2002 (Montoliu, 2002). Then, in 2007, prion protein gene (*PRNP*) was knocked-out in cattle via SCNT which may be a useful model for prion research and could provide industrial bovine products free of prion proteins (Gurgul and Słota, 2007). Later, gene targeting techniques with genome altering frameworks such as transcription activator-like effector nucleases (TALEN) and clustered regularly interspaced short palindromic repeats (CRISPR) significantly improved productivity to create transgenic animal' models for biomedical and agricultural purposes. Notably, many laboratories adopted this technique for targeting the specific gene using CRISPR-Cas9 targeted multiple sites as a genome editor in livestock species such as cattle, pig and goat and other laboratory animals (Jabbar et al., 2021). The applications of CRISPR in future will make significant innovations in health and medicine to increase knowledge about the disease and discovering new treatments. Thus, CRISPR has made significant innovations in genetic engineering, and

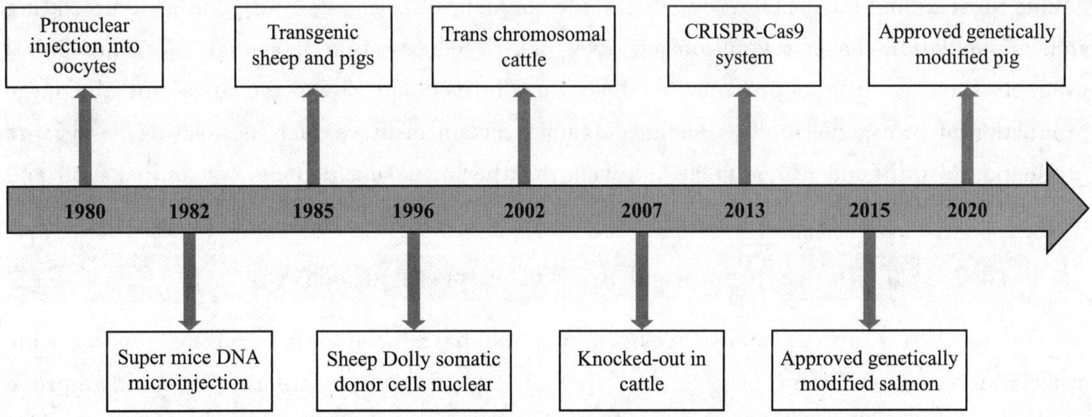

Fig. 10-1 Point and application of transgenic animal technology

it is now considered the most widely used genome modification technique in molecular biology. On the other way, transgenic animal technologies as a huge industry have been shown to have great commercial value in medicine and livestock around the world especially in some mainly scientific powerhouse.

In November 2015, the USA Food and Drug Administration (FDA) approved the entry of genetically modified salmon which, grown rapidly to the market, became the first genetically modified animal in the world. And then came the second genetically modified pig referred to as GalSafe pigs in December 2020. This intentional genomic alteration resulted in no detectable α-gal sugar on the surface of the pigs' cells. The approval covered both food use and potential medical use of tissues from the pigs. The GalSafe pig-derived tissues may potentially provide a source of porcine-based materials to produce human medical products that are free of detectable α-gal sugar (Katiyar et al., 2021). Therefore, genetically modified animal technology is changing the status-quo of research and development in biomedicine, agricultural production, environmental protection, biomaterials, and even the entire life sciences.

10.1.1 Application of transgenic animal technology in livestock

In livestock, transgenic animal technology can exactly modify the genome of the animal and remarkably promote animal traits such as growth, meat, milk yield and so on. The traditional animal breeding are revolutionized and joined at a higher level of stage breeding by molecular writing (BMW) which brings great opportunities for species improvement or animal modification in future (Liu et al., 2018). On this basis, China provided about 24 billion yuan for the project of "National Major Special Fund for New Cultivar Breeding of Genetically Modified Organisms" to breed new varieties of transgenic animals and plants as a national medium- and long-term development plan in last 15 years. Up to now, Remarkable achievements have been made in China's implementation of major transgenic projects in the animal field such as anti-tuberculosis cattle, "human emulsified" cows, "super-pigs" with double muscles and buttocks. Altogether, the sophistication and revolution in genome editing and manipulation have revolutionized livestock by harvesting essential biotechnological products such as drugs, proteins, and serum. It has laid down the areas for the large production of transgenic food, resistance against certain diseases such as mastitis, and large production of milk and leaner meat. Nowadays, the increasing demand for animal food and protein is fulfilled using transgenic animal technology (Table 10-1).

10.1.1.1 Improving animal production performance

In the 21st century, many livestock genome have been sequenced by genome-wide association study and some target genes are found to play important role or could improve production performance. However, to obtain excellent offspring, traditional animal breeding needs several generations which requires a considerable expenditure of time and energy.

Table 10-1 A selective list of gene-edited livestock for different purposes

Species	Transgenic trait	Altered genes	Gene transfer method
Mouse	Cerebral amyloidosis, dystrophic synaptic boutons and robust gliosis	*Aβ42* knock-in	Co-injected into male pronuclei of oocytes
	Mice on a calorie-restricted diet	*SIRT1* knock-in	
	Weight loss as well as virus replication in the lungs infected with SARS-CoV-2	*hACE2* knock-in	CRISPR-Cas9
Pig	tg-FⅨ contains no detectable Neu5Gc	Recombinant human factor Ⅸ (*tg-FⅨ*)	DNA microinjection
	To study progressive growth of cysts in kidneys	*PKD2*	Somatic cell nuclear transfer
	Protecting the pigs from porcine reproductive and respiratory syndrome virus	*CD163*	CRISPR-Cas9
	Appropriate for xenotransplantation	*α1,3GT*	CRISPR-Cas9
Goat	Human G-CSF is up to 50 μg/mL in transgenic goat milk	Human granulocyte colony stimulating factor (*G-CSF*)	DNA microinjection
	Secrete recombinant human lactoferrin (rhLf) up to 0.765 mg/mL	*hLf*	DNA microinjection
	Biallelic alterations in the *MSTN* helps in the development of more muscular tissues	*GDF8*	CRISPR-Cas9
	Production of skeletal muscles enhanced	*FGF*, *GDF8*	CRISPR-Cas9
Cattle	Recombinant lysozyme is up to 25.96 mg/L, as measured by radioimmunoassay	Recombinant human lysozyme	Somatic cell nuclear transfer
	Recombinant HSA is expressed in the milk at about 4-8 mg/mL	Recombinant human serum albumin (*HSA*)	Somatic cell nuclear transfer
	Resistance to tuberculosis increases	*NMRAP1*	CRISPR-Cas9
	99.9% *MSTN* mutation rate in offsprings with a double-muscling phenotype	*MSTN* mutation	CRISPR-Cas9
Chicken	High concentration of human erythropoietin (hEPO) in the ovalbumin	*hEPO*	Pseudotyped lentivirus vector microinjection
	hEC-SOD protein is expressed in the egg white and showed antioxidant activity	Human extracellular superoxide dismutase (*EC-SOD*)	Recombinant lentiviruses injected
	Deletions of MSTN	*MSTN* knock-out	CRISPR-Cas9

Transgenic animal technology can directly transfer required excellent genes and obtain excellent transgenic traits such as growth rate, reproduction rate, lean meat rate, gross yield, etc., and finally cultivate new varieties (lines) with excellent quality that meet people's needs. For example, in pig, myostatin gene (*MSTN*) restricts the growth of skeletal muscles. When the *MSTN* knocked-out, Er hua lian pigs with improving growth performance exhibited partial double-muscled phenotype such as prominent muscular protrusion, wider back and hip compared with wild-type piglets. In cattle, the main targets on the genetic modifications increased the generation (e.g., knocking out of *MSTN* improved growth), elimination of substances that cause allergy (e.g., knocking out of beta-globulin gene) and improved health (e.g., polled or having no horns) characteristics (Jabbar et al., 2021). In goat, inserting the thymosin beta 4 gene (*Tβ4*) into the gene position of chemokine receptor 5 of cashmere goats using CRISPR-Cas9 increased cashmere production without influencing the quality and fines (Li et al., 2019).

10.1.1.2 Disease resistance

Livestock disease occurrences adversely affect the production and animal welfare, and have an impact on both human health and public perception of food-animals production. So, efforts are seriously made to continuously explore the effective disease control approaches for the production of safe animal-originated food. Furthermore, there is a wide range of applications of transgenic animal technology by introducing the genomic marker responsible for disease resistance or immunocompetence in livestock that includes the production of resilience against disease. *M. bovis* infected more than 5 million cattle around the globe. The incidence rate of bovine tuberculosis raised vastly, which caused movement restriction of 10% cattle and slaughtering of animal's costs of £91 million. However, when the *NRAMP1* gene was inserted in the cattle's genome, cattle became resistant to bovine tuber-culosis caused by *M. bovis* (Gao et al., 2017). In pigs, the knock-out of CD163 receptors produced genetically engineered pigs that were resistant to porcine reproductive and respiratory syndrome virus (PRRSV) through the CRISPR-Cas9 technique. All the livestock with CD163 receptors knocked-out were protected against PRRSV and never showed symptoms from this infection (Jabbar et al., 2021). Therefore, the technical ease and low cost of transgenic animal technology will enhance the techniques application for producing disease resistance livestock.

10.1.1.3 Environmental protection

As we know, farms can lead to emissions of greenhouse gases (GHG) and ammonia (NH_3) in air, nitrates and antibiotics pollution in water bodies, and phosphorus, antibiotics and heavy metals in the soil. In order to have a sustainable agricultural environment, new methods must be placed in the highest priority in the livestock industry. Nitrogen and phosphorus emissions pollution produced by animals is a very serious problem in agriculture.

Canadian and Chinese scientists integrated the *E. coli* gene that can produce phytase with the murine gene that can control the secretion of salivary glands in part or required enzymes (mannanase, xylanase and phytase) in their breeding of a transgenic pig (somatic cell nuclear transfer technology) which leads to a 75% or 46.2% reduction in phosphorus content in feces, respectively. Therefore, the phytase produced by its saliva can effectively digest plant phosphorus, which is of great benefit to environmental protection, hence the so-called "environmental protection pig".

10.1.2 Application of transgenic animal technology in biology

10.1.2.1 Gene function annotation

There are numerous tools available to modify the genetic makeup of experimental animal. Of those animals, mouse is a powerful and indispensable experimental model in fundamental research via introduced DNA sequences (e.g., genes) of interest into the germ line genome. Briefly, the DNA sequences can be integrated into the genome randomly or into a specific locus by homologous recombination, in order to: ①delete or insert mutations into genes of interest to determine their function, e.g., T cell receptor-β (TCR-β) knock-out mice that lack T cells; ②introduce human or other species genes into the genome of mice to generate animal models enabling study of specific genes and diseases, e.g., mice susceptible to infections by human-specific pathogens of interest; ③introduce individual genes or genomes of pathogens (such as viruses) in order to examine the contributions of such genes to the pathogenesis of the parent pathogens; ④introduce reporter genes (green fluorescent protein or bringing new molecular probe) that allow monitoring *in vivo* or *ex vivo* the expression of genes of interest (Jabbar et al., 2021). Furthermore, the use of recombination systems, such as Cre/LoxP or FRT/FLP, enables conditional induction or suppression of gene expression of interest in a restricted period of mouse's lifetime, in a particular cell type, or in a specific tissue.

10.1.2.2 Protection of endangered animals and germplasm resources

The human footprint on our planet is currently threatening biological diversity across habitats which lead to the rate of extinction in the last century conservatively estimated to be 22 times faster than the historical baseline rate. Today, genome-scale data will improve the capacity of resource managers to protect species. To wild animals, the primary application of genetic engineering to wild species involves cloning involved in transgenic animal technology. This technology could be applied to either extinct or endangered species; for example, there have been plans to clone the extinct thylacine and the woolly mammoth. To the farm animals, China has rich livestock germplasm resources such as Erhualian pig, Hu sheep, Nanyang cattle and so on. However, those valuable germplasm resources are threatened by foreign commercialized varieties. Therefore, transgenic animal technology will play important role in this field.

10. 1. 3 Application of transgenic animal technology in pharmaceuticals

10. 1. 3. 1 Bioreactor to produce expensive drugs and biological materials

Animal bioreactors are genetically modified animal systems that have the potential to reduce production cost, and improve production efficiency, of pharmaceutically relevant recombinant proteins. Several species including goats, cattle, rabbits, and hens have been genetically modified to secrete target proteins into milk, egg whites, blood, or other bodily fluids. Take the transgenic bovine as bioreactors for example, dairy cows produce 1-14 g/L of heterologous protein in milk daily for a 305-day lactation cycle each year, more than 10 g/L protein compared to mammalian cell culture in 10-12 d in Chinese hamster ovary cell (CHO cell) which needs more completion condition with large-scale, serum-free, suspension batch-fed cultures. Furthermore, the recombinant protein production is more readily purified in milk than in mammalian cell culture and no specific conditions, media, or cells are required (Monzani et al., 2016). In recent year, two biopharmaceuticals produced in the mammary gland of transgenic animals have been released and are commercially available for treating human disease: antithrombin (ATryn®) treating hereditary antithrombin deficiency, produced by GTC Biotherapeutics UK Limited, and C1-esterase inhibitor (Ruconest®) treating hereditary angioedema, produced by the Pharming Group. Other biopharmaceutical companies are working to produce diverse biopharmaceuticals in the milk of transgenic animals. However, recombinant protein production is initially limited due to the longer interval from birth to the first lactation in the domestic animals and lower efficiencies of transgenesis techniques than in avians. Because ①avians bioreactor have more efficiency with a short generation time, quick establishment and high egg production; ② transgenic avian systems allow for appropriate post-translational modification with the use of constitutive strong promoters or the tissue-specific ovalbumin promoter; ③recombinant proteins can be incorporated into egg whites and easily collected from the sterile environment of the egg (Rachel et al., 2018).

On the other hand, there are transgenic animals for the development of unique biological materials like polymers based on spider silks that may be useful as suture or plastic materials in facial and orthopedic restorative surgery. For example, Spider silk not only has excellent mechanical properties, but also has the characteristics of corrosion resistance, low temperature resistance, and anti-enzymatic solution, and is one of the toughest and most elastic natural animal fibers known today. Researchers have recombined spider silk genes into the genome of silkworms, and specifically express spider silk proteins in the silk glands of silkworms, so that silk contains spider silk components which will have broad application prospects in medical, military and building materials, aerospace and navigation, etc.

10. 1. 3. 2 Genetically modified pigs for xenotransplantation

For a long time, many patients have been on waiting lists for organ transplantation and some patients die because of failure to obtain a suitable organ (Brennan et al., 2018). Facing the background of a worldwide shortage of donor organs, the use of tissues and organs via transgenic animal technology from lower mammalian species as pigs which are similar to humans in body size and in the structure and metabolism of the internal organs could provide a near-term solution for the human organ shortage, which is even considered as the next great medical revolution. At first, pigs that can accept human cells and tissues must be produced by immune tolerance, e. g. , expressing human hematopoietic cytokine receptors in pigs are likely to enhance the establishment of porcine hematopoietic chimerism or introducing human *CD47* into porcine *HCs* reduces their phagocytosis by human macrophages by transgenic animal technology. And then, xenotransplantation must overcome another hurdle which is the risk of transmission of zoonotic pig microorganisms especially porcine endogenous retroviruses (PERVs). To remove this possibility, the researchers used a fusion of cultured regularly interspaced short palindromic repeats-Cas9 system also CRISPR-Cas9 technique, apoptosis inactivators to produce porcine endogenous retroviruses-nonfunctional basic pig cell lines. With these chains of cells, porcine generated endogenous retroviruses-nonfunctional fertilized ova, fetus, and alive pigs (Denner, 2021). Anyway, pigs in which grafted human cells can flourish are essential for the studies of the production of human organs in the pig and for the verification of the efficacy of cells and tissues of human origin for use in regenerative therapy.

10. 1. 3. 3 Establish animal models of human diseases

Transgenic animal technology allows researchers to probe and decipher aspects of human diseases especially the etiology and pathophysiology of diseases, which is critical for effective discovery and development of novel therapeutics for rare diseases. Since the first transgenic mice were generated in 1982, transgenic animal models (mice, rats, zebrafish, monkeys and rabbits) have been used extensively to investigate biomedical important mechanisms underlying a variety of diseases, to develop and evaluate new therapies. For example, in cancer diseases studies, oncomouse was first transgenic animal to be patented in Harvard. Its germ cells and somatic cells contained an activated human oncogene sequence introduced into the animal at an early embryonic stage to ensure that the oncogene was present in all the animal cells. Then, more and more specific transgenic mouse models were obtained to study the mechanisms for tumor progression, tumor immunity and metastasis. In cardiovascular diseases, transgenic models of heart failure and hypertrophy like gene overexpression of calmodulin, gene mutation of alpha cardiac myosin heavy chain and knockout gene model of transforming growth factor were developed (Gim et al. , 2021). In recent year, severe acute respiratory syndrome coronavirus 2 (SARS-CoV-2) was the etiological agent of coronavirus disease 2019 (COVID-19), an emerging respiratory infection caused by introduction of a novel coronavirus into humans in late 2019. As of 5 June 2022, over 529 million confirmed cases and over 6 million deaths have been reported globally. To study human disease pathogenesis and test new candidate COVID-19 therapeutics and expedite drug repurposing, researchers have been developing COVID-19 animal models by modifying transgenic mice to express human angiotensin-converting enzyme 2 (hACE2) via expression of a tissue-specific promoter (e. g. , for epithelial cells). In addition to transgenic mice, the other animal models using transgenic animal technology have been explored for COVID-19 studies such as minks (*Neovison vison*), cats, and pigs (Fan et al. , 2022).

10.2 Main technical links of transgenic animal technology

Considering so important technology, the complete method including: gene isolation of interest, expression vector construction, gene receptor (early embryo or cell) acquisition, gene introduction, transgenic embryo transfer, transgenic integrated expression, detection and genetic stability testing of transgenes. Among them, the construction and introduction of gene of interest is a key step in the implementation of transgenic technology (Fig. 10-2).

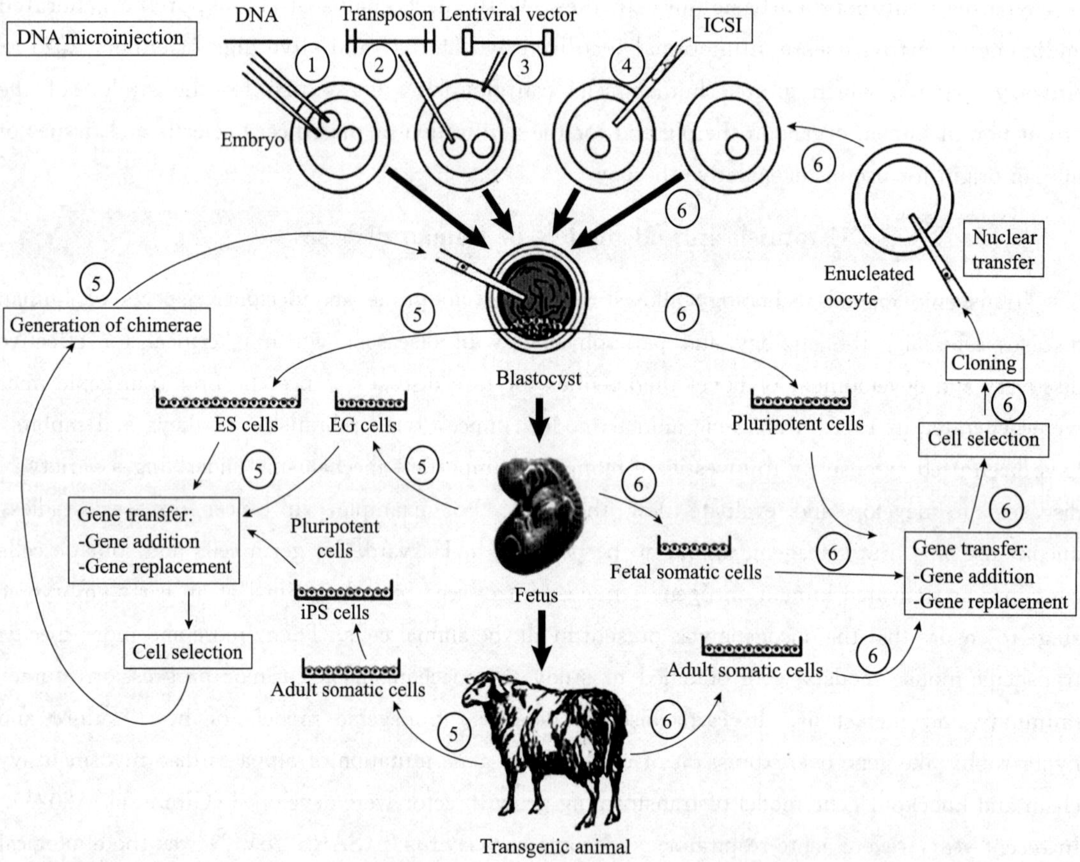

①DNA microinjection method ②Transposon mediated method ③Retroviral infection method
④Spermatozoa-mediated method ⑤Embryonic stem cell-mediated method
⑥Somatic cell nuclear transfer (SCNT) method

Fig. 10-2 Different methods to generate transgenic animals (Houdebine, 2014)

10.2.1 Targeting gene selection

In the short time frame since 2005, the development of high-throughput-next generation sequencing (HT-NGS) technologies with advancement of modern bioinformatics tools sequenced individual genomes of living organism at unprecedented pace with low cost and short time. Therefore, many draft genomes from other organisms have been published such as pig, cattle, chicken, and donkey. Based on the in-depth genome sequence information and analysis for most of the mammals and livestock, DNA microarray (ChIP-chip) or sequencing (ChIP-seq), RNA sequencing (RNA-seq), whole genome genotyping, *de novo* assembling and re-assembling of genome, genome wide structural variation, mutation detection and carrier screening, DNA library preparation would help to obtain useful targeting gene in animal genome. Next, the systematic sequencing of genomes and expressed sequence tags (EST) has facilitated cloning and investigating the function of most abundantly expressed genes (Fig. 10-3). For example, correlations between the sequence of microsatellites and other genetic markers in the same family members with cystic fibrosis led to the identification of the cystic fibrosis transmembrane conductance regulator (CFTR) gene and mutations that are causing the disease in question. At last, a given DNA fragment, after verification of correct cloning and confirmation of the nucleotide sequence, is purified through several steps such as isolation by electrophoresis through agarose, running the DNA on a column of DEAE-Sephacel, or glass bead adsorption to ensure that the targeting gene DNA fragment does not contain nicks or strand breaks and is as pure as possible.

10.2.2 Gene modification and expression vector construction

Once the target gene is obtained, this exogenous gene needs to be modified and constructed into a vector. The main modification is to add a long coding sequence to express a fluorescent reporter protein, Cre recombinase, or conditional allele (floxed gene). Furthermore, numerous types of specialized modification can be introduced based on different purposes. For example, DNA donor needed preventing CRISPR-Cas9 cleavage of the chromosome and insertion of restriction enzyme sites unique to the donor, to simplify downstream genotyping in CRISPR-Cas9 technology. One thing to note, the longer the insertion would led to the less efficient (Lanigan et al., 2020).

About the expression vector, it is usually containing the side sequence of the regulatory element, the structural gene sequence and the transcriptional termination signal in order to improve the function and specific and efficient expression of the product. Then, we need to insert the reporter gene and the natural promoter or tissue-specific promoters to enhance target gene expression. Naturally, the regulatory sequence is upstream in the expression vector, the enhancer is downstream, and the intron is retained in the target gene. However, to accurately control the high expression efficiency of foreign gene coding sequence, matrix attachment region (MAR), internal ribosome entry site (IRES), insulator and Kozak

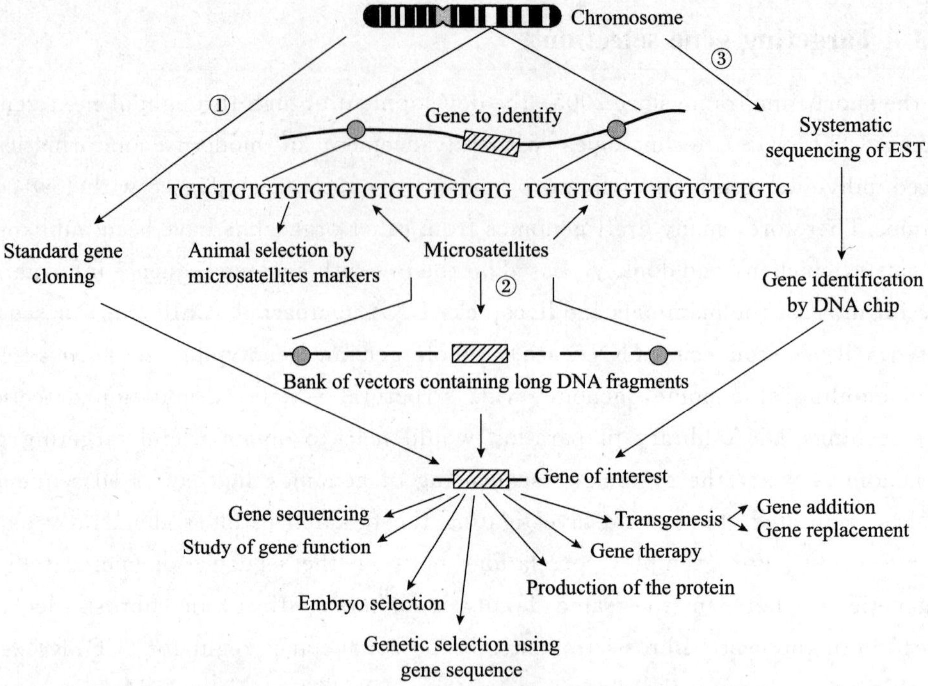

①Already known gene isolated (cleared background and function)
②Unknown gene isolated via microsatellites
③Unknown gene isolated via expressed sequence tag (EST)

Fig. 10-3 Different methods to isolate genes and the contribution of transgenesis in gene study and use (Houdebine, 2014)

sequence are added to improve expression vector. Then, various reporter genes, such as chloramphenicol acetyltransferase gene, fluorescent enzyme gene and alkaline phosphatase gene are also used in the design of transgenic vectors. For example, growth hormone (GH), a polypeptide hormone secreted by the anterior pituitary, can promote fat decomposition and improve meat quality in animal development and livestock production. When preparing GH transgenic pigs, Tet-On gene expression system (Tet-On) tetracycline-controlled is introduced and modified to achieve safe and controllable expression of exogenous *GH*. And then, the regulatory system rtTA and inducible tre expression vector are constructed on the vector which greatly improved the operation efficiency. Therefore, *GH* is inserted between the inducible expression vector system PtRe and the regulatory system rtTA to obtain the controllable expression of *GH*.

10.2.3 Introduction of exogenous genes

Introduction of exogenous genes into the cell, is the key step of generating transgenic animals. From the beginning to now, DNA microinjection has been the most widely applied method for gene transfer in mammals. However, variety of technologies building on DNA microinjection and related techniques are to improve relative efficiencies. Here, we will describe this methodology's principles and requirements of the technology and then introduce other methods.

10. 2. 3. 1 DNA microinjection method

Today, DNA microinjection still has utility in a variety of applications. In 1982, the world's first integrated GH mouse was obtained using a microscope to inject foreign genes directly into the prokaryotic nucleus of a single-cell fertilized egg with the help of a microscope manipulator. Briefly, DNA microinjection contained main steps as follows: superovulation of animal, production of eggs from superovulated females, *in vitro* culture of superovulated eggs, harvesting superovulated eggs, microinjection needles and slides, microforge shaping of pipettes, microinjection slide for eggs. Systems for controlling microinjection of DNA and holding eggs are grouped into either air-driven systems (e. g. , Carl Zeiss, Thornwood, NY, USA; Eppendorf, Hauppauge, NY, USA; Brinkman, Westbury, NY, USA; Harvard Apparatus, Holliston, MA, USA) or oil-driven hydraulic systems and DNA microinjection into cytoplasm (Fig. 10-4). The air-driven systems can also be attached to electronically controlled delivery systems that allow preprogrammed regulation of fluid delivery volume, time, and pressure. In a word, those steps involved in producing and evaluating transgenic mice derived by DNA microinjection are straightforward, although mastering the steps can be very time-consuming with 5 and 12 months of dedicated effort to become proficient at the outlined techniques. Therefore, microinjection practice will lead to the production of transgenic animal but need continued commitment to develop proficiency at these techniques.

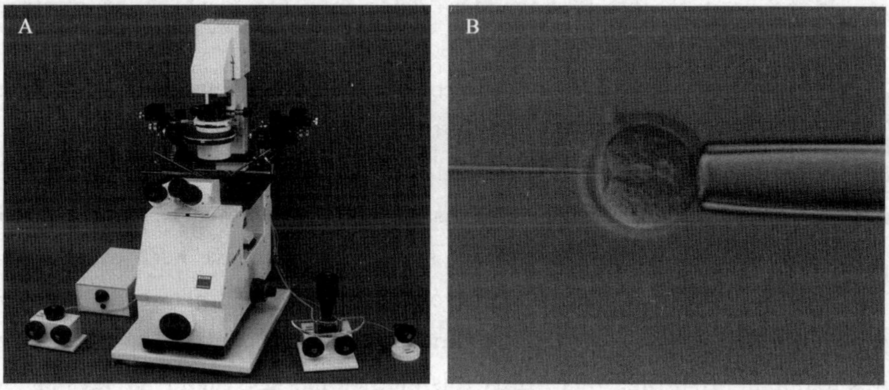

Fig. 10-4 Microscopes and micromanipulators (A) and DNA microinjection into cytoplasm (B)
(Pinkert and Polites, 2014)

10. 2. 3. 2 Retroviral infection method

Retroviruses can be used as vectors in the introduction of exogenous genes. When these viruses infect a host cell, the viral RNA is reversely transcribed in the cytoplasm which is transported into the host cell nucleus and integrated into a chromosome directly. In retroviruses, C-type retroviruses are extensively studied and therefore well characterized for use in vector construction such as murine sarcoma virus (MSV), gibbon ape leukemia virus (GaLV), avian leucosis virus (ALV), reticuloendotheliosis virus (REV), and Rous sarcoma virus (RSV). Non-SIN vectors are most common having the structure LTR-cDNA of gene 1-internal promoter-cDNA of gene 2-LTR and two exogenous genes are transferred simultaneously and expression of gene 2 is controllable by choosing an appropriate internal promoter. In general, the retrovirus vector contains exogenous genes that are injected into the early embryo, or to poison the culture cells of the animal and co-culture with the embryo so that the virus infects the embryo and obtains the transgenic animal. Compared to DNA microinjection method, retroviral infections are easy to operate without special instrument, higher integration efficiency, stable integration into the recipient genome. Although retrovirus vectors have several advantages, some drawbacks include size limitation that an insertion of permissible size (around 8 kb) can only code for a protein as big as 300 kDa, high recombination rate, potential pathogenicity and carcinogenicity. Therefore, the researchers have developed novel vector lentiviruses, including human immuno-deficiency virus type 1 (HIV-1), feline immunodeficiency virus (FIV), and equine infectious anemia virus (EIAV), which are convenient for the insertion of foreign genes into the animal chromosomes.

10.2.3.3 Embryonic stem cell mediated method

Embryonic stem cell (ESC), derived from the outgrowth of the inner cell mass of blastocyst, can be used to inject exogenous genes into blastocyst after electroporation, calcium phosphate precipitation or retrovirus infection (Miao et al., 2013). ESCs will participate in the formation of host embryo, construct chimera until species chimerism, and transmit the extrinsic gene to the next generation and produce transgenic animals (Song et al., 2016). For example, a mouse model with hypoxanthine-guanine phosphoribosyl transferase (HGPRT) knockout was established using ESC-mediated method. In 2012, Chinese scientists used the haploid embryonic stem cells that have been transgenic to prove that the artificially constructed haploid embryonic stem cells can also exist stably *in vitro*, and they carried out transgenic operations, and replaced the "fertilization" between sperm and oocytes to obtain offspring. In a word, this method has several advantages including that it is easy to identify the stable expression level, the copy number of foreign genes inserted into ESC and more controllable than other methods. However, the main drawback of this method is the difficulty of establishment of ESCs in culture and the chimera can cause reproductive disorders and infertility (Li et al., 2012).

10.2.3.4 Somatic cell nuclear transfer (SCNT) method

Based on the successful development of cloning technology for somatic cells, the researcher transfers somatic cell nuclear as nuclear donors who has been stably integrated into exogenous genes to produce transgenic animals. This method using preselected transgenic cells as a source of nuclei for SCNT increases the efficiency of transgenic animal production to 100% and overcomes the problem of mosaicism observed after pronuclear microinjection. Furthermore, SCNT avoids the limitation of the reproductive cycle and the separation of genetic traits brought about by sexual reproduction; the separation of gene integration and embryo production is conducive to fixed-point integration and gene expression. Therefore, the method has been successful in animals such as sheep, goats, cattle and pigs. For example, McCreath et al. described the efficient and reproducible gene targeting in fetal fibroblast cells to insert the therapeutic AATC2 transgene that encodes the recombinant human alpha 1-antitrypsin (rhAAT) at the ovine α_1(I) procollagen (COL1A1) locus and the production of live sheep by nuclear transfer. Recombinant human alpha 1-antitrypsin was detected at a concentration 650 μg/mL from milk samples (McCreath et al., 2000; Miao et al., 2013).

In addition to the above methods, there are also transposon-mediated method, sperm-mediated method, herpes simplex virus infection and so on.

10.2.4 Gene integration and expression analysis

During transgenic animal technology, there are two questions that need to be concerned, that is, whether gene integration and expression occur in transgenic animal. The integration event of DNA transgenes has been a black box without the following analyses.

10.2.4.1 Gene integration analysis

Due to the lack of the mechanism of DNA integration, many strategies have been evaluated and reported including adding specific transcriptional regulatory elements, matrix-attachment regions, the use of genomic-type constructs, the use of chromosomal-type transgenes, and inserting the desired transgene under the control of the endogenous locus whose expression pattern is required.

On the other way, the easiest and most commonly used method to analyze the integration of transgenes in the host genomes is PCR or real-time PCR approaches. Furthermore, several caveats need to know including that ①make sure positive and negative controls are taken, ②the confirmation of genomic DNA with good quality, ③using suitable concentration of genomic DNA as 1/10 and/or 1/50 diluted.

The third way, transgene integration detection using Southern blot approach which denatured single stranded DNA fragments in gel are adsorbed in situ onto nitrocellulose or nylon membrane by capillary action of a solution with a certain concentration of salt ions.

The fourth way, transgene integration detection using the slot-blot approach in which transgene-specific DNA band is used to hybridize equivalent amounts of genomic DNA immobilized onto a membrane concentrated in a small area.

The fifth way, Northern blots hybridization analysis can be used to detect RNA of exogenous genes and can reveal the presence of multiple transcripts that differ from one another by greater than 50-100 bases.

Recently, more sophisticated methods including the use of microarrays and a next-generation sequencing step are devised for the analysis of the integrity and the integration site of transgenes in mice.

10. 2. 4. 2 Gene expression analysis

The target gene can be expressed as a protein in transgenic animals and in specific tissues, which is the ultimate measure of the success of transgenic research. Common methods for measuring protein levels expressed in target genes are ELISA, RIA, and Western blotting. In addition, fluorescent antibodies conjugated with fluorescein or rhodamine can be used to detect single cells from transgenic animals under a fluorescence microscope or by fluorescence-activated cell sorting(FACS) and flow cytometry. Considering the offspring of transgene-positive animals, their offspring by natural mating or artificial insemination are tested for the integration, transcription, and expression of the offspring's foreign DNA.

10. 2. 5 New progress in the research of animal transgenic technology

In recent years, there have been new advances in animal transgenic technology with higher efficiency and more convenience, including transgenic technology mediated by conditional gene targeting, RNA interference (RNAi) and genome editing technology.

10. 2. 5. 1 Conditional gene targeting

In earlier studies, the expression of exogenous genes after being transferred into the animal genome could not be exactly controlled such as expression only in a certain tissue or developmental stage. The Cre recombinase/locus of X-over P1 (Cre/LoxP) system was induced to achieve the spatiotemporal specificity in gene targeting. Briefly, at first, the target gene was added to between the two loxP sites with reporter flag such as GFP to produce transgenic mouse named as Loxp mouse. And then, other transgenic mice named as Cre mouse strain carrying the Cre recombinase gene was produced. At last, Cre recombinase recognized 34 bp loxP (locus of recombination) elements and catalyzed recombination between the two loxP sites when these two mice were mated (Lanigan et al., 2020). Therefore, deletion of the critical exon causes a premature termination codon to occur in the mRNA transcript which achieves a local knockout of the target gene in specific tissues or organs. Meanwhile, Cre/LoxP system is also to achieve knock-in mouse (Fig. 10-5). Although Cre/LoxP system has some advantages of high efficiency, strong specificity, wide range of applications, it has difficulty in big animals. In 2014, researchers have successfully constructed the world's first large animal model (pig) of Rosa 26 site, which was widely used as a safe site for foreign gene expression knock-in of Cre recombinase reporter gene as mTomato, mGFP dual fluorescent by using primary mesenchymal stem cells, and animals generated by nuclear transfer (Li et al., 2014; Jabbar et al., 2021). Furthermore, other site-specific recombinases, such as FLP, Dre, and Vika that work on the same principle have also been applied to mouse models.

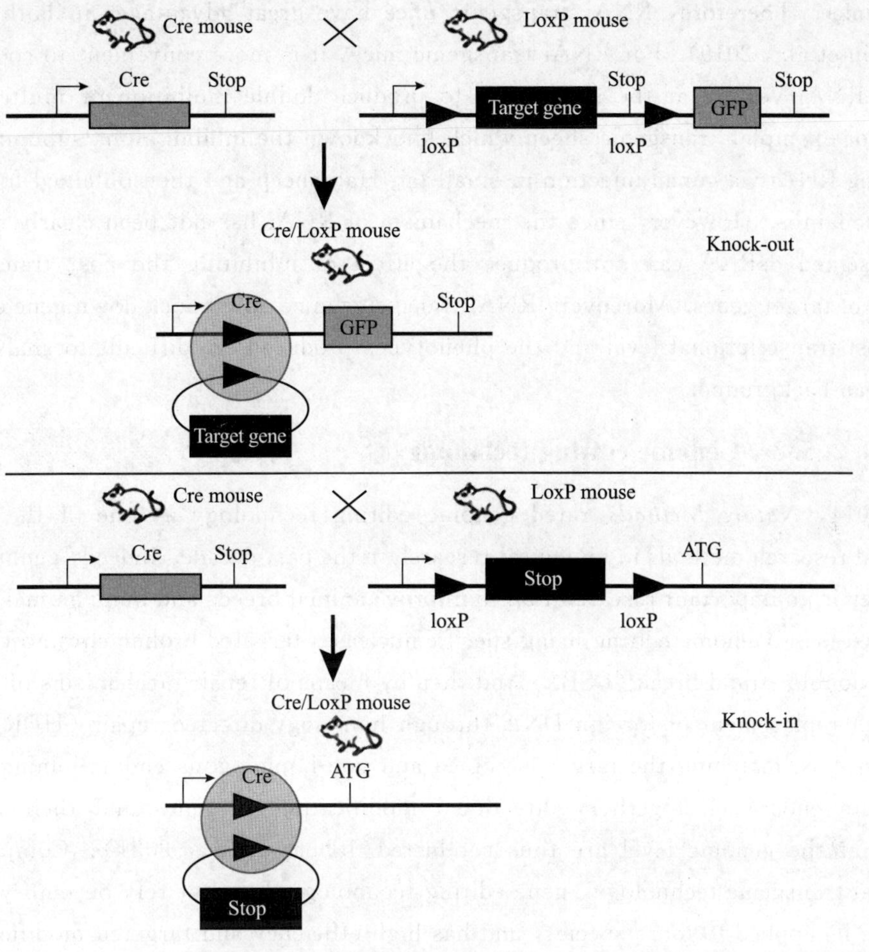

Fig. 10-5 Diagrammatic representation for conditional gene targeting by Cre/LoxP recombination
(Miao, 2013)

10.2.5.2 RNAi-mediated method

RNAi is the silencing of specific gene expression mediated by the formation of double-stranded RNA that results in the inhibition of gene expression by degrading mRNA. In 2001, scientists found that the introduction of synthetic 21-23 base double stranded RNA into mammalian cells could cause specific gene silencing and would not trigger the cell resistance response. Therefore, RNAi-mediated methods were applied in transgenic animals. The successful design of RNAi transgenic vector is the most critical step in the process of constructing RNAi transgenic mouse model. The general strategy is to use the H1 or U6 atriovens of RNA polymerase Ⅲ to start the transcription of siRNA, followed by a specific sequence that can form hairpin like dsRNA, followed by polyT. Some vectors also add fluorescent reporter genes downstream. Compared to traditional transgenic animals, RNAi transgenic mice only need three steps: vector construction, microinjection and F_1 generation founder mice. Therefore, RNAi transgenic mice have great advantages in both time and steps (Kin et al., 2019). For RNAi transgenic mice, it is more convenient to co-inject two or more RNAi vectors at the same time to produce double inhibition or multi-inhibition effect. For example, transgenic sheep which knockdown the inhibin alpha subunit (INHA) were using RNAi via gonad-injection in small-tail Han sheep and then obtained four of nine transgenic lambs. However, since the mechanism of RNAi has not been clearly explained, many designed dsRNA can not produce the effect of inhibiting the post transcriptional silencing of target genes. Moreover, RNAi transgenic mice only knock down gene expression at the post transcriptional level and the phenotypes produced are difficult to analyze due to the unclean background.

10.2.5.3 Genome editing technology

In 2014, *Nature Methods* rated genome editing technology as one of the ten most influential research methods in biological research in the past decade. Indeed, genome editing technology is an important research tool to improve animal breeds and build animal models of human diseases. Genome editing using specific nucleases targeted broken chromosomal DNA caused a double-strand break (DSB). And then by means of repair mechanisms of insertion, deletion or replacement of foreign DNA through homology-directed repair (HDR) where a donor sequence matching the target is copied and non-homologous ending-joining (NHEJ) which puts ends back together, directional modification of genes and their regulatory elements at the genome level are thus conducted (Jabbar et al., 2021). Compared with traditional transgenic technology, gene editing technology does not rely on embryonic stem cells, can be applied to more species, and has high efficiency and targeted modification. To date, the specific nucleases successfully applied to genome editing include zinc finger nuclease (ZFN), TALENs and CRISPR-Cas, and have formed ZFN, TALEN and CRISPR-Cas genome editing technology systems, respectively. Among the three technologies, CRISPR-Cas technology is simpler, more efficient and less costly, and the representative technology of which CRISPR-Cas9 has become a widely used genome editing technology (Jabbar et al., 2021).

ZFN is a kind of synthetic restriction endonuclease which consists of a zinc finger DNA binding domain responsible for the specific recognition sequence and a DNA cleavage domain for nonspecific restriction endonuclease cleavage. Three zinc fingers can be combined to specifically recognize nine base pairs on one DNA strand and then ZFN heterodimers play the role to break sequence-specific chromosome base on FokI restriction endonuclease, and exogenous genes are fused to the DNA-cutting domain of each zinc finger. Researchers can process and modify the zinc finger DNA binding domain of ZFN to target different DNA sequences, so that ZFN can bind to the target sequence in the complex genome and perform specific cleavage by the DNA cleavagedomain (Lanigan et al., 2020). At present, ZFN technology has been widely used in the mutation of targeted genes among zebrafish, frogs, rats, mice, cattle and other species. By artificially modifying genome information, new species with modified genetic background can be generated. This technology is also of great value in the medical field. It has potential significance for gene therapy of diseases and has a very broad application prospect.

The development of TALENs followed after ZFN technology. Tal effector (TALE) was first discovered in a plant pathogen named *Xanthomonas* sp. as an invasive strategy for bacterial infection of plants. These Thales were injected into plant cells through the bacterial type Ⅲ secretion system, and regulated transcription by targeting effector specific gene promoters to promote bacterial colony formation. Because tale had sequence with specific binding ability, researchers formed a powerful tool with specific genome editing function, talen, by connecting *Fok* Ⅰ nuclease with an artificial tale. TALENs were made up of tandem repeats of 34 amino acids. The central amino acids at positions 12 and 13, named repeat variable di-residues (RVDs), determined the base to which the repeat will bind. To achieve a specific chromosomal break, 15 TALE repeats assembled and fused to the *Fok* Ⅰ endonuclease domain (TALEN monomer) were required. Thus, one TALEN monomer bound to 15 base pairs on one DNA strand, and a second TALEN monomer to bases on the opposite strand. When the *Fok* Ⅰ endonuclease domains were brought together, a double-stranded DNA break occurred. In this way, a TALEN heterodimer can be used to cause a sequence-specific chromosome break. It was estimated that, within the entire genome, TALENs had potential target cleavage sites every 35 bp. Compared with ZFNs, TALENs are easier to construct with publicly available resources, and TALENs have been adopted for use in mouse and rat genome engineering in several laboratories that have produced knockout and knock in animal models (Jabbar et al., 2021).

In analogy with TALENs, the CRISPR-Cas system has been described initial in bacteria and archaea as their acquired "immunity" mechanism against viruses. In the bacterial CRISPR/Cas9 system, the mature crRNA binds to tracrRNA to form a tracrRNA: crRNA complex that partially binds to the Cas9 and guides the entire system to its target site (Rosenbaum et al., 2019). The crRNA-specific hydrolysis of both DNA strands via CRISPR-Cas9 requires the presence of a complementary target sequence in the DNA, followed immediately by a protospacer adjacent motif (PAM) sequence. After the complementary connection of the crRNA molecule with the target site within one DNA strand, it is hydrolyzed by the Cas9 HNH nuclease domain, while the hydrolysis of the complementary DNA strand to the recognized target site is carried out by the Cas9 RuvC nuclease-like domain (Jabbar et al., 2021). This creates a two-strand DNA break three nucleotides downstream of the PAM sequence (Golkar, 2020) (Fig. 10-6). The CRISPR/Cas9 system was first used in genetic engineering in 2013. Editing the genome through CRISPR/Cas9 system depends on the formation of a DSB and the subsequent cellular DNA repair processes based on HDR and NHEJ homologous recombination mechanism which delivers appropriately designed donor DNA. Computational analysis of genomic sequences of several organisms, including bacteriophages, led researchers to notice the importance of CRISPR repeat and spacer sequences. Conspicuously, CRISPR had been identified as an atypical bacterial DNA repeat element for many years before it was reported as the bacterial immune system and subsequently introduced as a potent re-programmable gene targeting tool kit (Golkar, 2020)(Fig. 10-6B). Compared to TALENs, CRISPR-Cas system is a highly efficient and robust instrument for genome alteration in the cultured cells and the embryos of species. Using CRISPR-Cas9 technology, knockout of single gene or multiple genes, point mutation, deletion and inversion of large chromosome fragments, gene silencing, etc. have been realized in the animal field. Furthermore, more and more new techniques of CRISPR-Cas system have been formed (Golkar, 2020). For example, obtaining the whole chromosome knockout technology of F_0 generation completely knockout primates in one step CRISPR-Cas system greatly shortens the time to obtain homozygous primates. This technology realizes the complete knockout of target genes, truly "cutting without residue", improves the efficiency of gene editing, and the accuracy of gene editing tools. In livestock, there is a wide range of applications of CRISPR-Cas9 including the production of resilience against disease, improvement of reproductive traits, and production of animals.

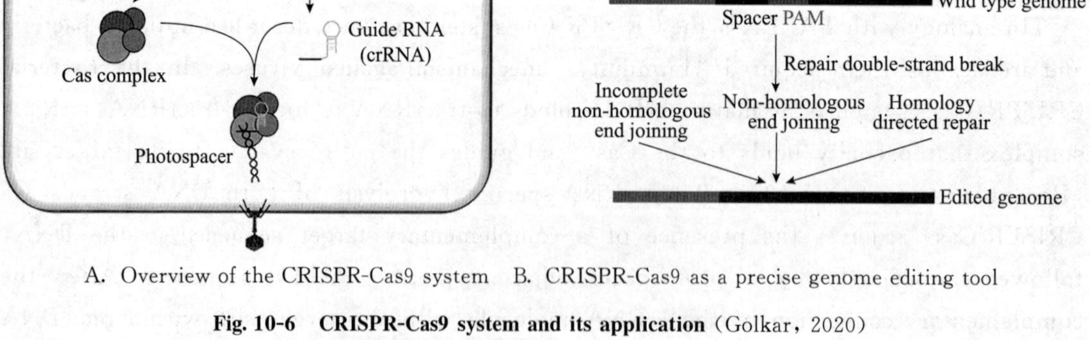

A. Overview of the CRISPR-Cas9 system B. CRISPR-Cas9 as a precise genome editing tool

Fig. 10-6 CRISPR-Cas9 system and its application (Golkar, 2020)

10.3 Main questions and developmental trend in transgenic animal technology

10.3.1 Main question in transgenic animal technology

Although the field of genetically modified animals has developed rapidly, there are still some problems which need to pay attention to and address. We will discuss as follows:

10.3.1.1 Improvement of transgenic animal technology

The main limiting factors in transgenic technique are low success rates and survival rates of transgenic animals. Then, the integration efficiency of extrinsic genes at the target site is low and unstable. It affects the intrinsic gene, damages the host's genome, or activates the closed gene in normal conditions to express and subsequently produces abnormalities.

10.3.1.2 Safety of transgenic animal technology

Since the first transgenic mice obtained in 1982, there were only two transgenic animals in the market. Attention has been focused on the safety of transgenic animal.

The first is food safety. These foods are made by inserting genes of other species into their DNA. In intact, DNAs in food are degraded during cooking and processing, but others remain via exogenous DNA and horizontal gene transfer (HGT). Consumed DNA is largely hydrolyzed during digestion. Several studies documented the survival of DNA in food/feed from transgenic plant (genetically modified crops) throughout the gastrointestinal tract in pigs (*cry1Ab* and *cry9C*), piglets (*cry1Ab*) and human intestinal microflora with low levels of the exogenous gene (Zhang and Shi, 2011; Bawa et al., 2013).

The second is environmental safety. Consequences on a global scale have become a significant factor in domestic calculations. When new hybrids are created, raising questions about environmental impacts, such issues are not limited by national borders. This will be a factor with transgenic animals, when considering their environmental impact or when concerns arise about the proliferation of disease. In addition, transgenic animal may contaminate other genes, and produce a variety of threats to ecological balance and species diversity.

The third is product safety. Several species including goats, cattle, rabbits, and hens have been genetically modified to secrete target proteins into milk, egg whites, blood, or other bodily fluids. Therefore, for example, the predominant transgenic host species has been mouse, the genomes of rats, rabbits, chickens, and cows have also been modified to express human antibodies. The creation of transgenic animal platforms expressing human antibody repertoires has revolutionized therapeutic antibody drug discovery. The safety of non-human primate's antibodies must be evaluated in a preclinical setting prior to human trials or highlighting their impact on human risk assessment. This important step of drug development usually necessitates that the clinical candidate cross-reacts to the appropriate preclinical animal species chosen for toxicology.

10.3.1.3 Consumer's attitude and ethics

Consumer acceptance is conditioned by the risk that they perceive from introducing food into their consumption habits processed through technology that they hardly understand. Consumers misunderstand the nature of genetic engineering or do not appreciate its prevalence or both. In a study conducted in Canada, 44% Canadians are confused about health effects of genetically engineered foods and ingredients. In total, 40% believe that there is not significant testing on genetically engineered food to protect consumers. In total, 52% are uncertain on their consumption of genetically engineered food, despite its prominence in the agrifood marketplace. Scientific literacy of respondents on genetic engineering is low (Einsiedel, 2005). On the other way, ethics procedures especially of animal welfare need to be strictly performed based on 3R (replacement, reduction, refinement). In addition, the statement that it is irresponsible to proceed with any clinical use of germline editing unless and until the relevant safety and efficacy issues have been resolved based on appropriate understanding and balancing of risks, potential benefits, and alternatives was announced at the First International Summit on Human Gene Editing in 2015. However, indeed, the resistance HIV baby via CRISPR-Cas9 system violated several ethical norms including international consensus guidelines, national regulations and well-established principles of bioethics in 2018.

10.3.2 Developmental trend in transgenic animal technology

Undoubtedly, it is foreseeable that with the continuous development of transgenic animal technology, especially genome editing technology, transgenic animal research (especially in the field of biomedicine) will have broad and beautiful prospects. However, the developments of transgenic animal technology need more new technology.

10.3.2.1 Development of reproduction

In recent years, more animal genome sequencing and the advance of molecular biology such as epigenetics would help make clear the mechanism of exogenous genes transferred into the animal genome.

10.3.2.2 New technique application

Interdisciplinary approach will become the great tool in transgenic animal technology.

1. Nanoscience and nanotechnology

Nanoscience and nanotechnology are used in vector construction, gene delivery and creating stem cell niches. For instance, a variety of nano-vectors has been constructed which help gene transfection with their properties to be selectively internalized into target cells and high efficiencies. Nanovectors used both for *in vivo* and *in vitro* gene transfection are of two types—viral vectors and non-viral vectors (Deb et al. , 2012).

2. Induced pluripotent stem cells (iPSCs) technique

In 2006, Yamanaka developed an alternative method of generating embryonic-like stem cells from adult cells which contained most of the advantages of embryonic stem cells as can be self-renewal and differentiation various cells. It was envisaged that iPSCs would facilitate the generation of transgenic animals for use as preclinical models and production of animals with valuable traits through using chimeric or nuclear transfer technologies (Takahashi and Yamanaka, 2006). However, to date, no germline-competent-iPSC line has been established for large animals.

3. Organoids technology

Organoids technology using 3D culture technology allows embryonic and adult mammalian stem cells to exhibit their remarkable self-organizing properties, and the resulting organoids reflect key structural and functional properties of organs such as kidney, lung, gut, brain and retina. Researchers have developed iPSC-derived cerebral organoids as a model for recombinant adeno-associated virus (rAAV) capsid selection and for testing efficacy of AAV-based transgenes in a human context. In a word, transgenic animal technology will play very important role in future.

Assignments

1. Please describe an exact case about transgenic animal technology application in livestock.
2. What are the main steps in transgenic animal technology?
3. What are the methods of exogenous genes transfer into the cell?
4. What are the mechanism and application of CRISPR-Cas9 system?
5. Please introduce the new advances in transgenic animal technology.

(Zhang Wei)

Chapter 11 Hormone Immunization and Genetic Immunization

Summary

Code 11-1

Hormone immunization was developed in the 1970s which applies hormones to stimulate the body to produce specific antibodies, or directly applies antibodies against hormones to neutralize the endogenous hormones, resulting in the alteration of the concentration and biological activities of the endogenous hormones. Genetic immunization was discovered in early 1990s, and is a simple and promising method for inducing both humoral and cellular immune response. Vaccines produced by this technology are called gene vaccines, which have been considered as the third-generation vaccines. Gene vaccines are useful for humans and animals in the fields of cancer, infectious diseases, and allergies. For animals, hormone and genetic immunization have been investigated for various applications including improving the fertility and immunocastration. This chapter mainly introduces the principles, methods and effects of hormone and genetic immunization in the regulation of animal reproductivity.

Hormone immunization and genetic immunization have been applied in the control of animal reproductivity and immunocastration, such as the regulation of the spermatogenesis in males, and the estrus induction, superovulation, fertilization and embryo development in females, by neutralization of the endogenous or exogenous hormone, or specific bioactive molecules.

11.1 Hormone immunization

The hormone neutralization technique developed in the 1970s has provided a theoretical basis for the study of reproductivity regulation. A variety of hormones can be applied to improve animal fecundity by immune neutralization, for example, neutralization of gonad-derived steroid hormones to reduce their negative feedback effect on hypothalamic GnRH, thus promoting gonadotrophic hormones (GTH) secretion, eutralization of gonad-derived INH and/or AMH to attenuate the inhibition of pituitary FSH secretion, or improve the response of preantral/small follicle to FSH, eutralization of hypothalamic GnIH to remove its inhibitory effect on GnRH, FSH and LH secretion (Yang et al., 2019).

So far, remarkable effects have been achieved in steroid vaccines used to increase the twinning rate of sheep. INH and other hormone immunization have also been applied in various animals and increased the ovulation rates, but no commercial vaccine is available yet. Although active or passive immunization of reproductive hormones can improve animal fertility, the physiological activities of reproductive endocrine regulation is very complicated, and the endocrine physiology differs among species or individuals, so the application of reproductive hormone immunization to increase animal reproductivity remains limited. For example, the steroid immunization is effective in increasing the lambing rate in sheep, while it has little effect on cattle, indicating the immunization technique still needs further study.

11.1.1 Steroid hormone immunization

The use of steroid hormone immunization to increase the twining rate of sheep was first successfully developed in Australia. CSLRO scientists found that immunized with hormone antigens prepared by steroids such as estrone, androstenadione or testosterone could improve the ovulation rate in ewes. In 1983, cooperated with Glaxo, they successfully developed the world's first commercial twining vaccine, Fecundin, which is composed of androstenedione-7 α-carboxyethyl thioether-human serum albumin with diethylamine ethyl glucan as adjuvant. After years of large-scale experiments in Australia, New Zealand and other countries, Fecundin has been proved to significantly improve the lambing rate of sheep by 18%-40%. After that, numerous tests were carried out in other countries with similar principle and method of Fecundin. Taking Fecundin as an example, the basic principle, method and effect of this technique are introduced here.

11.1.1.1 Basic principle

It is believed that follicle growth and development are mainly regulated by pituitary GTH. In mono-ovulated animals such as sheep, estrogen secreted by dominant follicles can exert feedback effect on the hypothalamus and pituitary to control the secretion of FSH and LH at certain levels. Therefore, the remaining follicles can not develop into mature follicles and ovulate due to insufficient GTH. Steroids such as progesterone and androgen are the precursors of estrogen. Immunization with estrogen or its precursors in ewes can induce a large number of antibodies against steroids, which neutralize the endogenous steroid hormones. By this way, the decrease in the concentration of synthetic estrogen or its precursors downregulates the concentration of peripheral estrogen and relieves the inhibition of hypothalamus and pituitary function, which increases pituitary GTH secretion, stimulates follicular development, and eventually leads to increased ovulate rate.

11.1.1.2 Methods and effects

1. Active immunization

Take the androstenedione-derived complex (androstenedione-7α-carboxyethyl sulfide-human serum albumin) as an example. This specific complex (1.2 mg) is dissolved in physiological saline (2 mL) containing 5% DEAE glucan and subcutaneously injected into the neck of ewes 5-7 weeks before breeding. Booster is given 3-5 weeks later (sheep immunized in the previous year need not be immunized for the first time in the second year, but only needed to be vaccinated once more). 2-3 weeks later, ewes in estrus are mated. According to the experiment in Australia, the twining rate of ewes for different breeds are increased by immunization with Fecundin, but the response is different. Similar studies and experiments have been carried out in China. Fecundin is also successively developed by Nanjing Agricultural University, Lanzhou Institute of Animal Science, Shanghai Institute of Biochemistry and Xinjiang Institute of Chemistry, Chinese Academy of Sciences. The main ingredients are testosterone-3-carboxymethyl oxime, estrone-3-half succinate, testosterone-3-carboxyethyl sulfide or 11α-androstenedione. Fecundin has been tested and popularized in Xinjiang fine-wool sheep or its hybrid sheep on a large scale and has achieved the increase in lambing rate by 5%-25%.

2. Passive immunization

The basic principle of passive immunization is similar to active immunization in improving the lambing rate of ewes. The difference is that hormonal antibodies (antisera) are directly injected into ewes by passive immunization which is more convenient than active immunization. It was reported that immunization of Xinjiang fine-wool sheep and the crossbred sheep with monoclonal antibody against androstenedione achieved the increase in double ovulated ewes (30% vs. 0) and twins' rate (27.5% vs. 7%).

11.1.2 Inhibin immunization

Inhibin (INH), a kind of glycoprotein hormone with α and β subunits, is secreted mainly by ovarian granulosa cells in females and testicular Sertoli cells in males. It cooperates with activin to regulate the synthesis and secretion of pituitary FSH. INH immunization, whether active or passive, can promote follicular development and ovulation in females, increase semen production in male animals, as well as increase litter size and even improve embryo quality.

11.1.2.1 Basic principle

As INH has a specific inhibitory effect on the synthesis and secretion of FSH, INH can be used to immunize animals to produce specific antibodies to neutralize endogenous INH, reduce the level of circulating INH, and thus weaken the inhibitory effect of INH on the secretion of FSH, so as to improve endogenous FSH level, promote follicular development and increase the ovulation number.

11.1.2.2 Methods and effects

1. Active immunization

There are three main sources of INH used for the preparation of antigen, including natural, synthetic peptide and recombinant INH. Natural INH can be extracted from the follicular fluid or semen plasma. Generally, INH in the follicular fluid or semen plasma can be extracted by immunoaffinity chromatography with specific antibody. Synthetic INH peptides can be chemically synthesized based on specific amino acid sequence of its subunit. There are two main types of INH peptides, C-terminal peptide and N-terminal peptide of alpha subunit. Recombinant INH can be prepared by genetic engineering technology. The natural INH extracted from follicular fluid is a kind of macromolecule protein with excellent immunogenicity, and the immunogenicity of heterogeneous follicular fluid extract is better than that of homologous. However, this kind of INH is difficult to purify and the prepared antibody has a large cross-reaction. Synthetic peptides must be coupled with carrier proteins (such as serum albumin and ovalbumin, etc.) to acquire good immunogenicity due to the weaker immunogenicity with low molecular weight. The recombinant INHα subunit fusion protein has complete antigenicity and can be directly used in animal immunization.

It was reported that immunization of INH purified from bovine follicular fluid resulted in increases of INH antibody titer and GTH levels accompanied by an increase of 57% in the ovulation rate in New Zealand Romney sheep. Immunization of a synthetic INH fragment with N-terminus 25 amino acid of sheep INHα resulted in an increase of 107% in the ovulation rate of sheep. Immunization of recombinant INHα subunit resulted in an increase of 209%-335% in the ovulation rate. Immunization of porcine semen INH increased the estrus rate by 21.4%, ovulation rate by 27.1%, and twinning rate by 26.3% in cattle. Active immunization in goats with INH increased the follicle number and ovulation rate. Similarly, immunization of recombinant INH also improved the large follicle number, ovulation rate and embryo recovery in Boer goats. Inhibin α subunit immunization also enhanced the reproductive performance during non-breeding season in Beetal goats when estrus was induced using a progestin-based treatment regimen (Rehman et al., 2021). Immunization of recombinant INH in Holstein cattle prior to superovulation increased the number of transplantable embryos and embryos with grade A, as well as the blood levels of FSH, estradiol, activin, and the ratio of activin/follistatin. Immunization against inhibin α subunit plus P_4 treatment enhanced ovarian follicle and the subsequent early embryo developments that help to greatly improve the fertility of Holstein dairy cows (Chen et al., 2022).

2. Passive immunization

Immunization with INH antiserum also increased the ovulation rate. Dairy cattle immunized with bovine INH antiserum had increased follicle number. Immunization in goats with INH antiserum also increased the developing follicle number and ovulation rate.

Supplement of oocytes cultured *in vitro* with INH antibody increased the maturation rate and cleavage rate after fertilization. In addition, INH passive immunization increased the hormone-induced testicular spermatogenic cell number, and testicular blood flow in goats (Samir et al., 2020).

11.1.3 Gonadotropin inhibiting hormone immunization

Gonadotropin inhibiting hormone (GnIH), an important reproductive suppressor secreted by the hypothalamus, has the opposite biological effect on GnRH. GnIH mainly inhibits the secretion of FSH and LH from the anterior pituitary and directly participates in the regulation of ovarian function by paracrine or autocrine manner. It was reported that immunization in Gansu alpine fine-wool sheep with synthetic GnIH epitope polypeptide increased the blood FSH and LH levels (Zhang et al., 2016).

11.1.4 Anti-Müllerian hormone immunization

Anti-Müllerian hormone (AMH), also known as the Müllerian inhibiting substance (MIS) or Müllerian inhibiting factor (MIF), belongs to the TGF-β superfamily. AMH is secreted by granulosa cells of growing follicles in the ovary and Sertoli cells in the testis. In females, AMH has been shown to suppress the primordial follicle initiation, and inhibits FSH-dependent increase in aromatase activity, LH receptor expression, and COC *in vitro* maturation (Yu et al., 2022).

Using sheep as a model, Campbell et al. (2012) first reported that active immunization with AMH could decrease the population of gonadotropin-responsive preantral and small antral follicles and increase in both the number of gonadotropin-dependent antral follicles and ovulation rate. Chen et al. (2020) demonstrated that vaccination against AMH enhanced ovarian follicular development and increased clutch sizes by 1-2 eggs in 2 consecutive laying incubation cycles in geese. Zhang et al. (2021) further indicated that active immunization against INH-a, AMH, and prolactin fused protein could decrease the broodiness rate and improve egg laying performance in geese.

11.1.5 Immunocastration

Traditional castration involves surgical removal or chemical destruction of animal gonads, depriving the sexual activity to reduce the impact of sexual activity on the growth and development, meat quality and behavior. However, traditional methods not only affect animal growth and development, but also affect animal welfare. Immunocastration suppresses or eliminates sexual activity by neutralizing endogenous reproductive hormones, and promotes animal growth, and it's beneficial for the management and can improve the quality of animal products.

11.1.5.1 Overview of immunocastration

Hormone immunocastration mainly targets the reproductive hormones (such as GnRH, FSH and LH) of the hypothalamus-pituitary-gonad (HPG) axis. The GnRH vaccine has been studied the most and has made the fastest progress in immunocastration. There have been various GnRH vaccines in clinical trials and on the market (e.g., Improvac® and Bopriva). The *KISS1* encodes a family of neuropeptides called kisspeptins, which are produced by neurons of the hypothalamus anterior zone and are a principal regulator of reproduction processes. Kisspeptin is a major stimulus for the secretion of GnRH and gonadotrophins through binding to GPR54 and activating the downstream signaling pathway to switch reproductive activity, which may be another target of immunocastration. It was reported that a dual immunogen combining GnRH and kisspeptin are more efficacious through targeting two levels of the axis which were confirmed in prepubertal, pubertal and as adult male and female rats (Junco et al., 2021).

11.1.5.2 Basic principle of immunocastration

The reproductive hormones of the HPG axis are targets for castration vaccines, which induce antibodies to neutralize endogenous hormones, leading to the decline of gonadal hormone levels, sexual organ atrophy. Take GnRH as an example: specific anti-GnRH antibodies can be induced by immunization with the conjugation of GnRH and appropriate macromolecular carriers. Antigen-antibody complexes formed with endogenous GnRH can inactivate most of GnRH, reduce the response of pituitary cells, and specifically cause decreases in LH and FSH etc, resulting in degenerative changes of gonads. It was reported that immunization of GnRH can also reduce the number of pituitary GnRH receptors, thereby affecting the binding of endogenous GnRH to its receptors. Other studies found that GnRH immunization can destroy GnRH-secreting cells in the hypothalamus, causing permanent changes in reproductive endocrinology. GnRH immunization induces sharp decrease in LH, but slight decrease in FSH, indicating the secretion of LH is mainly directly regulated by GnRH in the hypothalamus, while FSH is not only regulated by GnRH, but also by other factors such as INH.

11.1.5.3 Methods of immunocastration

The methods of traditional immunocastration based on proteins include passive immunization, synthetic peptide, and genetic engineering.

1. Passive immunization

GnRH passive immunization can neutralize endogenous GnRH, inhibit FSH, LH and testosterone levels, and therefore inhibit animal sexual behavior. It was reported that GnRH passive immunization in adult rams rapidly reduced testosterone levels and significantly inhibited sexual behavior after booster immunization.

2. Synthetic peptide vaccine

There are four kinds of synthetic peptide for GnRH vaccines, namely GnRH monosomes, crossosomes, crossosomes dimers and GnRH analogues. All of them need to be coupled with carrier proteins. The immunogenicity of a single GnRH molecule conjugated with the carrier protein is poor and the immune dosage should be high. Fusing two GnRH molecules together to form a peptide with 20 amino acid residues (tandem) enhances the immunogenicity. Based on this, the dimerization of GnRH tandem to form GnRH tandem dimer can further enhance the immunogenicity. Improvac, a commercial vaccine developed by Pfizer Inc., through conjugating the synthetic GnRH analogues to the carrier proteins, has been used for the castration and effectively inhibited the endocrine function and sexual behavior, and lightened the smell of boars. This vaccine has been widely used in New Zealand, Australia and Brazil etc. It was reported that Improvac® also reduced the testicular volumes in colts (Birrell et al., 2021), inhibited breeding behaviors in male sea lions (Martinez-Giménez et al., 2021), enhanced the contraceptive effect in female lions (Shiihara et al., 2021), and represented an animal-friendly, non-invasive and reliable way to avoid early pregnancy in heifers as well as the slaughter of pregnant cattle (Schütz et al., 2021).

3. Genetic engineering vaccine

It was reported that mice immunized with the fusion protein ova-GnRH expressed by *E. coli* significantly reduced the ovarian and uterine weight. Female mice immunized with the fusion protein GnRH3-hinge-MVP induced higher level of anti-GnRH antibody. A fusion protein encompassing dog ZP3 fragment and two copies of GnRH with appropriate promiscuous T cell epitopes (dZP3-GnRH2) immunization reduced pregnancies as well as the number of pups born (Gupta et al., 2022). Recombinant vaccine composed of GnRH-Ⅰ and GnRH-Ⅱ (GnRH Ⅰ+Ⅱ) lowered the testosterone levels and spermatocytes in males, the progesterone level and luteal number in females, and the pregnancy rate and offspring number in mice (Ahn et al., 2022).

11.2 Genetic immunization

Genetic immunization (GI), usually called nucleic acid immunization, DNA immunization, is a technology that transforms eukaryotic expression vector with a target gene into animal host by a certain route, and the encoded target antigens can be expressed *in vivo*, causing antigen-specific humoral and cellular immune responses. Vaccines produced by this technology are called gene vaccines, naked gene vaccines, and DNA vaccines. In fact, mRNA molecules can also be translated into the target antigen and developed as nucleic acid vaccine, such as COVID-19 mRNA vaccines produced by BioNTech and Moderna (FDA, 2021). However, mRNA vaccine regulating reproductive performance are rare.

Genetic immunization began with the work of Wolff and his colleagues (1990) demonstrating that murine skeletal muscle cells could capture plasmids containing foreign genes and induced sustained protein expression after inoculation. In 1992, Tang et al. observed that mice given plasmids encoding *hGH* by gene gun induced anti-GH antibodies, and the antibody level was significantly increased after a booster immunization, indicating that the exogenous gene expression product was able to induce an immune response *in vivo*.

So far, gene vaccines are investigated for various applications including therapy of cancer, human immunodeficiency virus type 1, allergies, autoimmune, and infectious diseases. In the USA, at the moment over 500 clinical trials that focus on DNA vaccination are registered, targeting especially viral infections and cancer. Notably, four DNA vaccines have been approved for veterinary use, including West Nile virus DNA vaccine, infectious hematopoietic necrosis (IHN) virus DNA vaccine, canine melanoma DNA vaccine, and growth hormone releasing hormone (GHRH) DNA vaccine. In addition, a variety of DNA vaccines against hormone genes have been developed to regulate the growth, lactation, and reproductive traits of animals, including somatostatin, inhibin, kisspeptin, AMH, and GnIH DNA vaccines.

11.2.1 Principle of genetic immunization

In contrast to vaccines that employ recombinant bacteria or viruses, genetic vaccines consist only of DNA (as plasmids) or RNA (as mRNA), which is taken up by cells (such as myocytes or keratinocytes) and translated into protein in its native form. DNA vaccines can efficiently express antigens in transfected host cells, thus encoding endogenous antigens, which are more effective in inducing T cell responses, resulting in producing both CD4+ and CD8+ T cell responses in parallel. Apart from the cellular immune response, a humoral immune response is also stimulated once the B cell receptor recognizes the target antigen. Considering the ability of DNA immunization inducing humoral immune responses, DNA immunization has been regarded as a technology platform for monoclonal antibody induction (Liu et al., 2016). That DNA vaccines induce adaptive immune responses has been shown in Fig. 11-1 (Hobernik et al., 2018).

11.2.2 Methods of genetic immunization

Generally, there are two crucial steps in the process of genetic immunization. The first is to design or construct plasmid of gene vaccines using vector and target gene, followed by plasmid extraction and *in vitro* expression of DNA vaccine antigen. The second is to deliver gene vaccines, including selection of animal models, immunization schedule, and choosing the delivery methods. The conditional routs of genetic immunization are needle injection via either intramuscular (IM) or intradermal (ID) routes. Given that DNA vaccine is immunogenic in animal models, it has overall low immunogenicity in humans or large animals. Lots of methods have been developed to enhance immunogenicity of DNA vaccines, including optimized DNA vaccine design and development of new DNA vaccine delivery systems. Among them, the use of physical delivery approaches, including gene gun or

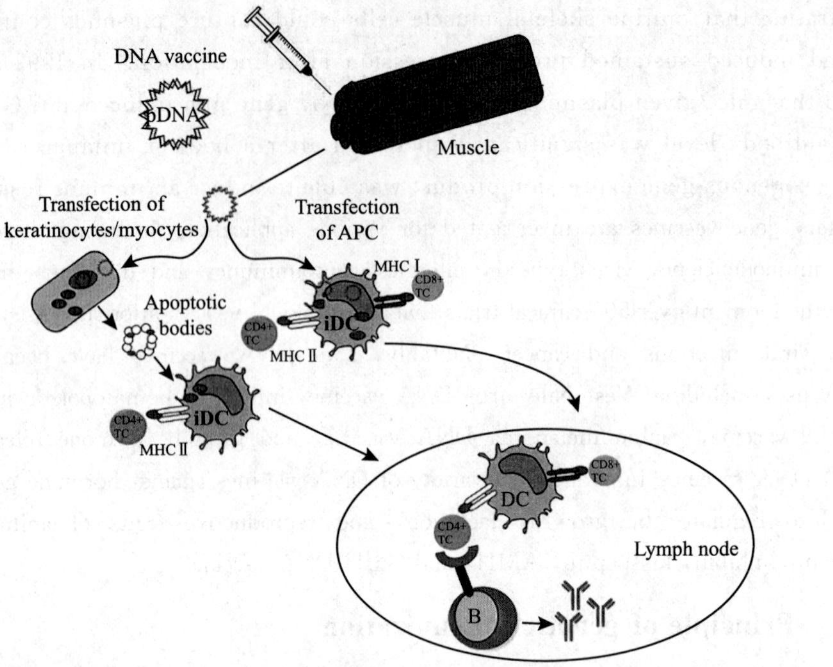

Fig. 11-1 DNA vaccines induce adaptive immune responses (Hobernik et al. , 2018)

electroporation (EP), which have their unique advantages and have greatly improved the immunogenicity of DNA vaccines in the humans or livestocks. Another attractive strategy is to perform "heterologous prime-boost vaccination", in which the hosts are first immunized with a DNA vaccine, followed by boost immunizations with the different types of vaccines containing the same antigens, which are more effective than homologous prime-boost using multiple times of the same vaccines.

11. 2. 3 Advantages and safety of genetic immunization

Gene vaccination provides several advantages over the traditional protein/peptide-based vaccines. The most obvious advantages seem to be the purity, stability, low cost, and easy to manufacture for use, which facilitate the application of gene vaccines. In the host cell, vaccine antigens are expressed in their native form, thus optimizing their presentation and processing to the immune system.

The gene vaccine is relative safe to the host due to its plasmids non-live, non-replicating and non-spreading, leading to a little risk of either reversion to a disease-causing form or secondary infection. However, there are still some safety issues that deserve attention, such as the potential to integrate into host genome DNA leading to insertional mutation and possible tumorigenesis, the development of autoimmune responses including anti-DNA antibodies, and the possibility of antibiotic resistance. It must be mentioned that there is no evidence for chromosomal integration of injected DNA. With regards to antibiotic resistance issue, the antibiotic-free plasmid selection system is explored and used.

11.2.4 Genetic immunization for regulating animal reproduction

11.2.4.1 Inhibin DNA immunization

Genetic immunization with inhibin against animals can induce the specific anti-INH antibody, which further neutralizes the endogenous INH and reduces the negative feedback regulation of INH on gonadotropins, resulting in the increase of the secretion and release of FSH, ultimately improving the reproductive performance of animals (Table 11-1).

Table 11-1 Immune response of inhibin DNA vaccines

Vaccines	Animals	Routes	Positive ratio/% (P/N value)	References
PCIS	Rats	IM	55.6(20/36)	Mao et al., 2003
PCIS	Rats	IM	38.9(21/54)	Han et al., 2008
PCISI	Rats	IM	61.1(33/54)	Wang et al., 2012
C500/pXAIS	Mice	Oral	NA	Han et al., 2014
C500/pXAIS	Rats	IM	60(6/10)	Hui et al., 2014
p-SINH/SRFRP	Mice	Elect	$80(8/10)^{I}/60(6/10)^{R}$	Dan et al., 2016a
p-TPA-SINH/TPA-SRFRP	Mice	Elect	$100(10/10)^{I}/90(9/10)^{R}$	Dan et al., 2016a
PCIS plasmid	Sheep	IM	46.7(14/30)	Han et al., 2008
PCISI plasmid	Chickens	IM	57 (17/30)	Mao et al., 2016
p-TPA-SINH/TPA-SRFRP	Sheep	IM	$3.288 \pm 0.477^{I}/3.119 \pm 0.225^{R}$	Dan et al., 2016b
C500/pXAIS	Buffalo	IM	90.24	Liu et al., 2016
C500/pXAIS	Buffalo	Nasal	94.12	Liu et al., 2017

Notes: IM=intramuscular injection; Elect=electroporation; I means Inhibin; R means RFRP; NA means not applicable.

Inhibin genetic immunization study began with Jiang et al., in which INH α (1-32) fragment was inserted into pCDNA 3.1 vector for generating eukaryotic expression plasmid (pINH). After immunizing mice with pINH, antibody-positive mice reached to 30%, whereas there was no significant difference of the litter size and birth weight between positive mice and negative mice groups. Given the poor immune response of pINH plasmid, the hepatitis B surface antigen gene (*HBsAg*) was subsequently used as particulate carrier to enhance the immunogenicity of inhibin α subunit, named pCIS (one copy of inhibin gene) and pCISI (two copies of inhibin gene) plasmids. The immune response of pCIS DNA vaccine has been tested in mice, and results showed that rats vaccinated with pCIS plasmid obtained a greater number of mature follicles (Han et al., 2008; Fig. 11-2). Regarding pCISI DNA vaccine, Wang et al. (2012) reported that pCISI plasmid was able to induce humoral immune response, raise FSH and estradiol concentrations, and increase the number of placentas and litter size.

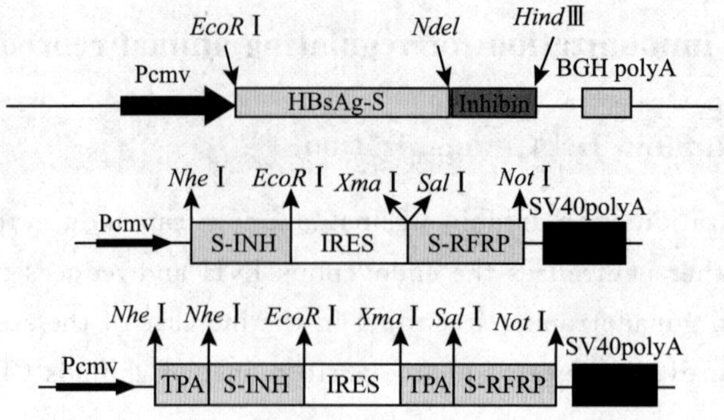

Fig. 11-2 The simple map of partial inhibin vaccines

(Han et al. , 2008, Dan et al. , 2016a)

On the basis of the pCIS plasmid, Han et al. (2014) further induced *asd* labeled balanced lethal system to develop crp^-/asd^- C500/pVAX-asd-IS (pXAIS) vaccine, in which *asd* rather than the kanamycin gene was used as selection marker (Han et al. , 2014) . Meanwhile, the immune response and the effects on reproductive traits in mice were assessed, and results showed that the novel vaccine can efficiently induce anti-INH antibodies and increase in number of large follicle population in mice. Likewise, Hui et al. (2014) found that rats in C500/pXAIS treated group exhibited more corpora lutea and higher estradiol and progesterone contents, suggesting this inhibin DNA vaccine may stimulate ovarian folliculogenesis and increase the ovulation in rats. Moreover, the efficacy of C500/pXAIS vaccine was also evaluated in rhesus monkeys with obvious immune response and increased the concentrations of E_2 and P_4 (Chen et al. , 2018).

To further improve fertility of animals, synthesized SINH and SRFRP [*INH* and *RFRP* were separately fused to the C-terminus of the hepatitis B virus gene (*HBV-S*)] fragments were inserted into multiple cloning site of pIRES vector, named p-SINH/SRFRP (Dan et al. , 2016a). Furthermore, the tissue plasminogen activator (TPA) signal sequence was subsequently inserted into p-SINH/SRFRP plasmid, eventually producing p-TPA-SINH/TPA-SRFRP recombinant plasmid. Mice immunized with p-TPA-SINH/TPA-SFRFP plasmid showed the higher positive rates (100% for anti-INH, 90% for anti-RFRP, respectively), resulting in more litter size (Dan et al. , 2016a; Fig. 11-2; Table 11-1). Considering the lower expression of downstream ORF in pIRES vector, Tang (2020) constructed a novel INH and RFRP dual expression DNA vaccine by using 2A peptide (pVAX-SINH-2A-SRFRP-asd), which presents the similar expression between the upstream ORF and downstream ORF. Tang (2020) further observed that mice treated with pVAX-SINH-2A-SRFRP-asd vaccine obtained higher litter size when compared to control group.

Besides the experimental animals, the effects of the above mentioned INH vaccines have been investigated in large animals, including cattle, sheep, and chickens. In sheep, the twinning rate in pCIS immunization groups were significantly higher than that in control groups (39.2% vs. 10%). Similar results were reported by Dan et al. (2016b) that maximum twinning rate of ewes were found in the p-TPA-SINH/TPA-SRFRP (37.5%) and p-SINH/SRFRP (37.5%) treated groups. In hens, Mao et al. (2016) observed that vaccination with pCISI promoted egg lay performance, increased the amounts of dominant, small yellow and large white follicles in the ovary, whereas it had no effect on egg quality and hatching rate.

In 2016 and 2017, Liu et al. combined inhibin genetic immunization with synchronization and timed artificial insemination (TAI) programs, developed and applied two different inhibin immunization procedures in crossbred buffalo (Fig. 11-3). Results indicated that the number and diameter of large follicles (≥10 mm) as well as ovulatory follicles in C500/pXAIS vaccine group were significantly greater than non-inhibin immunization group, resulting in a greater conception rate in buffalos with positive antibodies (51.02 % vs. 30.00%).

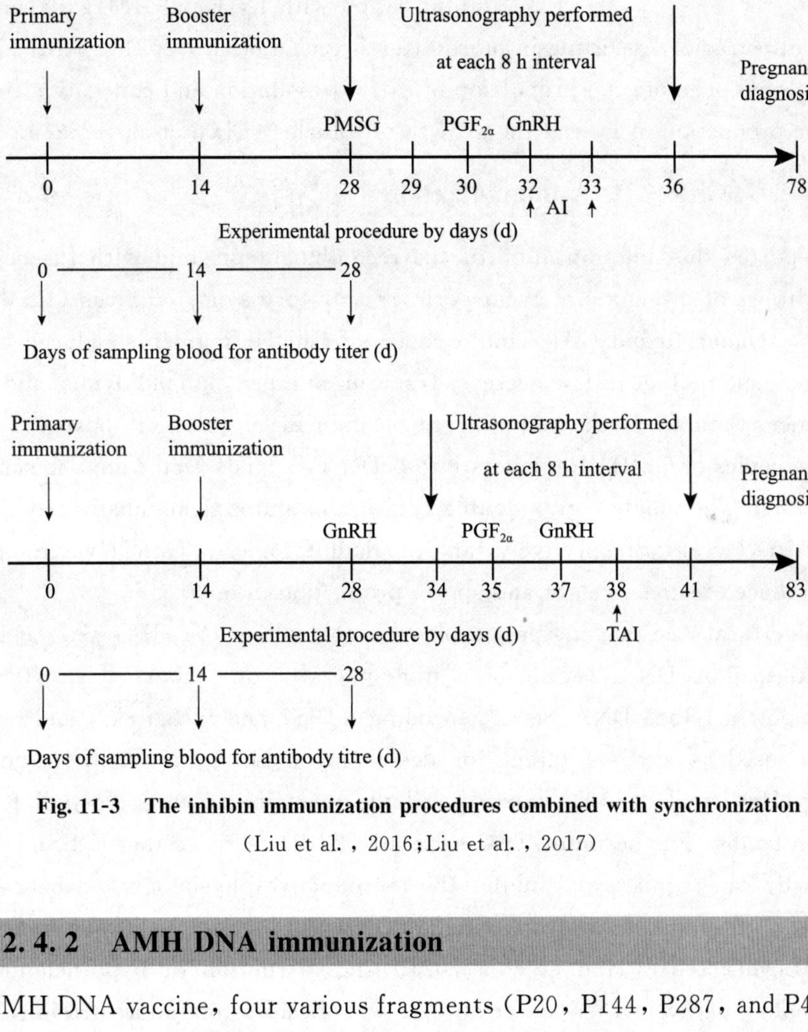

Fig. 11-3 The inhibin immunization procedures combined with synchronization
(Liu et al., 2016; Liu et al., 2017)

11.2.4.2 AMH DNA immunization

For AMH DNA vaccine, four various fragments (P20, P144, P287, and P450) of AMH were employed to construct AMH eukaryotic expression plasmids (pVAX-tPA-SAMH-asd, Fig. 11-4), and it was reported that the novel AMH DNA vaccine could increase the numbers of secondary follicles, resulting in higher litter size (Jiang et al., 2018). Based on

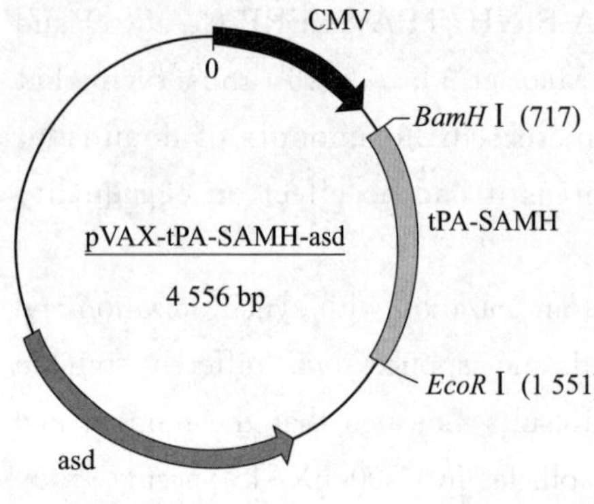

Fig. 11-4 AMH eukaryotic expression plasmid (Jiang et al. ,2018)

this AMH DNA vaccine, Tang et al. (2020) constructed serials AMH co-expression plasmids, including AMH-RFRP dual co-expression plasmid (pVAX-SAMH-2A-SRFRP-asd) and AMH-RFRP-INH triple co-expression plasmid (pVAX-SAMH-2A-SINH-2A-SRFRP-asd), and found that both double-expression and triple-expression AMH vaccines increased the number of offspring in mice. Among them, the triple-expression AMH-INH-RFRP vaccine showed the highest litter size (15.93 ± 1.90).

In addition, a recent study showed that co-immunization with INH and AMH plasmids induced higher levels of estradiol, resulting in more litter size (Yu et al. , 2021). AMH-INH-RFRP DNA vaccine could promote the proportion of estrus, ovulation and conception in buffalo by promoting the production of E_2 and the growth of follicles (Chen et al. , 2023).

11. 2. 4. 3 GnRH DNA immunization

It was reported that immunization of the recombinant plasmid with fused genes of T helper cell epitopes of plasmodium cyclosporin, respiratory syncytial virus (RSV), measles virus (MVP), tetanus toxoid (TT) and 5 copies of GnRH fragments reduced testosterone level, inhibited spermatogenesis, decreased sperm number in epididymis and implanted embryos in mice. Similarity, the recombinant plasmid fused genes of above 4 T helper cell epitopes and 4 copies of GnRH fragments into pcDNAV5-HisB vector got the same effect of castration in rats. The single copy, double copies and amino acid-substituted *GnRH* were cloned into pVAX1 vector, respectively, and the double copies of *GnRH* vaccine pVAX-S2G significantly reduced testicle weight and sperm production in mice.

Given the critical role that kisspeptin plays in controlling the HPG axis, it is proposed to develop kisspeptin DNA vaccine for immunocastration. Han et al. (2015) first constructed a potent KISS1 DNA vaccine encoding *KISS1* and *HBsAg-S*, and reported that KISS1 can be used as a novel target for developing a DNA immunocastration vaccine. Subsequently, Han et al. (2018) investigated the reversibility and safety of KISS1 DNA vaccine in ram lambs. Furthermore, Tesema et al. (2019) reported that KISS1 DNA vaccine (C500/pKS-asd) can regulate and inhibit the reproductive physiology and behavior of ram lambs. Notably, the suppressor effect of KISS1 DNA vaccine may be related to the apoptosis of testicular Leydig cells (Han et al. , 2021) and disruption of hypothalamic-pituitary-testicular signaling pathway (Wassie et al. , 2019). Besides immunocastration, the KISS1 DNA vaccine have also been used in female rats and results showed that the oral KISS1-54 DNA vaccine with fused tPA signal peptide was more effective than that KISS1-10 DNA vaccine in decreasing the litter size of female rats (Han et al. , 2018).

As mentioned above, Junco et al. (2021) developed a dual immunogen combing GnRH and kisspeptin, and indicated that kisspeptin-GnRH immunogen is efficacious in producing immunoneutralising antibodies and infertility in rats. To this point, it is worth developing GnRH-kisspeptin dual expression DNA vaccine to inhibit the reproductive capacity.

Assignments

1. What hormones can be used by immunological methods to enhance animal fecundity?
2. What hormones can be used by immunological methods to reduce animal fecundity or as immunocastration antigens?
3. What are the forms of hormonal immunogens?
4. What are the concept and advantages of genetic immunization?
5. What are the safety issues of genetic vaccines?
6. What is the progress of inhibin gene vaccine?
7. What are the problems that need to pay attention to in hormone genetic immunization?
8. Based on the principle of hormone genetic immunization, could you please design a DNA vaccine to improve bovine fertility except INH, AMH, and GnIH DNA vaccines?

(Mao Dagan, Liang Aixin)

Chapter 12 Avian Reproductive Biotechnology

Summary

Different from mammals, birds are very special animals easy to handle and treat. Avian reproductive biotechnology is an important part and branch of animal reproductive biotechnology. This chapter mainly introduces the mechanism and useful protocols of avian seasonal reproductive activity control, embryo manipulation, chimeric technology, sex control, and transgenic technology.

The germ cell is the only cell type capable of transmitting genetic information to the next generation. Therefore, the manipulation of primordial germ cells (PGCs), progenitor cells for gametes (spermatozoa and ova), is directly linked to the manipulation of individuals, making PGCs attractive target objects in the field of biotechnology. However, the application of avian biotechnologies is difficult in accessing and manipulating the large yolky zygotes. As an alternative, a unique biological property of avian PGCs is that they circulate temporarily in the vasculature during early development, which allows accessing and manipulating the avian germline. We introduce the principles and procedures of avian germline chimera production and their application for germline modification as well as genetic resource conservation. The fundamental avian reproductive technologies are also described, including seasonal reproduction control, artificial insemination and embryo culture, and chimera production in avian species.

The impact of sex on economic benefits is particularly prominent in poultry production. In the intensive production of laying hens, people hope to get as many female chicks as possible. Sex reversal and sex identification are two common phenomena in the poultry production. The original factor of sex differentiation in birds, which is similar to the function of *SRY* in mammals, is likely to control gender differentiation by regulating the expression of aromatase gene, as well as the relative levels of estrogen and androgen in embryos. The research on transgenic avian is of great significance to improve the disease resistance and product quality, especially in the research of bioreactor. The common transgenic technology of avian includes the method of sperm vector, technique of microinjection for fertilized egg, transfection of embryonic placental cells using retroviral vectors, transfection of blastoderm cells with lipofection packaging vector, and the method of blastoderm cell and primordial germ cell vectors.

12.1 Techniques to control avian seasonal reproduction

Seasonal reproduction is the phenomenon of reproduction in a particular season in order to adapt the natural selection. Light is the main regulation factor of seasonal reproduction. The seasonal reproduction responded to changes in photoperiod involves many complex physiological processes, including the perception and conversion of light signal, the regulation of hypothalamus-pituitary-gonad (HPG) axis and secretion and releasing of various reproductive hormones. In contrast to mammals, birds have extra-ocular photoreceptors in the pineal organ and the deep brain. In birds, the eyes are not necessary for seasonal reproduction, and light is instead directly perceived by deep brain photoreceptors.

12.1.1 Mechanism of avian reproductive seasonality

Among all the environmental factors, light has the most significant effect on the seasonal reproduction of birds. Changes of light are perceived mainly through photoreceptors in the hypothalamus and accordingly, adapt their reproductive activities. The key factor of seasonal variation of reproductive activity in birds is the change of gonadotropin-releasing hormone (GnRH) neuron activity in hypothalamus. Seasonal reproductive activity is mainly regulated by the HPG axis. The secretion of GnRH and vasoactive intestinal peptide (VIP) in hypothalamus is regulated by light, which regulates the seasonal reproductive activities of birds through the secretion of luteinizing hormone (LH) and prolactin (PRL), respectively. Light acts on GnRH neurons through the pituitary portal system to secrete follicle-stimulating hormone (FSH) and LH. Then FSH and LH stimulate small follicles to produce estrogen (E_2) and progesterone (P_4) by binding to receptors on theca cells and granulosa cells.

The initiation of seasonal reproduction in birds is related to the secretion of PRL with increasing temperature. The occurrence of broody cycle and the transition between breeding and non-breeding seasons are regulated by the alternating secretion of PRL and LH. PRL inhibits reproduction through two aspects: on the one hand, it inhibits the secretion of pituitary gonadotropin in hypothalamus; on the other hand, it directly inhibits the expression of steroid hormone synthesis gene through PRL receptor binding to gonads, thus inhibiting follicular development and leading to follicular atresia.

In addition, melatonin regulates the reproduction of birds through a similar way to TH-mediated process of regulating GnRH release in mammals, and may also indirectly inhibit the release of gonadotropin by regulating the synthesis and secretion of GnIH in hypothalamus, thus reducing the activity of HPG axis.

When birds are exposed to prolonged light, the phenomenon known as photo refractoriness occurs, in which hypothalamus photoreceptors are less able to convert light energy into nerve impulses and can not maintain high levels of gonadotropin secretion. The secretion level of PRL increases continuously during the reproductive period with prolonged illumination, and finally inhibits the secretion of GnRH in the hypothalamus when it reaches a certain threshold, thus terminating the reproductive activity and the breeding season. This change is actually an adaptive mechanism that restricts the birds to continue to breed for a certain period of time, so that they can have well physiological conditions for the next breeding cycle.

12.1.2 Procedures to control avian seasonal reproduction

Seasonal reproductive activities of avian are regulated by the HPG axis system. The seasonal variation of reproductive hormones secreted by HPG is regulated by the photology cycle. The possible physiological pathways of seasonal reproduction are: light stimulation acts on the retina and is transformed into nerve impulses, which act on the pineal gland through the hypothalamus and the suprachiasmatic nucleus (SCN), the main circadian clock in birds, and the pineal gland secretes melatonin. Melatonin regulates the impulsive release of GnRH in hypothalamus through complex neuroendocrine processes. It stimulates the pituitary gland to secrete and release gonadotropins (such as FSH and LH), and then initiates reproductive activities such as gonadal development and ovulation in birds. This process is regulated by a variety of factors, such as positive and negative feedback regulation of steroid hormones, seasonal nutrition regulation and survival pressure.

A retrograde mechanism discovered in recent years may be involved in the indirect regulation of neuro-endocrine in hypothalamus by melatonin. Melatonin from pineal gland, as seasonal signal, does not directly act on GnRH cells, but by acting on anterior pituitary gland, which releases thyroid stimulating hormone (TSH) and then regulates the synthetic release of hypothalamus reproductive hormone.

12.2 Avian embryo manipulation technique

The prominent characteristic of embryo development in birds is ovulation, which is different from that in mammals. The embryo goes through two developmental stages from zygote to chick, the womb stage *in vivo* and the shell stage *in vitro*. The embryo development *in vitro* is independent on the mother, which is conducive to the operation of the embryo, but the relatively firm and closed eggshell becomes a major obstacle. *In vitro* culture technology refers to the fertilized egg or early embryo of poultry in the substitution shell or artificial container composed of relatively open *in vitro* culture system incubation to obtain individuals. Using *in vitro* culture technology can meet the needs of the operation of embryos in all stages of development, and has the advantages of easy observation. Poultry embryos are good experimental material of developmental biology and play an increasingly important role in study in the development science and technology, embryology, developmental biology, genetics, transgenic bioreactor, toxicology, etc. Therefore, the poultry embryos *in vitro* culture technique also has the very strong practical significance.

12.2.1 Artificial insemination and embryo transplantation

Artificial insemination has been used in particular for the turkey, chicken and guinea fowls due to the simplification of the insemination process through the establishment of methods for semen collection and vaginal insemination. Especially, artificial insemination is becoming common in breeding broilers in accordance with the recent advancement of geneticselection. Artificial insemination has also been applied to test the possibility of inter specific hybridization.

Generally, after continuous observation, the hen is killed by cervical dislocation about 30 min after laying, and the egg is obtained by operation in the funnel of the fallopian tube or in the abdominal cavity. Each egg is placed in a modified version of Ringer's solution. Fresh chicken semen is collected, diluted and washed with the above solution, and the sperm solution is dripped onto the placenta of the ovum, and then placed in a 5.0% CO_2 incubator at 41 ℃ for 15 min for fertilization. After fertilization, the sperm is washed twice with culture medium to remove excess sperm.

12.2.2 Culture of avian embryos in the early stage

Currently, most of the methodologies used in avian biotechnology involve the accession to and manipulation of developing embryos. After continuous exploration, great progress has been made in the research on *in vitro* culture of poultry embryos, and the culture system of each development stage has been preliminarily established. There are three main methods of accessing developing avian embryos: the non-eggshell vessel culture, the surrogate shell system and the three-stage culture system.

12.2.2.1 Early non-eggshell vessel culture

Initially, people use ceramic dishes, plastic bags and other containers to add artificial synthetic culture medium to cultivate poultry embryos. Embryonic day 1 chicken and duck embryos are incubated in solid medium and the embryos can survive *in vitro*. Chicken embryos cultured in the above containers die about 10 d after incubation. Non-eggshell vessel culture has not yet hatched individuals, the longest survival up to 21 d. The embryos cultured by the above methods all show the lack of calcium, magnesium and other mineral nutrients, and the embryo development is delayed, so it is impossible to get hatched chicks. At present, the non-eggshell vessel culture method has rarely been adopted.

12.2.2.2 Surrogate shell system

Eggshells play important roles of embryo development in poultry, not only protecting embryos, but also providing adequate mineral nutrition for embryo development. With the analysis of the symptoms of embryo retarding and severe calcium deficiency caused by vessel culture and the deepening of the understanding of the close relationship between embryo development and eggshell, result shows the irreplaceable role of eggshell in embryo development. And then the system of using substitute shell system for embryo culture has been established. There were 3% of quails hatched by transferred quail embryos incubated for 2.5 d into chicken egg shells. By transferring chicken embryos incubated for 3 d, the rate of hatchability was 20.0% in turkey eggshells and 21.7% in chicken double-yolked eggshells, respectively. The hatched chicks have been achieved for the first time by using the surrogate shell system and demonstrated that embryos could use albumen and eggshells from other species.

12.2.2.3 Three-stage culture system

The egg of poultry is large with much yolk, and fallopian tube is located in the deep abdominal cavity, so the operation of fertilized egg transplantation into the recipient is difficult. In order to apply microinjection technology to transgenic poultry, researchers have made unremitting efforts to improve the substitute shell system and develop a complete *in vitro* culture system. Chicken embryo development is divided into three stages for the purpose of *in vitro* culture: fertilization to blastoderm formation (stage Ⅰ) lasts for 1 d, embryogenesis (stage Ⅱ) lasts for 3 d, and embryonic growth (stage Ⅲ) lasts 18 d. Then the embryos are cultured in three different culture systems in corresponding to different three stages.

The first culture system which is normally in the oviduct is designed to provide an environment similar to that *in vivo*. To extend the course of development into embryogenesis, transfer of a cultured blastoderm to a host eggshell is required (system Ⅱ). After transferring the contents of a glass jar in system Ⅰ to a surrogate eggshell prepared from a freshly laid egg by cutting off the harp end, the shell is filled with thin albumin, and sealed with cling-film secured by plastic rings and rubber bands. The reconstituted egg is incubated for about 3 d at 38 ℃ and 50%-60% relative humidity, with 90° tilting in hourly cycles. For further development, the embryo cultured in system Ⅱ is transferred again to a larger host eggshell containing an air space to expose the vascular systems of the extra-embryonic membranes to the atmosphere (system Ⅲ). There were 7% chicks hatched, thus the three-stage culture system was successfully established and improvements were developed on this basis since then. Early cleavage embryos (stage Ⅱ) were obtained without killing hens by injection of vasopressin to produce soft-shell eggs. And 22.5% hatchability was obtained after three stages of culture. By further improved the method, removing concentrated proteins in the second stage of culture, about 35% hatchability was yielded in chicken. Therefore, *ex-vivo* culture system enables easy access and precise manipulation of the developing avian embryos.

12.3 Avian chimeric technology

Chimeric technique is a good model for studying developmental biology and cell differentiation. Chimera production technologies are now essential for avian biotechnologies. Donor cells can be cultured *in vitro*, and donor PGCs or their progenitors can be differentiated into functional gametes in chimeric gonads to obtain chimeras. Genetic modification and selection of donor PGCs, transfection of exogenous genes and preparation of chimeras can obtain progenies containing exogenous genes, which is one of the effective ways to study transgenic poultry. In addition, chimeras of applied lines can also be used to preserve the genetic resources of the same species or even other species with high reproductive performance and a large number of poultry, so as to achieve species conservation and rapid propagation.

12.3.1 Classification of avian chimeras

There are two main methods to prepare poultry chimeras according to donor cell types: blastodermic chimeras and primordial germ cell chimeras. According to the difference of the biological classification of donor and recipient, it can also be divided into intra-species chimeras and inter-species chimeras. There are two types of germline chimeras and somatic chimera according to the site of cell implantation.

12.3.1.1 Chimera of blastocysts

The blastodermal cells are the cells in the stage X of the newly fertilized egg. At this time, the blastodermal cell consists of approximately 60 000 cells. The blastodermal cell consists of two cell types, somatic cells and PGCs, which develop into extra-embryonic tissue, embryonic body and germ cells respectively. When transplanted into the central bright region, blastocytes can differentiate into epithelial tissue and mesenchymal tissue in the recipient, indicating that blastocytes are a kind of pluripotent cells. Blastoderm cells from White Leghorn chickens are injected into the blastoderm of Rock chickens, and chimeric embryos are determined from the phenotype of feather color. The blastodisc stage is the earliest *in vitro* development of chicken embryos, which is easy to acquire and manipulate.

For a long time, chicken, quail and turkey have been used as models in chimerism studies. Chimera of transverse Rock and White Leghorn chickens were obtained by transplantation of blastocysts. Phenotypic chimera was obtained from feather color judgment of hatchling individuals, and the offspring was found to be germinal chimera. DNA fingerprinting analysis of blood and sperm also confirmed that donor cells were chimera in blood and gonadal tissues. These results indicated that PGCs in blastodisc cells had entered the receptor gonads and differentiated into functional gametes. The stage X blastocyst can enter germline of chimera, hematopoietic organs and epidermal tissues after being injected into the corresponding recipients, so it is a kind of multi-differentiation potential cell.

Since then, studies on blastodisc cells have gradually increased. Chimeras of Rock chickens and White Leghorn chickens cultured *in vitro* were prepared. The chimerism between different varieties of quails was prepared by injecting wild color quails' blastoderm cells into the inferior cavity of recessive white feather quails' blastoderm, and the phenotypic chimerism rate was 14%. The chimeras of different breeds of chickens were prepared from the frozen and thawed germ disc cells. Female and male chimeras were interbred with each other, and 9% of the offspring were offspring from the donor chicken. This was the first report on the restoration of the donor by chimera technology in birds.

The establishment of a long-term culture system of blastocyst will undoubtedly bring great convenience to the operation of blastocyst *in vitro*. The successful isolation of mammalian ES cells has greatly promoted the research of avian stem cells. The blastocyst cells of chickens were cultured *in vitro* using STO as feeding layer, and many kinds of growth promoting and differentiation inhibiting growth factors were added, such as bFGF, IGF-I, LIF, IL-1 and SCF, etc. It was found that blastocyst cells could divide and expand fast on STO feeding layer. Further tests after continuous passage showed that the cultured blastodisc cells had the characteristics of mouse ES cells. If the cells had alkaline phosphatase activity, they were tested positive by mouse ES cell-specific antigen antibodies ECMA-7, SSEA-1, EMA-1, etc. The cells had strong telomerase activity, and after removing the factors inhibiting differentiation *in vitro*, they could form embryoid bodies containing three different cell types of germ layers. The cultured and passed blastocytes were transplanted into recipient blastoderm to prepare chimeras, 50% of the hatched individuals were phenotypic chimeras, and germline chimeras were also obtained. In conclusion, cultured chicken blastocyst cells have similar characteristics to mouse ES cells.

12. 3. 1. 2 Chimera of primordial germ cells

The most important aspect of avian biotechnologies is that manipulated cells can contribute to the germline and form particular functional gametes. There are two types of cells in the early stage of blastocyst: somatic cells and PGCs, so somatic and germline chimeras can be prepared by transplantation of blastocyst. The number of PGCs in the early embryo is quite few, and the proportion and degree of chimerism in the germline are low. The ultimate goal of chimerism is to obtain a high proportion of offspring from donor cells, whether as a way to transgene or as a measure to preserve genetic resources.

PGCs is the ancestor cell of germ cell lineage, and the direct operation of PGCs to prepare chimera will undoubtedly be beneficial to improving the ratio of germline chimera. Poultry PGCs originate in the center of the blastoderm at the freshly ovulation stage and then migrate to primitive germinal ridge at a certain stage of embryonic development. Different from mammals, the migration mode of PGCs is not only inter-tissue migration, but also passive migration with blood circulation, which brings great convenience to the operation of poultry PGCs. Following the development of micromanipulation of PGCs in chickens, germline chimeras are produced through the transplantation of PGCs to the appropriate sites of developing recipient embryos. There are generally two ways of producing germline chimeras: transplantation of donor PGCs into the sub germinal cavity at the blastoderm stage or into the blood vessel at the somitogenesis stage in recipients. According to the different sources of donor PGCs, there are three approaches as follows: PGCs from germinal crescent, blood circulation system, or primary gonads.

It has been a long process to transplant PGCs into chimeras. When PGCs isolated from the turkey germinal crescent are injected into the blood vessels of early chicken embryos, turkey PGCs are found in the gonads of the recipient, which makes the first successful turkey-chicken interspecific chimerism. It is proved that PGCs in the germinal crescent region of the donor can differentiate into functional gametes in the gonads of the recipient.

PGCs migrating with the bloodstream can be obtained by collecting early embryonic blood, which contains only PGCs and red blood cells. But the low number of PGCs in the blood makes it difficult to gain enough PGCs to operate. After continuous improvement, the purity of separation PGCs reaches more than 50%. The improvement of separation method has greatly promoted the study of PGCs chimeras. Currently, the proportion of offspring from donor PGCs can be up to 97%. Such high germline transmission indicates that chimeric technology is feasible for conservation of poultry genetic resources and transgenic research.

Although PGCs can be easily isolated in blood and germinal crescent, the amount of PGCs increases rapidly after migrating to the primitive germinal ridge and entering mitotic stage, and the number reaches 2 000 at stage 26, accounting for 5.0% of total cells. Therefore, primary gonads are favorable sources of PGCs. Primordial PGCs of 5-day-old chicken are transplanted into the 2.5-day-old recipient blood of after 5 d culture *in vitro*, and it is found that PGCs transplanted after proliferation and division in settled gonad can still migrate to the recipient gonads. The successful preparation of gonadal PGCs chimeras indicates that PGCs with different development stages still have the ability of migration and differentiation in the recipient, thus opening up another important source for the collection of PGCs.

12.3.1.3 Interspecific germline chimera

The intraspecific chimeras have been successful produced by transplanting of PGCs from quail, chicken and turkey with a high proportion of the offspring from the donor. Exogenous PGCs can enter the gonads of the recipient and differentiate into functional gametes to prepare interspecies chimeras. This method will make it possible for wild endangered birds to recover in poultry embryos, providing a new way for breeding and protection of wild endangered birds.

Germline chimeras have been produced by germ cell transplantation to the same species (allogeneic) and in cross-species situations (xenogeneic). The preparation of interspecific chimeras depends on two critical factors: whether the donor PGCs can migrate to the gonads of the interspecific recipient, and whether the donor PGCs can differentiate into functional gametes. The xenogeneic fish germline chimeras are produced successfully by transplantation of PGCs or spermatogonial stem cells (SSCs) of salmonid into rainbow trout and functional donor-derived gametes are obtained. Xenogeneic germline chimeras in mammal can be produced by microinjection of ES or iPS cells into blastoderm, or by transplantation of SSCs into seminiferous tubules of the recipient. Transplantation of SSCs obtained from rodents such as rats leads to spermatogenesis in the testes of immunodeficient mice. In avian species, interspecific chimeras of chicken and quail have been used commonly to trace cell lineages. However, no donor-derived offspring have been produced from recipients when quail blastoderm cells or PGCs are transplanted into chickens. The key ability of xenotransplanted PGCs to migrate to the gonads is well conserved among avian species. To date, there are a series of reports on the production of viable progeny derived from xenotransplanted PGCs through testcross analysis of recipients in males. Although additional histological evidence is required to confirm xenogeneic spermatogenesis, the report provides new insights into the possibility of spermatogenesis of avian species. It is also reported that histological observations of the existence of oogonia in primary follicles derives from female chicken PGCs in the ovaries of quail recipients. The xenotransplanted female PGCs are capable of entering recipient gonads and subsequently starting follicle development, but most interspecific follicles either disappear before maturation, or their formation is suppressed. Transplantation of donor PGCs carrying a reporter gene to interspecific recipients would provide important information on the ability of xenogeneic gametogenesis through sequential histological analysis. In the future, the fate of xenotransplanted PGCs in recipient gonads needs further detail investigation.

The chemotaxis of embryogenic gonad anlag plays an important role in the directional migration of PGCs. Both *in vivo* and *in vitro* experiments have shown that the gonad anlag also plays a chemotaxis role in heterologous PGCs. Chicken embryo progenies have chemotactic effects on PGCs of quails, turkeys, ducks and even mice, and there may be a common pattern of PGCs migration beyond species boundary barrier. This provides a theoretical basis for the preparation of interspecific chimera by transferring heterologous PGCs through bloodstream. Studies on interspecific chimerism began in the 1990s when quail PGCs were transferred into the blood vessel of chicken embryos at stage 14 and it was found that quail PGCs could migrate to the primary gonads of chicken embryos. Quail PGCs accounted for 14.2% of the gonads of chimeric chicken embryos at stage 29, and there were chicken and quail spermatogonial cells in the testes of hatched cocks. Differentiation of PGCs into gametes is a very complex process, involving the interaction between PGCs and gonadal stromal cells and the regulation of reproductive endocrine hormones at all levels. It is not clear whether these effects are different between species, and whether exogenous PGCs causes immune rejection in the recipient gonads. Therefore, the survival, development and differentiation of exogenous PGCs in recipient gonads need to be further studied.

12.3.2 Production of avian chimeras

12.3.2.1 Subgerminal cavity injection

The preparation of blastodermal chimeras is a process of transferring donor cells into the subgerminal cavity of recipient which is a small space between the blastoderm. Injection of blastoderm cells into the subgerminal cavity is a useful method for assaying the contributions of transferred cells to all tissues including somatic as well as the germline in recipients. The operation of blastodermal chimeras is relatively simple, and somatic as well as germline chimeras can be produced by transferring blastoderm cells into the subgerminal cavity of recipient embryos at stage X. The transplanted blastodermal cells will contribute to three germ layers (ectoderm, mesoderm, and endoderm) and germline in recipients. For the production of germline chimeras, donor cells should be prepared from the central part of the area pellucida of stage X blastoderm where PGCs are localized. Three stages culture systems allow the precise manipulation of recipient embryos, and thereby facilitates the production of both somatic and germline chimeras. The ratio of germline chimeras is relatively low. Therefore, a fairly large number of chimeras should be produced to ensure germline chimeras.

Long-term cultured chicken blastoderm cells are maintained in an undifferentiated state which can contribute to the three germ layers both *in vivo* and *in vitro*, but germline competency seems to be lost rapidly in culture systems. Although isolated PGCs can repopulate the gonads after transplantation into the subgerminal cavity, chicken PGCs in this culture system do not survive. Instead, the remaining epiblast-derived cells, which are already committed to somatic lineages, propagate for the long-term and maintain their abilities to contribute to all three germ layers.

12.3.2.2 Intravascular injection

In avian, PGCs are transported to the gonadal ridge by the blood circulation. This unique biological property and accessibility of avian PGCs during early development provides an opportunity to genetic manipulation. Germline chimeras can be produced by intravascular injection of PGCs. Because the blood circulation of avian PGCs is transitory, PGCs transplantation is feasible only during a short circulation period. Firstly, the entrance of chicken PGCs from the germinal crescent region to the vascular system starts at stage 10 and completes at stage 13. Then, the migration of circulating PGCs to the gonadal ridge starts at stage 15 and completes at stage 17. However, access to the developing vasculature is technically difficult until a chicken embryo reaches stage 13. In addition, intravascular transplantation of chicken PGCs to stage 17 and later embryos leads to the failure of germline chermiras. Taking these into consideration, the appropriate time for transplanting chicken PGCs by bloodstream is in embryonic stage 13-16.

Avian PGCs obtained from different sites such as the germinal crescent region and primary gonads can migrate to the gonads and then differentiate into functional gametes after intravascular transplantation. However, the germinal crescent region is the least useful source for germline chimera production, given the low number of PGCs during this period. Both male and female PGCs obtained from stage 27 to stage 31 chicken primary gonads could be used as donor cells for the production of germline chimeras. SSCs obtained from adult chicken testes also have the ability to migrate to the recipient gonads, and subsequently differentiate into functional spermatozoa after transplantation. Thus, these researches confirmed that avian germ cells have developmental plasticity. In contrast to spermatogenesis in chickens, oogenesis can not complete by transplantation because the germ cells in female gonads enter meiosis and stop at the diplotene stage after hatching. In germ cell transplantation, only the mitotic germ cells would contribute to repopulating the recipient gonads. Germ cell meiosis in female chicken embryos occurs at embryonic 15. 5 d of incubation, and meiotic germ cells are abundant by embryonic 18. 5 d. Although a few female germ cells at embryonic 20 d still retain the ability to repopulate gonads after intravascular transplantation, PGCs obtained from earlier stages in development are much more efficient for chimera production.

12. 3. 3 Identification of chimeras in poultry

The identification of chimeras refers to the detection of the distribution, development and differentiation of donor cells at each developmental stage of chimeras. The natural or artificial markers are mainly used to track and detect donor cells. For example, according to the difference of donor and recipient cells in surface antigen, monoclonal antibody can be prepared and detected by immunohistochemistry. The other is to use molecular biological methods to find donor-specific DNA fragments as probes according to the genetic material differences between donor and recipient. *In situ* hybridization or PCR of DNA probes can well display the distribution of donor cells in the recipient. Artificial labeling is to label donor cells with fluorescent reagents with a long half-life and make tissue sections for fluorescence microscopy detection. In addition, intraspecies chimeras are generally used as a means of transgenic chickens, so the donor cells of chimeras are generally genetically manipulated, so the target gene, reporter gene or their expression products can be used as markers for detection.

Individual phenotype such as feather color, bodily form can also be used. However, sexual chimeras can not be judged by phenotype alone. The experiment shows that phenotypic chimerism is not necessarily sexual chimerism, and there is no proportional relationship between the degree of sexual chimerism and phenotypic chimerism. The identification of germline chimeras is generally done by offspring detection method. The obtained chimeras are backcrossed with the donor or interbred with male and female chimeras, and the traits of offspring are observed to judge. A common genetic phenotypic marker for intra specific chimeras is feather color, while inter specific chimeras are individuals born from the same species as the donor.

12.4 Sex control technology of poultry

In livestock production, many economic traits are directly related to the sex of animals, such as lactation of dairy cows, antler production of bucks and sow pregnancy. The impact of sex on economic benefits is particularly prominent in poultry production. In the intensive production of laying hens, people hope to get as many female chicks as possible. In broiler production, cocks with higher growth rate and feed conversion rate are widely needed. Therefore, the research on identifying and controlling the sex has practical value in the poultry production. There are two different theories about sex determination in birds, including the theory of sex chromosome determination and the theory of autosomal balance. Many factors are related to sex differentiation. The original factor of sex differentiation in birds, which is similar to the function of *SRY* in mammals, is still unclear. The similar gene is likely to control gender differentiation by regulating the expression of aromatase gene, as well as the relative levels of estrogen and androgen in embryos. Sex reversal and sex identification, two common production phenomena, which is closely related to production benefits.

12.4.1 Sex reversal

Sex differentiation is a process from genetic sex to phenotypic sex. Sex reversal is the phenotypic sex transition regulated by genes and their expression products. Up to now, aromatase, sexual hormone and AMH are known to play critical role in sex differentiation. Most methods of artificial sex reversal in birds are realized by regulating the levels of the above substances. Exogenous hormones can be artificially applied in the early stage of embryo, mainly including estrogen, androgen, anti-estrogen, anti-androgen, aromatase inhibitor and so on. For example, it is shown that injecting a dose of estrogen (estradiol, diethylstilbestrol) and androgens (testosterone and testosterone propionate) into 7-day-old chicken embryos can induce sexual reversal in varying degrees. In addition, the treatment of aromatase inhibitor on the embryo in the undifferentiated stage can inhibit the activity of aromatase and reduce the level of estrogen, along with the increased androgen, which leads female embryos into males. Similarly, the researchers treated chicken embryos with aromatase inhibitors at the undifferentiated stage of gonads, resulting in 50% of females in sex reversal. The left gonad of female chicks with sex reversal can easily develop into ovary, while the left gonad often develops into testis. If the left ovary is removed in the early stage of embryonic development, the right one will develop into an intact ovary. However, if the left ovary is removed at hatching, the right one will develop into testis. The cock with sexual reversal has two testicles in small size, which have the reproductive ability of male. After sexual maturity, their spermatogenesis is normal, while the number of sperms is smaller than normal. Besides, sex reversal cocks can produce W sperms that do not exist in nature, but they have no fertilization ability.

The study of sex reversal is still in the laboratory stage, and practical application is still a long way off. Moreover, some sex reversal does not develop completely, along with intersexual individuals or reversal back. As we all know, sex differentiation is related to genes. It is of great value to control the direction of sex differentiation by regulating the expression of related genes at DNA level. It is revealed that new candidate genes for gonad differentiation of chicks provide evidence of the preferential expression of BMPs in the developing ovary and inhibin/activin subunits in the testis.

12. 4. 2 Sex identification

The phenotypic sex differentiation of birds is later than mammals. It is difficult to distinguish the sex even in some adult chicks. Early identification of poultry sex can improve the efficiency of production. The following approaches are mainly used for sex identification of birds. According to the differences in the anatomical structure of cloaca between newborn male and female chicks, vent sexing is widely used for identification. However, this method requires long-term training for operators. In addition, there is an approach associated with genetic identification, which is based on the differences between sex chromosomes, including karyotype analysis (karyotype Ping) and PCR amplification of specific DNA fragments on W chromosome. Karyotype analysis is first used, but the process is cumbersome. By means of breeding, the phenotypic genes with sex inheritance are used to establish special male and female lines, such as fast and slow feather lines, which have been widely used and have achieved good production benefits.

12. 4. 2. 1 Experimental principle

The difference in sex chromosomes between male and female avian is W chromosome, which is the theoretical basis for sex identification. Molecular identification of sex can be achieved by detecting the structural differences in the composition of W and Z chromosomes. Based on the target fragment used, the two methods can be classified as follows.

1. W chromosome DNA repeat sequence

The W chromosome of avian contains a large amount of heterochromatin, which accounts for 70% of the entire. The heterochromatin consists of highly repetitive DNA sequences that can be obtained by digestion of genomic DNA with different restriction endonucleases. The W chromosome of avian mainly consists of enzyme cleavage sequences of Xbol (0. 7 kb and 1. 1 kb) and of *EcoR* I (1. 2 kb). 0. 7 kb and 1. 1 kb Xbol fragments are repeated for approximately 14 000 and 6 000, accounting for 60% of the DNA on the W chromosome. The repetitive sequences of the two families reside in different regions on chromosome W. The enzyme cleavage sequence of 0.7 kb Xbol has 68% homology with that of 1.2 kb *EcoR* Ⅰ. A 0.4-kb *Pst* Ⅰ cleavage sequence with up to 10 000 repeats was also found in the W chromosome of turkey, while a 0.5-kb *Taq* Ⅰ cleavage sequence was found in pheasant. In previous studies, no identical repetitive sequences were found in the genomes of male avian. It is believed that the above repetitive sequences are specific to the W chromosome in the female genome, which were the first sequences used for molecular identification of avian sex. Later, the weak positive bands of enzyme cleavage sequence were found in the amplification products based on the genome of male chicks, especially when the number of amplification cycles was large. It is speculated that the same sequence with low copy also exists in the genome of male chicks. In addition, the variation of repeat sequences is less affected by environmental selection, so the conservation between species is poor. A set of primers is only applicable to one species or other species with close biological distance. Therefore, different detection schemes for the different avian species should be designed.

2. Specific genes on W chromosome

With the development of gene localization and DNA sequencing technology, the structure of W chromosome is gradually clear. Three genes have been identified, including *CHD1W*, *ATP5A1W* and *PKC1W*. Copies which are homologous to the above genes, have also been identified in the Z chromosome. Therefore, no specific gene on W chromosome has been found. *CHD* (chromo-helicase-DNA) is probably a transcriptional regulator at the chromatin level. The *CHD1W* gene on the avian W chromosome is similar to that of human. The *CHD1W* gene is located on the W chromosome of most non-ratite birds.

However, a *CHD* gene, named *CHD1Z*, was subsequently found to be localized outside the pseudoautosome region of Z chromosome. Therefore, the *CHD1Z* gene is located in both male and female (ZW/ZZ) genomes. The extron of *CHD1W* and *CHD1Z* genes are highly homologous, while the differences between their introns are usually large. Sex can be identified based on the difference in the size of PCR products by amplifying fragments of different introns. According to the difference of intron size between *CHD1W* and *CHD1Z* genes, Fridolfsson designed primers of 50 species of 11 orders of non-ratite birds and distinguished the sex of them. Among the 50 species, 47 were successfully identified, and successful results were also obtained in the other 3 species by using another pair of primers. It is believed that with the implementation of the chicken genome mapping project, more specific markers will be found on the W chromosome for molecular identification of sex.

12.4.2.2 Experimental method

Although there are various methods for molecular identification of sex, most of them are conducted by the use of PCR and have several steps including target sequence selection, primer design, preparation of genomic template DNA and PCR amplification detection. The application of 2550F/2718R primers for the molecular identification of chicken sex is described below (Fig. 12-1).

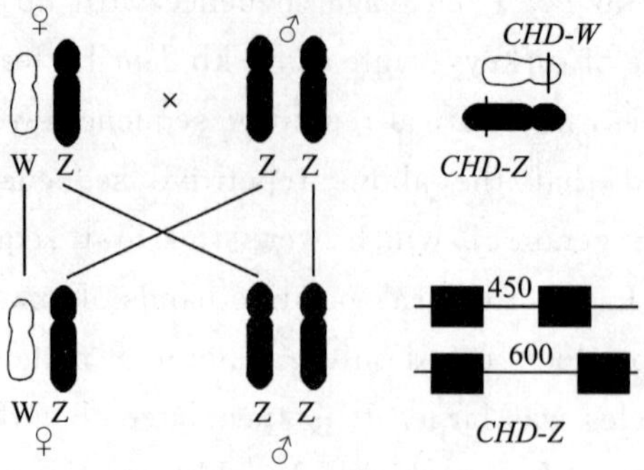

Fig. 12-1 Schematic diagram of gender molecular identification based on intron differences of *CHD1W* and *CHD1Z* genes (Shamim and Mason, 1999)

1. Primer selection

2550F: 5′-GTTACTGATTCGTCTACGAGA-3′;

2718R: 5′-ATTGAAAT GATCCAGTGCTTG-3′.

2. Extraction of avian genomic DNA

(1) Nucleated whole blood (birds, reptiles)

① Transfer 50 μL whole blood to a 1. 5-mL microfuge tube, add PBS solution and wash by centrifugation, remove the supernatant.

② Add 50 μL of digestion solution (10 mmol/L Tris, pH 8, 10 mmol/L ethylenediaminetetraacetic acid, pH 8, 1% SDS, 100 μg/mL proteinase K, pH 7. 6) and perform them at 45 ℃ overnight with gentle shaking.

③ Inactivate proteinase K at 95 ℃ for 15 min, cool to 4 ℃, centrifuge at 10 000*g* for 10 min. Take the supernatant and dilute it to 500 μL, and use 3-5 μL for PCR reaction.

(2) Extraction of genomic DNA from early embryonic disc cells

① Use pipette pler device to make micro glass needle with 30-50 μm outer diameter from capillary borosilicate glass tubes (OD 1 mm).

② Open a 6-8-mm diameter hole in the equatorial part of the egg, and adjust the position to expose the embryo disc. Aspirate 5 μL of PBS solution with a prepared micro glass needle. Puncture the embryo disc and aspirate a small amount of embryo disc cells, usually 300-800 cells can be collected.

③ Place the collected embryonic disc cells into 50 μL of cell lysis solution (10 mmol/L Tris, pH 8, 10 mmol/L ethylenediaminetetraacetic acid, pH 8, 0. 2% SDS, 100 μg/mL proteinase K, pH 7. 6), and perform them at 58 ℃ overnight with gentle shaking.

④ Inactivate proteinase K at 95 ℃ for 15 min, then cool it to 4 ℃. Take the supernatant and dilute it to 500 μL, 5 μL of which is used for PCR reaction.

(3) Extraction of genomic DNA from feathers

① Take and shear the base of a chest feather, and put the tips into a 1. 5-mL microfuge tube.

② Add 1 mL digestion solution (20 mmol/L Tris, pH 8, 20 mmol/L ethylenediaminetetraacetic acid, pH 8, 4% SDS, 10 mg/mL proteinase K, pH 7.6) into the tube. Water bath and shock the tube at 58 ℃ overnight. Then, inactivate proteinase K at 95 ℃ for 15 min.

③ Because there are many impurities in feathers, the obtained cleavage products must be extracted with phenol chloroform, precipitated with ethanol and dissolved in TE solution.

④ Take the supernatant and dilute it to 500 μL, and use 3-5 μL for PCR reaction.

3. PCR reaction to amplify the target fragment

(1) Template and primer Recommended amounts of DNA template for a 10-μL reaction system are as follows: 50 ng genomic, 2550F and 2718R 1 μmol/L.

(2) *Taq* DNA polymerase concentration *Taq* DNA polymerase may range from 5-50 units/mL (0.25-2.5 units/50 μL reaction) in specialized applications. The final concentration of dNTPs is typically 40 μmol/L of each deoxynucleotide.

(3) Denaturation An initial denaturation of 30 s at 94 ℃ is sufficient for most amplicons from pure DNA templates. The time for the annealing step is 30 s.

(4) Annealing Annealing temperature is based on the T_m of the primer pair and is typically 45-68 ℃. Annealing temperatures can be optimized by doing a temperature gradient PCR, which starts at 5 ℃ below the calculated T_m.

(5) Extension The recommended extension temperature is 72 ℃. Extension times are generally 1 min/kb. A final extension of 5 min at 68 ℃ is recommended.

(6) Cycle number Generally, 25-35 cycles yield sufficient product. Up to 45 cycles may be required to detect low-copy-number targets.

(7) Agarose gel electrophoresis After the reaction, 10 μL of the product is taken to conduct electrophoresis with 2% agarose gel electrophoresis, and observe placed under UV detector. The size of the amplification products varies from species to species. Generally, males show a band of *CHD1Z* with 600-650 bp. Females show a band of *CHD1W* with 400-450 bp, apart from the band with 600-650 bp. The size of the reaction products of *CHD1W* and *CHD1Z* are 450 bp and 600 bp, respectively.

12.4.3 Problems of sex control

Although sex differentiation is closely related to factors during embryonic development, the fundamental mechanism of sex determination has not been clearly revealed. Many challenges in the sex control of poultry remain. First of all, there is no approach to achieve sex control at the gamete stage. The phenotypic reversal can be realized to a certain extent by using chemicals before sex differentiation, but the effect is unstable. Secondly, checking the anus and cultivating sex-related genetic phenotypic breeds are two important methods for sex control of commercial laying hens in the intensive production. However, there is a lack of methods for sex identification of other special avian species. Although sex identification can be achieved at embryo stage by using molecular identification, the operation will bring harm to the embryo and affect the hatching rate. It is not suitable for the needs of intensive production. There are still some limitations of the technology, which is far away from practical application. With the progress of gene chip technology, the identification of avian sex becomes more rapid and accurate. The determination of physical profile of avian sex chromosome is helpful to finding a new way for control sex at the molecular level.

12.5 Transgenic technology of avian

Due to the unique reproductive physiology, the transgenic technology widely used in mammals can not be directly used in avian. Nevertheless, scientists are still trying to overcome the difficulties of embryo operation, exploring and establishing gene transfer strategies suitable for avian. The research of transgenic avian is of great significance to improve the disease resistance and the product quality, especially in the research of bioreactor. Normal egg white contains 3.5-4 g protein, more than half of which derive from one coding gene (ovalbumin). Using gene expression to recombine protein can obtain more than 1 g protein in normal eggs. Therefore, people try to produce medicinal protein by changing the chicken genome. Many avian transgenic companies have been established in the 20th century with great technological progress. The common transgenic technology of avian includes the method of sperm vector, technique of Microinjection for fertilized egg, transfection of embryonic placental cells using retroviral vectors, transfection of blastoderm cells with lipofectin packaging vector, and the method of blastoderm cell and primordial germ cell vectors.

Birds produce pole tip megalecithal egg, which are large and opaque. During fertilization, multiple sperms can enter the egg to form multiple male pronucleus. The male pronucleus fused with the female pronucleus is difficult to identify. The oviduct is located in the deep part of the abdominal cavity. The fertilization happens in the magnum of the oviduct. Therefore, we can only use the method of planning to collect eggs. At the same time, it is difficult to transplant the operated single-cell fertilized eggs into the body. The success rate of *in vitro* hatching of fertilized eggs is not high. The fertilized egg quickly enters the cleavage stage in the oviduct, which is surrounded by albumin, shell membrane and eggshell. When the egg is produced from the cloaca, the embryo has developed into the X stage placenta with 60 000 cells. The egg of pseudoinsemination has aged and lost its developmental potential. Also, the research background of avian embryology and genetics is far behind that of mammals, so the transgenic research of avian faces great difficulties.

12.5.1 Method of gamete vector

It is not easy to obtain the zygote of the avian. However, the gamete can be transduced with exogenous gene, then the transgenic zygote by fertilization can be obtained.

12.5.1.1 Method of sperm vector

The sperm of avian is small, and it has little cytoplasm. Only the nucleus of sperm enters the egg during fertilization, and then forms male prokaryotes. Therefore, it is achievable to transfer the exogenous gene into the nucleus of zygote by using sperm as the carrier. Furthermore, there is no capacitation of avian sperm, and artificial insemination technology has been widely used, so sperm vector method is easy to implement. The dead chicken sperm is treated with radiation for pseudogamy, and then fertilized with normal sperm. The result shows that some individuals of the offspring perform some phenotypic traits of dead sperm. After that, it is confirmed that the surface plasmon of chicken sperm could absorb exogenous DNA fragments, which can transfer genetic matter to offspring. In the initial stage of sperm vector research, the exogenous gene fragments are incubated together with sperm, and then use it into artificial insemination. The adsorption capacity of sperm lipid membrane to DNA fragments is instable, resulting into different phenotypic traits happening in a large number of experiments with this method. Therefore, people use various ways to make as many foreign genes stably enter the sperm or bind to the surface, so that the fertilized egg with the foreign gene is formed after fusion with the female pronucleus. Recently, lipofectins and electroporation have been used to transfer exogenous genes into human sperms. Following with this technical progress, researchers used lipofectin packaging transfection method to transfer exogenous genes into chick sperms. After fertilization, exogenous genes were detected in hatched chicks, which passed on genetic changes to their offspring. However, the exogenous gens disappeared in the offspring finally, indicating that the integration into the recipient genome is not stably. It is reported that transgenic mice can be obtained by injecting exogenous DNA directly into mouse testis with subsequent spermatozoa. However, we can not get transgenic chickens with the same method.

In sperm vector method, the normal fertilization process is used to complete the integration of foreign genes. It has the characteristics of low cost, simple operation, less damage to the egg prokaryote and mass production. Compared with others, people pay more attention to sperm vector method, which has obvious advantages in basic research of genetics and developmental biology. In recent years, great progress has been made, especially in promoting the development of avian transgenic field. However, the integration mechanism of foreign genes in the sperm and zygote genome is still not clear, so further research is needed.

12.5.1.2 Method of egg vector

Similar to egg vector method, the exogenous gene can be directly injected into the follicles of avian, and be transferred into the fertilized egg through fertilization. Injecting replication competent reticulointerstitial hyperplasia virus into the developing chick follicles can get 26% of the fertilized eggs with virus sequence after hatching. Due to the potential biological security risk of replication virus, replication deficient virus was used to repeat the above experiment, and exogenous DNA still existed in 8% of individuals. However, when 1-25 μg plasmid containing bovine growth hormone gene was injected into developing follicles, no exogenous genes were found in the progenies. The transferred foreign DNA may have been degraded in oocytes. Because the method requires surgery, and DNA is prone to degradation during follicular development, it is rarely used in practice.

12.5.2 Technique of microinjection for fertilized egg

Microinjection was invented by Gordon, an American scientist. Its basic process is as follows: firstly, inject the target gene into the prokaryote of the fertilized egg by a micromanipulation system directly; secondly, randomly integrate the gene into the chromosome according to the repair mechanism during DNA replication; after that, transfer the fertilized egg with exogenous gene into the uterus of a surrogate mother; finally, transgenic animals are produced.

The widely used prokaryotic microinjection is one of the relatively stable methods for producing transgenic animals. This approach has the advantage that the exogenous gene can be directly injected without vector component as well as length limitation. However, the structure of fertilized eggs and the development of avian embryo are special. It is difficult to carry out genetic operation on the fertilized egg directly. Furthermore, the chicken prokaryote is too small to directly inject for operators. In addition, we can only get one fertilized egg from one chicken through surgery. It is difficult to either transfer the egg back into the maternal oviduct, or to conduct *in vitro* incubation of fertilized eggs. Therefore, the early application of this method in chickens was restricted.

Studies have shown that chick egg is located in the middle and rear segment of the oviduct bulge site at 1.8-3.5 h after fertilization. At this time, the number of male prokaryotes near the reproductive disc of the zygote was 6-15, with a diameter of 7-17 μm. The prokaryotes were dispersed in the early stage, and then the male and female prokaryotes gradually occupied the central position. The other male prokaryotes migrated to the edge of the reproductive disc. The distance between male and female prokaryotes was 10 μm 3 h after fertilization. At 3.5 h, between male and female prokaryotes have been located in the center of the reproductive disc, 25 μm away from the outer surface, while the other male prokaryotes were 450 μm away from the center. Therefore, microinjection in fertilized eggs of avian is essentially cytoplasmic injection (Fig. 12-2). The commonly used transfer method is to inject foreign genes into the central plasma membrane at 50-200 μm under reproductive disc. The fertilized eggs after injection are hatched or re-transplanted to the recipient after a three-stage culture *in vitro*.

Whether the exogenous genes injected into the cytoplasm near the prokaryotes can enter the fused prokaryotes and integrate into the genome is a key factor for microinjection technique at the stage of fertilized eggs. Up to now, microinjection technique at the stage of

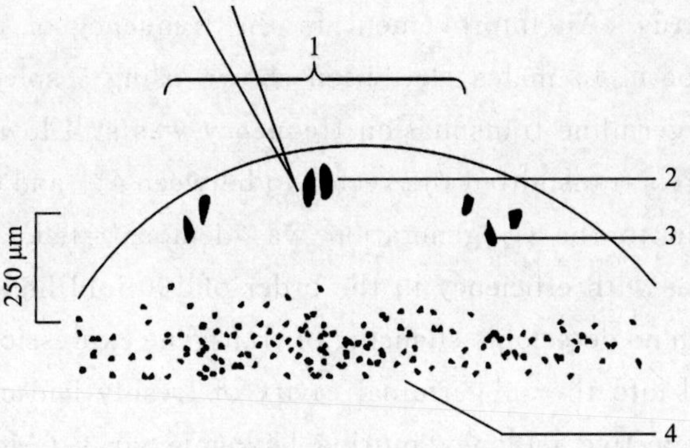

1. Genital disc 2. Male and female prokaryotes about to fuse 3. Other male pronucleus 4. Egg york

Fig. 12-2 Schematic diagram of chicken fertilized egg microinjection for gene transfer

(Shamim and Mason, 1999)

fertilized egg is a basic method for transgenic research in theavian field. Microinjection affects the hatching rate of embryos *in vitro* culture to a certain extent. Statistics show that the hatching rate of embryos after injection can reach 10%. The proportion of transgenic individuals reaches 0.2%-0.5%, but the results are still not stable. At present, many domestic researchers directly inject the exogenous genes into blastula. Compared with the traditional methods, direct injection is easy and concise, which can bring higher survival rate. However, we can only get chimeras with low efficiency of gene integration by using this method. It still needs hybridization and reselection to get transgenic chickens in the real sense.

The current research objective is to improve embryos survival rates and the integration efficiency of injected DNA on chromosomes, which can promote the production of transgenic chickens.

12.5.3 Transfection of embryonic placental cells using retroviral vectors

The exogenous gene is constructed into the sequence of the viral vector. With the infection of the virus, the exogenous gene and the viral sequence can be integrated into the host chromosome as part of the normal life cycle. Two types of vectors are currently used: replication vectors, which contain all the genes necessary for viral replication and allow for multiple rounds of repeated infections; replication-deficient vectors, which have no replication capacity and provide only the necessary factors for infection and integration.

Various strategies such as the viral infection of stage X embryos, microinjection of transgenes into fertilized eggs, and embryonic stem cells have been used to produce transgenic birds. Commonly used avian virus vectors are avian leucosis virus (ALV), rous sarcoma virus (RSV), moloney murine leukemia virus (MoMLV), spleen necrosis virus (SNV), equine infectious anemia virus (EIAV), recombinant retroviral vectors (RRV). The earliest methods developed were based on the use of avian retroviruses: replication-competent vectors derived from ALV and replication-defective vectors derived from reticuloendotheliosis virus. An improvement in the frequency of production of germline transgenic birds (to one in 15 males) has been shown using a spleen necrosis virus-based vector. However, the germline transmission frequency was still low (0.9%). It is proved that ten founder cockerels transmitted the vector to between 4% and 45% of their offspring, and stable transmission to the G_2 generation was demonstrated. EIAV can be used to generate transgenic lines with efficiency in the order of 100-fold higher than any previously published method, with no detectable silencing of transgene expression between generations. The researcher injected into the subgerminal cavity of freshly laid chicken eggs at stage X using the replication-defective moloney murine leukemia virus (MoMLV), a total of 38 chicks out of 573 virus-injected eggs, hatched and contained the *hupA* gene in their various body parts. The mRNA transcript of the *hupA* gene was present in various organs, including blood and egg, and was germline transmitted to the next generation. The researcher selected the reverse transcription expression vector pBABE-pour with MoMLV as the backbone and produced chimeric chickens expressing *EGFP* gene by *in vivo* infection of primordial germ cells according to the characteristics of stage X chicken embryos, and verified the successful integration of *EGFP* gene into the genome by retrovirus using PCR, and further analyzed the genomic integration using Southern blotting. The results showed that 66.7% of chicken embryos were positive, which proved that the recombinant retrovirus *in vivo* infection of chicken primordial germ cells was an effective method for the preparation of chimeric chickens.

Replicative vectors (ALV) were injected into embryos, and the frequency of integration of proto-viral sequences was detected in the progeny of hatched roosters was generally 1%-11%. However, the integration site is difficult to control, the size of the introduced exogenous gene fragment does not exceed 8 kb, and it is still doubtful whether this vector has carcinogenic risk and biological safety problems, and the application in production has certain risks, and the corresponding products are also resisted by consumers, after the 1990s, this method is gradually replaced by other methods, such as using primordial germ cells (PGCs) as vector. PGCs have been used widely in the generation of transgenic birds to overcome the limitation of low-efficiency germline transmission. Transgenes can be introduced into the cultured genomes of PGCs using transfection reagents to produce transgenic birds.

12.5.4 Transfection of blastoderm cells with lipofectin packaging vectors

Due to the safety issues with viral vectors, the use of non-viral vectors for transfection has been explored. Lipofectin-mediated transfection is one of them. This method involves the introduction of exogenous genes into the embryonic disc cells by packaging them in lipofectins, and then transfer them by injection into the subembryonic cavity.

The introduced exogenous genes can be expressed in the early embryo. Because most of the cells at the embryonic disc stage will develop into extra-embryonic tissues, the exogenous genes are mostly expressed in the extra-embryonic region of the transfected embryo. Therefore, transgenic avian obtained by this method often suffer from genetic instability. It was shown that positive results could be detected in chick embryos at 5 d of age after DNA and lipofectin complex injection. As the embryos develop, positive results disappear in the late embryonic stage. It is found that the exogenous *pEGFP-C2* gene could be integrated into the G_0 generation germline after lipofectin transfection into the embryonic disc cell of quails. The exogenous gene could be passed on to offspring, but the transfection rate out of the shell is less than 10%. The expression level of the transduced gene can be increased by means of mediation of nuclear matrix binding sequences, which is not prone to gene silencing. The ratio of lipofectins to DNA and the injection dosage can affect the transfection effect. In addition, exogenous genes can be used to generate transgenic embryos by transfecting primordial germ cells and spermatogonia, which does not require microscopic manipulation and avoids damage to the embryo.

12.5.5 Method of blastoderm cell and primordial germ cell vectors

The idea of preparing transgenic birds by chimera method also originated from in mammals' research. As ES and EG cell lines of mice and homologous recombination technology developed, the purposiveness and precision of the cell genome manipulation has been greatly enhanced. This method combing with gene targeting can be used for genetic manipulation of ES and EG cells, which overcomes the low integration rate, randomness and blindness of other transgenic methods. It can achieve the targeted insertion and specific modification of endogenous genes. Selection, subculture and amplification of genetically modified cells can be achieved *in vitro*, which can contribute to obtaining a large number of stable cell vectors incorporated with exogenous genes. Through the preparation of chimera, exogenous genes in the donor cell can be widely involved in the gonads, where they transfer into the differentiated gametes. Then, transgenic offspring with stable inheritance performance can be produced after mating. At present, chicken is the species with highest chimerism rate, which is also the earliest animal model used to obtain the offspring of donor cells. Therefore, chimerism is the most promising technique for transgenic avian research.

The chimera is produced through transfection of exogenous genes through lipofectins, electroporation or viral vectors into the blastodisk cell and PCGs isolated from the early embryo. The targeted exogenous genes can be detected in the gonads, which proves the accuracy of this technical route. The *PneoSVz/LacZ* vector covered by lipofectin was transfected into chick blastodisk cells *in vitro*, and cultured for 8 d after selection by Zeocin™. The expression of *LacZ* was found in 10% of the surviving blastoderm cell clones. Similarly, the *PneoSVz/LacZ* vector covered by lipofectin was transfected into quail placental cells *in vitro*. The expression of *LacZ* was observed after 52-53 h of cell proliferation. As for *LacZ* expression in surviving embryos, just one third exhibited it both embryonic and extraembryonic tissues. Another one third exhibited it only in extraembryonic tissues, and the remaining failed to show any effects. Because PGCs can transfer exogenous genes to the offspring directly, this method of transgenic technique shows a wide prospect. At present, there are three main ways to introduce exogenous genes into PGCs. The first is to transfect exogenous genes into the germinal crescent area of the early embryo directly by using a variety of cell transfection methods. The second way is to directly inject exogenous genes assembled by lipofectin into the blood of 14 day-old embryo to transfect the PGCs during the circulation period, through which the exogenous genes are also detected in the gonad. The third approach is to transfect PGCs and produce chimeras. The plasmid is introduced into gonadal PGCs of 6 day-old embryo by electroporation method. Then, cultures for 24 h and detect the gene expression of *LacZ*. It is found that electroporation method results in 80% positive X-gal of gonad PGCs, while the efficiency of lipofectin method is 17%. Furthermore, the expression of *LacZ* is detected in the gonads of embryos at 6 day-old and 10 day-old and chicks after hatching, with positive rates of 100%, 67%, and 41% respectively.

Presently, direct transfection of chicken blastoderm cells and PGCs *in vitro*, or chimeric transfection of exogenous genes after short-term culture, can only produce embryos or individuals with transient expression of exogenous genes. Therefore, it is difficult to obtain a large number of vector cells with stable integration of exogenous genes for producing highly chimeric individuals.

12.5.6 Problems and prospects

It can be seen that although there are many transgenic techniques for birds, most of them are in the exploratory stage. The efficiency of gene transfection, integration and expression are relatively low. We can only use the methods of virus vector and microinjection to obtain transgenic chicken successfully.

Gene targeting by homologous recombination or by sequence-specific nucleases allows the precise modification of genomes and genes to elucidate their functions. Researchers targeted the joining (J) gene segment of the chicken Ig heavy chain gene by homologous recombination in primordial germ cells to establish fully transgenic chickens carrying the knockout. In homozygous knockouts, Ig heavy chain production is eliminated, and no antibody response is elicited on immunization. The TALEN technique used in the chicken primordial germ cell line is a powerful strategy to create specific genome-edited chickens safely for practical applications. Researchers accomplished targeted knockout of the ovalbumin gene (*OV*) in the chicken primordial germ cells through the use of transcription activator-like effector nuclease (TALEN)-mediated gene targeting, and *OV* mutant offspring were generated through test-cross analysis. Similarly, the researcher also obtained *ddx4* gene knockout chickens through TALEN, resulting in infertility of adult female chickens.

With the development of transgenic technology, some scholars have made progress in the research of in chicken oviduct bioreactor. The human tissue kallikrein gene (*hKLK1*) cDNA-expression cassette was subcloned into an avian adeno-associated virus (AAAV) transfer vector, and recombinant viral particles were generated by transfection of AAV-293 cells with a 3-plasmid system. After intravenous injection of laying hens with the viral particles, long lasting and oviduct specific expression of rhK1 was demonstrated by reverse transcription (RT)-PCR and Western blotting. The high level of rhK1 was lastingly secreted into egg white. The oviduct specific lysozyme gene (*LYZ*) expression cassette was cloned into AAAV transfer vector, and high titer recombinant viruses were generated using the baculovirus/insect cell expression system. After intravenous injection of laying hens with the recombinant viruses, long-lasting and oviduct-specific expression of human lysozyme (rhLYZ) was obtained. The high level and long-lasting expression of rhLYZ in egg white indicate the usefulness of the method for expression of other recombinant proteins.

Nevertheless, there are still many problems as follows to be solved: ①constructing vectors with high efficiency and specific expression; ②improving the efficiency and accuracy of exogenous gene integration; ③establishing the *in vitro* culture system of avian ES and EG cells.

Assignments

1. What is the control mechanism of avian seasonal reproductive activities?
2. How many methods can be used to manipulate avian embryos?
3. Which method can be applied *in vitro* culture avian embryo?
4. How to classify the types of avian chimeras?
5. Which method can be applied to produce avian chimeras?
6. How to test chimerism in avian, especially the germline chimera?
7. Please briefly describe the test process of avian sex identification.
8. Master the main principles and methods of avian transgenic technology.

(Yu Minli, Lü Zengpeng)

Abbreviations

Abbreviation	Full name
6-DMAP	6-dimethylaminopurine
AAAV	avian adeno-associated virus
ACTH	adrenocorticotropic hormone
AI	artificial insemination
ALV	avian leucosis virus
AMH	anti-Müllerian hormone
AP	alkaline phosphatase
ART	assisted reproductive techniques
bMSC	bone marrow pluripotent stem cells
BMW	breeding by molecular writing
BSA	bovine serum albumin
CB	cytochalasin B
CFTR	cystic fibrosis transmembrane conductance regulator
CFU	colony forming unit
CFU-F	fibroblast colony forming unit
CHD	chromo-helicase-DNA
CHO cell	Chinese hamster ovary cell
CHX	cycloheximide
COC	cumulus oocyte complex
COVID -19	coronavirus disease 2019
CRISPR	clustered regularly interspaced short palindromic repeats
CRL	crown-rump length

Continued

Abbreviation	Full name
CS	cattle serum
CX	cycloheximide
DIA	differentiation inhibitory factors
DMSO	dimethyl sulfoxide
DSB	double-strand Break
DTT	dithiotherie
E_2	estrogen
EDTA	ethylene diamine tetraacetic acid
EIA	enzyme immunoassay
EIAV	equine infectious anemia virus
ELISA	enzyme-linked immunosorbent assay
EMG	electron microscopic grids
EP	electroporation
EPF	early pregnancy factor
ESC	embryonic stem cell
EST	expressed sequence tags
ET	embryo transfer
FACS	fluorescence-activated cell sorting
FAO	Food and Agriculture Organization
FCS	fetal bovine serum
FDA	Food and Drug Administration
FISH	fluorescence in situ hybridization
FIV	feline immunodeficiency virus
FM	fetal movement
FSH	follicle-stimulating hormone
FTET	fix time embryo transfer
GaLV	gibbon ape leukemia virus

Continued

Abbreviation	Full name
GH	growth hormone
GHG	greenhouse gases
GI	genetic immunization
GMP	glass micropipettes
GnIH	gonadotropin inhibiting hormone
GnRH	gonadotropin releasing hormone
GO	glass oviduct
GS	gestational sac
GTH	gonadotrophic hormones
GV	germinal vesicle
hACE2	human angiotensin-converting enzyme 2
HBsAg	hepatitis B surface antigen
hCG	human chorionic gonadotropin
HDR	homology-directed repair
HGF	hepatocyte growth factor
HGPRT	hypoxanthine-guanine phosphoribosyl transferase
HGT	horizontal gene transfer
HIV-1	human immuno-deficiency virus type 1
hKLK1	human tissue kallikrein gene
HMC	handmade cloning
HPG	hypothalamus-pituitary-gonad
HT-NGS	high-throughput-next generation sequencing
HTs	high-throughput screening
H-Y antigen	histocompatibility-Y antigen
ICSI	intracytoplasmic sperm injection
ID	intradermal
IGA	intentional genomic alteration

Continued

Abbreviation	Full name
IHN	infectious hematopoietic necrosis
IL-6	interleukin-6
IM	intramuscular
INH	inhibin
INHA	inhibin alpha subunit
iPS	induced pluripotent stem cells
ISCT	International Society for Cell Therapy
isEPF	immunosuppressive early pregnancy factor
IVC	*in vitro* culture
IVF	*in vitro* fertilization
IVM	*in vitro* maturaiton
KLFs	Krüppel-like factors
LAMP	loop-mediated isothermal amplification
LH	luteinizing hormone
LI	laparoscopic insemination
LIF	leukaemia inhibitory factor
LIF-R	LIF receptor
LSD1	lysine (K)-specific demethylase 1A
LYZ	lysozyme
MART	modern animal reproductive technology
MEL	melatonin
MF	microfertilization
MI	microinsemination
MIF	Müllerian inhibiting factor
MIS	Müllerian inhibiting substance
MoMLV	moloney murine leukemia virus
MSCs	mesenchymal stem cells

Continued

Abbreviation	Full name
MSTN	myostatin
MSV	murine sarcoma virus
MVP	measles virus
MyHC	myosin heavy chain
NCS	newborn bovine serum
NHEJ	non-homologous ending-joining
NT	nuchal translucency
NT	nuclear transfer
OPS	open pulled straw
OPU	ovum pick up
ORF	open reading frame
OV	ovalbumin
OXT	oxytocin
P_4	progesterone
PA	parthenogenetical activation
PAM	protospacer adjacent motif
PB1	first polar body
PBS	phosphate-buffered saline medium
PCR	polymerase chain reaction
PDT	population doubling time
PERVs	porcine endogenous retroviruses
PG	prostaglandin
PGCs	primordial germ cells
PGE	prostaglandin E
$PGF_{2\alpha}$	prostaglandin $F_{2\alpha}$
PMSG	pregnant mare serum gonadotropin
PRRSV	porcine reproductive and respiratory syndrome virus

Continued

Abbreviation	Full name
PZM-3	porcine zygote medium 3
rAAV	recombinant adeno-associated virus
REV	reticuloendotheliosis virus
rhAAT	recombinant human alpha 1-antitrypsin
RFPR	RFamide-valated peptide
RIA	radioimmunoassay
RIT	rosette inhibition test
RNAi	RNA interference
RRV	recombinant retroviral vectors
RSV	Rous sarcoma virus
SARS-CoV-2	severe acute respiratory syndrome coronavirus 2
SC	sex control
SCN	suprachiasmatic nucleus
SCNT	somatic cell nuclear transfer
SNV	spleen necrosis virus
SOD	superoxide dismutase
SSCs	spermatogonial stem cells
SSEA	stage-specific embryonic antigens
SUZI	subzonal insemination
TAI	timed artificial insemination
TALE	Tal effector
TALEN	transcription activator-like effector nuclease
TC	traditional cloning
TDF	testicular determinant factor
TPA	tissue plasminogen activator
TRA-1-60	tumor rejection antigen-1-60
TT	tetanus toxoid

Continued

Abbreviation	Full name
Tβ4	thymosin beta 4
USA	United States of America
VIP	vasoactive intestinal peptide
WOW	well-of-the-well
ZFN	zinc-finger nuclease
ZFY	zinc-finger Y
ZP	zona pellucida
β-HCG	human chorionic gonadotropin β subunit

References

[1]Abecia J A, Forcada F, Gonzalez-Bulnes A. Hormonal control of reproduction in small ruminants[J]. Anim Reprod Sci, 2012, 130(3-4): 173-179.

[2]Ahn H S, Park B J, Go H J, et al. Induction of immunocontraceptive effects in both male and female mice immunized with GnRH vaccine [J]. Vet Med Sci, 2021,7(5):1999-2007.

[3]Al-Dossary A A, Strehler E E, Martin-Deleon P A. Expression and secretion of plasma membrane Ca^{2+}-ATPase 4a (PMCA4a) during murine estrus: association with oviductal exosomes and uptake in sperm[J]. PLoS One, 2013, 8(11): e80181.

[4]Almiñana C, Caballero I, Heath P R, et al. The battle of the sexes starts in the oviduct: modulation of oviductal transcriptome by X and Y-bearing spermatozoa[J]. BMC Genomics, 2014, 15:293.

[5]Armant D R. Life and death responses to trophinin-mediated adhesion during blastocyst implantation [J]. Cell Cycle, 2011, 10(4):574-575.

[6]Birrell J R, Schulman M L, Botha A E, et al. Vaccination against GnRH as a prelude to surgical castration of horses[J]. Equine Vet J, 2021,53(6):1141-1149.

[7]Bouabe H, Okkenhaug K. Gene targeting in mice: A review[J]. Methods Mol Biol, 2013,1064: 315-336.

[8]Burnett T A, Kaur M, Polsky L, et al. Rumen-reticular temperature during estrus and ovulation using automated activity monitors in dairy cows[J]. Front Vet Sci, 2020, 7:597512.

[9]Cahuascanco B, Bahamonde J, Huaman O, et al. Bovine fetal mesenchymal stem cells exert antiproliferative effect against mastitis causing pathogen *Staphylococcus aureus* [J]. Vet Res, 2019,50(1):25.

[10]Cai Z G, An J H, Liu Y L, et al. Single layer centrifugation improves the quality of frozen-thawed sperm of giant panda (*Ailuropoda melanoleuca*)[J]. Anim Reprod Sci, 2018, 195:58-64.

[11]Calle A, López-Martín S, Monguió-Tortajada M, et al. Bovine endometrial MSC: mesenchymal to epithelial transition during luteolysis and tropism to implantation niche for immunomodulation [J]. Stem Cell Res Ther,2019,10(1):23.

[12]Campbell B K, Clinton M, Webb R. The role of anti-Müllerian hormone (AMH) during follicle development in a monovulatory species (sheep) [J]. Endocrinology, 2012, 153(9):4533-4543.

[13]Carré G A, Couty I, Hennequet-Antier C, et al. Gene expression profiling reveals new potential players of gonad differentiation in the chicken embryo[J]. PLoS One, 2011, 6(9): e23959.

[14]Chang L B, Peng S Y, Chou C J, et al. Therapeutic potential of amniotic fluid stem cells to treat bilateral ovarian dystrophy in dairy cows in a subtropical region [J]. Reprod Domest Anim, 2018,53(2):433-441.

[15]Chen C, Zhao X, An Z, et al. , Nasal immunization with AMH-INH-RFRP DNA vaccine for improving follicle development and fertility in buffaloes[J]. Front Endocrinol (Lausanne), 2023, 14: 1076404.

[16]Chen F, Lu J, Guo R, et al. Rectifying cow infertility under heat stress by immunization against inhibin and supplementation of progesterone[J]. Domest Anim Endocrinol, 2022, 80:106726.

[17]Chen R, Dai Z C, Zhu H X, et al. Active immunization against AMH reveals its inhibitory role in the

development of pre-ovulatory follicles in Zhedong White geese [J]. Theriogenology, 2020, 144:185-193.

[18]Chen W Z, Li Y M, Yu X, et al. The efficacy, biodistribution and safety of an inhibin DNA vaccine delivered by attenuated Salmonella choleraesuis[J]. Microb Biotechnol, 2018,11(1):248-256.

[19]Chung Y, Lee J, Oh S, et al. Automatic detection of cow's estrus in audio surveillance system[J]. Asian-Australas J Anim Sci, 2013, 26(7): 1030-1037.

[20]Curchoe C L, Swain J E. The changing culture of embryo culture: advances in the IVF culture system [M]. London: Cambridge University Press, 2021.

[21]Dan X, Han L, Riaz H, et al. Construction and evaluation of the novel DNA vaccine harboring the inhibin α (1-32) and the RF-amide related peptide-3 genes for improving fertility in mice[J]. Exp Anim, 2016a; 65(1):17-25.

[22]Dan X, Liu X, Han Y, et al. Effect of the novel DNA vaccine fusing inhibin α (1-32) and the RF-amide related peptide-3 genes on immune response, hormone levels and fertility in Tan sheep[J]. Anim Reprod Sci, 2016b, 164:105-110.

[23]Deb K D, Griffith M, Muinck E D, et al. Nanotechnology in stem cells research: advances and applications[J]. Front Biosci (Landmark Ed), 2012, 17(5): 1747-1760.

[24]Delhez P, Ho P N, Gengler N, et al. Diagnosing the pregnancy status of dairy cows: How useful is milk mid-infrared spectroscopy[J]. J Dairy Sci, 2020, 103(4):3264-3274.

[25]Denner J. Porcine endogenous retroviruses and xenotransplantation[J]. Viruses, 2021; 13(11): 2156.

[26]Dolgin E. Pig kidney transplant obscures value of engineered animals[J]. Science, 2021, 374(6568): 668-669.

[27]Eriksson S, Jonas E, Rydhmer L, et al. Invited review: breeding and ethical perspectives on genetically modified and genome edited cattle[J]. J Dairy Sci, 2018, 101(1):1-17.

[28]Fan C, Wu Y, Rui X, et al. Animal models for COVID-19: advances, gaps and perspectives[J]. Signal Transduct Target Ther, 2022,7(1):220.

[29]FDA Approves First COVID-19 Vaccine | FDA, https://www.fda.gov/news-events/press-announcements/fda-approves-first-covid-19-vaccine, 2021.

[30]FDA Authorizes Pfizer-BioNTech COVID-19 Vaccine for Emergency Use in Children 5 through 11 Years of Age, https://www.fda.gov/news-events/press-announcements/fda-authorizes-pfizer-biontech-covid-19-vaccine-emergency-use-children-5-through-11-years-age, 2021.

[31]Feng Z Q, Wang L, Zhao S J, et al. Application progress on Ovsynch and timed artificial insemination technology in beef cattle[J]. China Animal Husbandry & Veterinary Medicine, 2021, 48(8):2937-2946 (in Chinese).

[32]Gao F, Wu Y, Wen H, et al. Multilineage potential research on pancreatic mesenchymal stem cells of bovine [J]. Tissue Cell, 2019,56:60-70.

[33]Gao Y, Wu H, Wang Y, et al. Single Cas9 nickase induced generation of NRAMP1 knockin cattle with reduced off-target effects[J]. Genome Biol, 2017,18(1):13.

[34]Gim G M, Kwon D H, Ha J J, et al. Production of MSTN-mutated cattle without exogenous gene integration using CRISPR-Cas9[J]. Biotechnol J, 2021, e2100198.

[35]Goldman B D. Mammalian photoperiodic system: formal properties and neuroendocrine mechanisms of photoperiodic time measurement[J]. J Biol Rhythms, 2001, 16(4): 283-301.

[36]Golkar Z. CRISPR: a journey of gene-editing based medicine[J]. Genes Genomics, 2020, 42(12): 1369-1380.

[37]Gupta S K, Toor S, Minhas V, et al. Contraceptive efficacy of recombinant porcine zona proteins and

fusion protein encompassing canine ZP3 fragment and GnRH in female beagle dogs[J]. Am J Reprod Immunol, 2022, 87(6): e13536.

[38]Hafez B, Hafez E. Micromanipulation of gametes and embryos: *In vitro* fertilization and embryo transfer (IVF/ET): Hafez/reproduction in farm animals[M]. Lippincott Williams & Wilkins, 2016.

[39]Hammer R E, Pursel V G, Rexroad Jr. C E, et al. Production of transgenic rabbits, sheep and pigs by microinjection[J]. Nature, 1985, 315:680-683.

[40]Han J Y, Park Y H. Primordial germ cell-mediated transgenesis and genome editing in birds[J]. J Anim Sci Biotechnol, 2018, 9:19.

[41]Han L, Zhen Y H, Liang A X, et al. Oral vaccination with inhibin DNA delivered using attenuated Salmonella choleraesuis for improving reproductive traits in mice[J]. J Basic Microbiol, 2014, 54(9): 962-968.

[42]Han Y, Liu G, Jiang X, et al. KISS1 can be used as a novel target for developing a DNA immunocastration vaccine in ram lambs[J]. Vaccine, 2015,33(6):777-782.

[43]Han Y, Peng X, Li K, et al. Potent effect of KISS1-54 DNA vaccine compared with KISS1-10 DNA vaccine in inhibiting the fertility of female rats[J]. Vaccine, 2018,36(45):6631-6639.

[44]Han Y, Si W, Han Y, et al. Immunization with oral KISS1 DNA vaccine inhibits testicular Leydig cell proliferation mainly via the hypothalamic-pituitary-testicular axis and apoptosis-related genes in goats[J]. Anim Biotechnol, 2021, 32(3):395-399.

[45]Han Y G, Liu G Q, Jiang X P, et al. Reversibility and safety of KISS1 metastasis suppressor gene vaccine in immunocastration of ram lambs[J]. Asian-Australas J Anim Sci, 2018,31(6):835-841.

[46]Hickey D K, Fahey J V, Wira C R. Mouse estrous cycle regulation of vaginal versus uterine cytokines, chemokines, alpha-/beta-defensins and TLRs[J]. Innate Immun, 2013, 19(2):121-131.

[47]Hill A B, Hill J E, Bressan F F, et al. Derivation and differentiation of canine ovarian mesenchymal stem cells [J]. J Vis Exp,2018,142:e58163.

[48]Hill A B T, Bressan F F, Murphy B D, et al. Applications of mesenchymal stem cell technology in bovine species [J]. Stem Cell Res Ther, 2019,10(1):44.

[49]Hill A B T, Therrien J, Garcia J M, et al. Mesenchymal-like stem cells in canine ovary show high differentiation potential [J]. Cell Prolif, 2017,50:6.

[50]Hobernik D, Bros M. DNA vaccines—How far from clinical use? [J]. Int J Mol Sci, 2018, 19 (11):3605.

[51]Huaman O, Bahamonde J, Cahuascanco B, et al. Immunomodulatory and immunogenic properties of mesenchymal stem cells derived from bovine fetal bone marrow and adipose tissue [J]. Res Vet Sci, 2019,124:212-222.

[52]Huang G. Animal breeding [M]. Nanchang: Jiangxi Science and Technology Press, 2017.

[53]Huang Z, Li J, Li W, et al. Brief co-incubation of sperm and oocytes for *in vitro* fertilization techniques [M]. John Wiley & Sons, Ltd, 2011.

[54]Huo S, Yang J. Animal embryo engineering [M]. Beijing: China Agriculture Press, 2012 (in Chinese).

[55]Ioannidis J, Donadeu F X. Changes in circulating microRNA levels can be identified as early as day 8 of pregnancy in cattle[J]. PLoS One, 2017, 12:e0174892.

[56]Jabbar A, Zulfiqar F, Mahnoor M, et al. Advances and perspectives in the application of CRISPR-Cas9 in livestock[J]. Mol Biotechnol, 2021, 63(9):757-767.

[57]Jervis M, Huaman O, Cahuascanco B, et al. Comparative analysis of *in vitro* proliferative, migratory and pro-angiogenic potentials of bovine fetal mesenchymal stem cells derived from bone marrow and

adipose tissue [J]. Vet Res Commun, 2019,43(3):165-178.

[58]Jiang T T. Prediction of B epitopes of AMH protein and construction and identification of AMH DNA vaccines without antibiotic resistance gene[D]. Wuhan: Huazhong Agricultural University, 2018.

[59]Joe Bearden H, Fuquay J W, Willard S T. Applied animal reproduction[M]. 6th ed. London: Pearson Press, 2003.

[60]Johnston D, Malo Estepa L, Ebhardt H A, et al. Differences in the bovine milk whey proteome between early pregnancy and the estrous cycle [J]. Theriogenology, 2018, 114:301-307.

[61]Junco J A, Fuentes F, Millar R P. A dual kisspeptin-GnRH immunogen for reproductive immunosterilization[J]. Vaccine, 2021,39(43):6437-6448.

[62]Katiyar K S, Burrell J C, Laimo F A, et al. Biomanufacturing of axon-based tissue engineered nerve grafts using porcine GalSafe neurons[J]. Tissue Eng Part A, 2021, 27(19-20):1305-1320.

[63]Kawase K, Tomiyasu J, Ban K, et al. Contraceptive effect of a gonadotropin-releasing hormone vaccine on a captive female African lion (*Panthera leo*): A case study[J]. J Vet Med Sci, 2021, 83(9): 1454-1458.

[64]Khasawneh R R, Al Sharie A H, Abu-El Rub E, et al. Addressing the impact of different fetal bovine serum percentages on mesenchymal stem cells biological performance [J]. Mol Biol Rep, 2019, 46(4): 4437-4441.

[65]Khosla K, Bischof J, Varga Z M. Sperm cryopreservation, *in vitro* fertilization, and embryo freezing[M]. In Laboratory Fish in Biomedical Research. New York: Academic Press, 2022.

[66]Kiesner J, Eisenlohr-Moul T and Mendle J. Evolution, the menstrual cycle, and theoretical overreach [J]. Perspect Psychol Sci, 2020, 15(4):1113-1130.

[67]Knox R V. Recent advancements in the hormonal stimulation of ovulation in swine[J]. Vet Med (Auckl), 2015, 6:309-320.

[68]Laine A, Bastin C, Grelet C, et al. Assessing the effect of pregnancy stage on milk composition of dairy cows using mid infrared spectra [J]. J Dairy Sci, 2017, 100:2863-2876.

[69]Lamb G C, Dahlen C R, Larson J E, et al. Control of the estrous cycle to improve fertility for fixed-time artificial insemination in beef cattle: a review[J]. J Anim Sci, 2010, 88(13 Suppl):E181-192.

[70]Lara E, Rivera N, Rojas D, et al. Characterization of mesenchymal stem cells in bovine endometrium during follicular phase of oestrous cycle [J]. Reprod Domest Anim, 2017,52(5):707-714.

[71]Lee J, Byeon J S, Gu N Y, et al. Bovine tongue epithelium-derived cells: A new source of bovine mesenchymal stem cells [J]. Biosci Rep, 2020, 40(4).

[72]Lee S H, Gupta M K, Ho Y T, et al. Transgenic chickens expressing human urokinase-type plasminogen activator[J]. Poult Sci, 2013, 92(9):2396-2403.

[73]Ley D, Harraghy N, Le Fourn V, et al. MAR elements and transposons for improved transgene integration and expression[J]. PLoS One, 2013, 8(4):e62784.

[74]Li P S, Lei X Q, Xu T S, et al. Study on human tumor suppressor gene P53 in chicken[J]. Journal of Northwest A&F University (Natural Science Edition), 2019, 47(7):54-61(in Chinese).

[75]Li Q. Animal cell engineering and practice [M]. Beijing: Chemical Industry Press, 2005(in Chinese).

[76]Li S, Flisikowska T, Kurome M, et al. Dual fluorescent reporter pig for Cre recombination: Transgene placement at the *ROSA26* locus[J]. PLoS One, 2014, 9(7):e102455.

[77]Li W, Shuai L, Wan H, et al. Androgenetic haploid embryonic stem cells produce live transgenic mice[J]. Nature, 2012, 490(7420):407-411.

[78]Li X, Hao F, Hu X, et al. Generation of Tβ4 knock-in cashmere goat using CRISPR /Cas9[J]. Int J Biol

Sci, 2019, 15(8):1743-1754.

[79]Li Y, Wang X Y, Li Q F, et al. Transgenic quail carrying PEGFP-C2 gene was prepared by liposome transfection into blastodisk cells[J]. Chinese Journal of Veterinary Medicine, 2014, 34(03): 442-445(in Chinese).

[80]Li Z K, Wang L Y, Wang L B, et al. Generation of bimaternal and bipaternal mice from hypomethylated haploid ESCs with imprinting region deletions[J]. Cell Stem Cell, 2018, 23(5):665-676.

[81]Liu Q, Han L, Rehman Z U, et al. The efficacy of an inhibin DNA vaccine delivered by attenuated *Salmonella choleraesuis* on follicular development and ovulation responses in crossbred buffaloes[J]. Anim Reprod Sci, 2016 ,172:76-82.

[82]Liu Q, Rehman Z U, Liu J J, et al. Nasal immunization with inhibin DNA vaccine delivered by attenuated Salmonella choleraesuis for improving ovarian responses and fertility in cross-bred buffaloes[J]. Reprod Domest Anim, 2017,52(2):189-194.

[83]Liu Z C, He D J. Recognition method of cow estrus behavior based on convolutional neural network[J]. Nongye Jixie Xuebao, 2019, 50(7):8(in Chinese).

[84]Liu Z G, Wang BY, Mu YL, et al. Breeding by molecular writing (BMW): the future development of animal breeding[J]. Scientia Agricultura Sinica, 2018, 51(12):2398-409 (in Chinese).

[85]Łukaszewicz E, Jerysz A, Kowalczyk A. Effect of semen extenders on viability of ISA Brown and Hubbard Flex roosters' sperm stored for 24 h[J]. Poult Sci, 2020, 99:2766-2774.

[86]Manafi M. Artificial insemination in farm animals[M]. BoD-Books on Demand: 2011.

[87]Mao D, Bai W, Hui F, et al. Effect of inhibin gene immunization on antibody production and reproductive performance in Partridge Shank hens[J]. Theriogenology, 2016,85(6):1037-1044.

[88]Mapletoft R J. History and perspectives on bovine embryo transfer[J]. Anim Reprod, 2013,10:168-173.

[89]Martinez-Giménez J, Ferrero-Vicente L, Feltrer-Rambaud Y. Effect of a GnRH vaccine (IMPROVAC®) on Patagonian (*Otaria byronia*) and California sea lions (*Zalophus californianus*) [J]. J Zoo Wildl Med, 2021,52(2):721-725.

[90]Mayo L M, Silvia W J, Ray D L, et al. Automated estrous detection using multiple commercial precision dairy monitoring technologies in synchronized dairy cows[J]. J Dairy Sci, 2019, 102(3): 2645-2656.

[91]Ménoret S, Tesson L, Remy S, et al. Advances in transgenic animal models and techniques[J]. Transgenic Res, 2017, 26(5): 703-708.

[92]Monzani P S, Adona P R, Ohashi O M, et al. Transgenic bovine as bioreactors: Challenges and perspectives[J]. Bioengineered, 2016,7(3):123-131.

[93]Moreira A, Kahlenberg S, Hornsby P. Therapeutic potential of mesenchymal stem cells for diabetes [J]. J Mol Endocrinol, 2017,59(3):R109-120.

[94]Mozdziak P E, Borwornpinyo S, McCoy D W, et al. Development of transgenic chickens expressing bacterial beta-galactosidase[J]. Dev Dyn, 2003, 226(3):439-445.

[95]Nagy Z P, Varghese A C, Agarwal A. Practical manual of *in vitro* fertilization: Advanced methods and novel devices[M]. Springer New York, 2012.

[96]Nazari H, Shirazi A, Shams-Esfandabadi N, et al. The effect of amniotic membrane stem cells as donor nucleus on gene expression in reconstructed bovine oocytes [J]. Int J Dev Biol, 2016,60(4-6):95-102.

[97]Neethirajan S and Kemp B. Digital phenotyping in livestock farming [J]. Animals (Basel), 2021, 11(7).

[98]Nowicki A, Barański W, Baryczka A, et al. Ovsynch protocol and its modifications in the reproduction management of dairy cattle herds—An update[J]. J Vet Res, 2017,61(3):329-336.

[99]Okamura L H, Cordero P, Palomino J, et al. Myogenic differentiation potential of mesenchymal stem

cells derived from fetal bovine bone marrow [J]. Anim Biotechnol, 2018,29(1):1-11.

[100]Okutsu T, Shikina S, Kanno M, et al. Production of trout offspring from triploid salmon parents[J]. Science, 2007, 317(5844):1517.

[101]One H, Hoshino Y, Yasuo S, et al, Involvement of thyrotropin in photoperiodic signal transduction in mice[J]. Proc Natl Acad Sci U S A, 2008, 105(47):18238-18242.

[102]Osycka-Salut C E, Martinez-Leon E, Gervasi M G, et al. Fibronectin induces capacitation-associated events through the endocannabinoid system in bull sperm[J]. Theriogenology, 2020,153:91-101.

[103]Ouyang Q, Bao D, Lu Y, et al. A comparative study of libido in drakes: From phenotypes to molecules [J]. Poult Sci, 2021, 100:101503.

[104]Parrilla I, Perez-Patiño C, Li J, et al. Boar semen proteomics and sperm preservation [J]. Theriogenology, 2019,137:23-29.

[105]Peng S Y, Chou C W, Kuo Y H, et al. Potential differentiation of islet-like cells from pregnant cow-derived placental stem cells [J]. Taiwan J Obstet Gynecol, 2017,56(3):306-311.

[106]Pinkert C A, Polites H G. Transgenic animal technology: A laboratory handbook[M]. Elsevier, 2014.

[107]Poorna M R, Jayakumar R, Chen Jyh-P, et al. Hydrogels: A potential platform for induced pluripotent stem cell culture and differentiation[J]. Colloids Surf B Biointerfaces, 2021, 207:111991.

[108]Qin Y, Yang S, Xu J, et al. Deep insemination with sex-sorted cashmere goat sperm processed in the presence of antioxidants [J]. Reprod Domest Anim, 2018, 53:11-19.

[109]Rahbaran M, Razeghian E, Maashi M S, et al. Cloning and embryo splitting in mammalians: brief history, methods, and achievements[J]. Stem Cells Int, 2021:2347506.

[110]Rashid N A, Lalitkumar S, Lalitkumar P G, et al. Endometrial receptivity and human embryo implantation[J]. Am J Reprod Immunol, 2011,66(Suppl 1):23-30.

[111]Rehman A, Ahmad E, Arshad U, et al. Effects of immunization against inhibin α-subunit on ovarian structures, pregnancy rate, embryonic and fetal losses, and prolificacy rate in goats where estrus was induced during the non-breeding season[J]. Anim Reprod Sci, 2021, 224:106654.

[112]Rosenbaum L. The future of gene editing-toward scientific and social consensus[J]. N Engl J Med, 2019, 380(10):971-975.

[113]Ryczek N, Hryhorowicz M, Zeyland J, et al. CRISPR/Cas technology in pig-to-human xenotransplantation research[J]. Int J Mol Sci, 2021, 22(6):3196.

[114]Sakumoto R, Iga K, Hayashi K G, et al. Gene expression of CCL8 and CXCL10 in peripheral blood leukocytes during early pregnancy in cows[J]. J Anim Sci Biotechnol, 2018, 9:46.

[115]Samir H, El Sayed M A I, Nagaoka K, et al. Passive immunization against inhibin increases testicular blood flow in male goats[J]. Theriogenology, 2020,147:85-91.

[116]Sang R. Animal reproduction biotechnology [M]. Beijing: China Agriculture Press, 2002(in Chinese).

[117]Saragusty J, Arav A. Current progress in oocyte and embryo cryopreservation by slow freezing and vitrification[J]. Reproduction, 2011, 141(1):1-19.

[118]Schusser B, Collarini E J, Yi H, et al. Immunoglobulin knockout chickens via efficient homologous recombination in primordial germ cells[J]. Proc Natl Acad Sci USA, 2013, 110(50): 20170-20175.

[119]Schütz J, Rudolph J, Steiner A, et al. Immunization against Gonadotropin-releasing hormone in female beef calves to avoid pregnancy at time of slaughter[J]. Animals (Basel), 2021,11(7):2071.

[120]Sharma N, Do Luong Huynh S W K, Ghosh M, et al. A PiggyBac mediated approach for lactoferricin gene transfer in bovine mammary epithelial stem cells for management of bovine mastitis [J]. Oncotarget, 2017,8(61):104272.

[121]Skenandore C S, Cardoso F C. The effect of tail paint formulation and heifer behavior on estrus detection [J]. Int J Vet Sci Med, 2017, 5(2):113-120.

[122]Song H, Li H, Huang M, et al. Big animal cloning using transgenic induced pluripotent stem cells: a case study of goat transgenic induced pluripotent stem cells[J]. Cell Reprogram, 2016, 18(1):37-47.

[123]Steiger S P. *In vitro* fertilization[M]. New York: Nova Science Publishers, 2011.

[124]Takahashi K, Yamanaka S. Induction of pluripotent stem cells from mouse embryonic and adult fibroblast cultures by defined factors[J]. Cell, 2006, 126(4):663-676.

[125]Talovic M, Patel K, Schwartz M, et al. Decellularized extracellular matrix gelloids support mesenchymal stem cell growth and function *in vitro* [J]. J Tissue Eng Regen Med, 2019,13(10):1830-1842.

[126]Tang J M. Construction of AMH-INH-GnIH gene vaccine and Evaluation of its immunization effect and regulation of reproduction traits in mice[D]. Changsha: Hunan Agricultural University, 2020(in Chinese).

[127]Taylor L, Carlson D F, Nandi S, et al. Efficient TALEN-mediated gene targeting of chicken primordial germ cells[J]. Development, 2017, 144(5):928-934.

[128]Teijeiro J M, Cabada M O, Marini P E. Sperm binding glycoprotein (SBG) produces calcium and bicarbonate dependent alteration of acrosome morphology and protein tyrosine phosphorylation on boar sperm[J]. J Cell Biochem, 2008,103:1413-1423.

[129]Tesema B, Zhao J Y, Jiang X P, et al. Kisspeptin recombinant oral vaccine: A master gene vaccine inhibiting the reproductive physiology and behavior of ram lambs[J]. Vaccine, 2019, 37(32): 4630-4636.

[130]Tharasanit T, Thuwanut P. Oocyte cryopreservation in domestic animals and humans: principles, techniques and updated outcomes[J]. Animals (Basel), 2021, 11(10):2949.

[131]Toledo-Alvarado H, Vazquez A I, De Los Campos G, et al. Diagnosing pregnancy status using infrared spectra and milk composition in dairy cows[J]. J Dairy Sci, 2018, 101:2496-2505.

[132]Umehara T, Tsujita N, Shimada M. Activation of Toll-like receptor 7/8 encoded by the X chromosome alters sperm motility and provides a novel simple technology for sexing sperm[J]. PLoS Biol, 2019, 17: e3000398.

[133]Umezu K, Hiradate Y, Numabe T, et al. Effects on glycocalyx structures of frozen-thawed bovine sperm induced by flow cytometry and artificial capacitation[J]. J Reprod Dev, 2017, 63:473-480.

[134]Vajta G, Peura T T, Holm P, et al. New method for culture of zona-included or zona-free embryos: The Well of the Well (WOW) system[J]. Mol Reprod Dev, 2015, 55(3):256-264.

[135]Wakayama T, Perry AC, Zuccotti M, et al. Full-term development of mice from enucleated oocytes injected with cumulus cell nuclei[J]. Nature, 1998, 394(6691):369-374.

[136]Wang A P, Wang Y J, Wu S, et al. Study on the expression of human lysozyme in oviduct bioreactor mediated by recombinant avian adeno-associated virus[J]. Poult Sci, 2017, 96(7): 2447-2453.

[137]Wang D H, Liu Y L, Cai Z G, et al. Effects of extender type on the quality of post-thaw giant panda (*Ailuropoda melanoleuca*) semen[J]. Cryobiology, 2020, 94:95-99.

[138]Wang F. Animal reproduction[M]. Beijing: China Agricultural University Press, 2012(in Chinese).

[139]Wang S L, Han L, Ahmad S, et al. Effect of a DNA vaccine harboring two copies of inhibin α (1-32) fragments on immune response, hormone concentrations and reproductive performance in rats[J]. Theriogenology, 2012, 78(2):393-401.

[140]Wang X, Liu C. Livestock reproduction[M]. Yinchuan: Ningxia Sunshine Press, 2013(in Chinese).

[141]Wassie T, Liu G, Jiang X, et al. Immunization against Kisspeptin-54 perturb hypothalamic-pituitary-testicular signaling pathway in ram lambs[J]. Theriogenology, 2019,125:193-202.

[142]Wei L, Xi J, Zhang Y, et al. The use of RNAi technology to interfere with Zfx gene increases the male rates of red deer (*Cervus elaphus*) offspring[J]. Biomed Res Int. 2020:9549765.

[143]Wei Y, Yang C R, Zhao Z A. Viable offspring derived from single unfertilized mammalian oocytes[J]. Proc Natl Acad Sci USA, 2022, 119(12):e2115248119.

[144]Wernery U, Liu C, Baskar V, et al. Primordial germ cell-mediated chimera technology produces viable pure-line Houbara bustard offspring: Potential for repopulating an endangered species[J]. PLoS One, 2010, 5(12): e15824.

[145]Woodfint R M, Hamlin E, Lee K. Avian bioreactor systems: A review[J]. Mol Biotechnol, 2018, 60 (12):975-983.

[146]Xi J F, Wang X Z, Zhang Y S, et al. Sex control by Zfy siRNA in the dairy cattle[J]. Anim Reprod Sci, 2019, 200:1-6.

[147]Yamanaka S. Elite and stochastic models for induced pluripotent stem cell generation[J]. Nature,2009, 460(7251):49-52.

[148]Yang L G. New technology of animal breeding[M]. Beijing: China Agriculture Press, 2009 (in Chinese).

[149]Yang L G. Animal Reproduction[M]. Beijing: China Agriculture Press, 2019 (in Chinese).

[150]Yoshino H, Toji N, Sasaki K, et al. A predictive threshold value for the diagnosis of early pregnancy in cows using interferon stimulated genes in granulocytes [J]. Theriogenology, 2018, 107:188-193.

[151]Yu L, Zheng Y, Feng Y, et al. Role of *L*-selectin on leukocytes in the binding of sialic acids on sperm surface during the phagocytosis of sperm in female reproductive tract[J]. Me Hypotheses, 2018,120:4-6.

[152]Yu X, Li Z, Zhao X, et al. Anti-Müllerian hormone inhibits FSH-induced cumulus oocyte complex *in vitro* maturation and cumulus expansion in mice[J]. Animals (Basel), 2022,12(9):1209.

[153]Yu X, Qiao T, Hua L, et al. Synergistic regulatory effect of inhibin and anti-Müllerian hormone on fertility of mice[J]. Front Vet Sci, 2021, 8:747619.

[154]Zhang J, Lei C, Deng Y, et al. Hypoxia enhances mesenchymal characteristics maintenance of buffalo bone marrow-derived mesenchymal stem cells [J]. Cell Reprogram,2020,22(3):167-177.

[155]Zhang J, Qin X, Deng Y, et al. Transforming growth factor-beta 1 enhances mesenchymal characteristics of buffalo (*Bubalus bubalis*) bone marrow-derived mesenchymal stem cells [J]. Cellular Reprogramming,2021,23(2):127-138.

[156]Zhang L, Yang B, Yue Y, et al. Effects of active immunization with GnIH and INH epitopes peptide vaccine on reproductive hormones of Gansu alpine fine-wool sheep[J]. Chinese animal husbandry and veterinary medicine, 2016,43(4): 1039-1044(in Chinese).

[157]Zhang S, Lin H, Kong S, et al. Physiological and molecular determinants of embryo implantation[J]. Mol Aspects Med,2013,34(5):939-980.

[158]Zhang Y, Chen Z Y, An C, et al. Effect of active immunization with recombinant-derived goose INH-α, AMH, and PRL fusion protein on broodiness onset and egg production in geese (*Anser cygnoides*) [J]. Poult Sci, 2021, 100(11):101452.

[159]Zhu S. Animal reproductive physiology [M]. Beijing: China Agriculture Press, 2006(in Chinese).

[160]Zhu S. Livestock reproduction[M]. Beijing: China Agriculture Press, 2015(in Chinese).